THE OXFORD AUT

*General Editor: Frank K*

ANDREW MARVELL (1621–78) lived through some of the most momentous events in English history, but his poetry is detached and private, mostly lyrical, strikingly original in its combining of existing traditions, and seemingly untouched for the most part by what was happening around him. He left Cambridge on the death of his father and travelled on the Continent, possibly to avoid direct involvement in the civil wars. Later, back in England, he worked (like his friend John Milton) for the government of Oliver Cromwell, and still later, after the Restoration of Charles II, he strove in opposition as MP for Hull. This last period led to more direct involvement in political affairs. Most of his poems were published after his death in a volume (1681) edited by a woman claiming to be his widow. For a century and more after his death he was remembered primarily as a witty satirist in verse and a polemicist in prose.

FRANK KERMODE, retired King Edward VII Professor of English Literature at Cambridge, the General Editor of the Oxford Authors Series, and co-editor of this particular volume, is the author of many books, including *Romantic Image*, *The Sense of an Ending*, *The Classic*, *Genesis of Secrecy*, *Forms of Attention*, and *History and Value*.

KEITH WALKER is Senior Lecturer in English Language and Literature at University College London. He has written on eighteenth-century literature, and edited a selection of the works of Dryden for the Oxford Authors Series.

THE OXFORD AUTHORS

# ANDREW MARVELL

EDITED BY
FRANK KERMODE
AND
KEITH WALKER

Oxford New York
OXFORD UNIVERSITY PRESS
1990

*Oxford University Press, Walton Street, Oxford* OX2 6DP
*Oxford New York Toronto*
*Delhi Bombay Calcutta Madras Karachi*
*Petaling Jaya Singapore Hong Kong Tokyo*
*Nairobi Dar es Salaam Cape Town*
*Melbourne Auckland*
*and associated companies in*
*Berlin Ibadan*

*Oxford is a trade mark of Oxford University Press*

*Introduction, edited text, and editorial matter*
*© Frank Kermode and Keith Walker 1990*

*All rights reserved. No part of this publication may be reproduced, stored in a retrieval system, or transmitted, in any form or by any means, electronic, mechanical, photocopying, recording, or otherwise, without the prior permission of Oxford University Press*

*This book is sold subject to the condition that it shall not, by way of trade or otherwise, be lent, re-sold, hired out or otherwise circulated without the publisher's prior consent in any form of binding or cover other than that in which it is published and without a similar condition including this condition being imposed on the subsequent purchaser*

*British Library Cataloguing in Publication Data*
*Andrew Marvell.—(The Oxford authors).*
*1. Poetry in English. Marvell, Andrew, 1621–1678*
*I. Kermode, Frank, 1919– II. Walker, Keith, 1936–*
*821.4*
*ISBN 0–19–254183–8*
*ISBN 0–19–281347–1 (Pbk.)*

*Library of Congress Cataloging in Publication Data*
*Marvell, Andrew, 1621–1678.*
*[Selections. 1990]*
*Andrew Marvell/edited by Frank Kermode and Keith Walker.*
*p. cm.-- (The Oxford authors)*
*Includes bibliographical references.*
*I. Kermode, Frank, 1919– . II. Walker, Keith, 1936– .*
*III. Title. IV. Series.*
*PR3546.A6 1990 821'.4--dc20 90–34950*
*ISBN 0–19–254183–8*
*ISBN 0–19–281347–1 (Pbk.)*

*Typeset by Promenade Graphics, Cheltenham*
*Printed in Great Britain by*
*Richard Clay Ltd,*
*Bungay, Suffolk*

# CONTENTS

# INTRODUCTION

WHEN Marvell died in 1678, the government was considering an informer's report which named him as the author of *An Account of the Growth of Popery*. The report was correct, and the informer said nothing that was not widely known. At the time of the poet's death all the talk must have been of this and similar works which attacked the deviousness of the king, and his failure to prevent his brother from plotting a French papist takeover in England. Marvell was the satirist, the controversialist, expert in the joint territories of politics and religion; he was the classical republican, sketching in advance the principles of 1688, and an eighteenth-century man before his time. His epitaph in St Giles, London, grants him 'wit and learning, with a singular penetration and strength of judgment', but whereas we should use these words of the poet, the writer of the tribute thought of him exclusively as a patriot politician. When his *Miscellaneous Poems* appeared in 1681, they were read in dissenting circles but otherwise attracted little attention, though his satirical fame was great, and people thought him the author of a considerable body of anti-Establishment writing with which he had had nothing to do.

The consequences of having been born in 1621 were in some way like those of having been born in 1918: the course of a man's life must have been strongly affected by war and the aftermath of war. So, after 1642, it was with Marvell. There was nothing very unusual about his youth. Though of Cambridgeshire stock, he was born in Yorkshire, where his father was a parson of strong Calvinist convictions. Given this background it is not surprising that the poet went—at the age of twelve—from Hull Grammar School to Cambridge. Some said that as a student he was tempted by the Jesuits, but rescued from them by his father. When his father drowned in the Humber in 1641 Marvell left Trinity College and went to London. After that, as we know from Milton's letter recommending him to the President of the Council of State in 1653, Marvell travelled for some four years in Europe, possibly to avoid the Civil War, possibly as a tutor to some young gentleman, possibly even as some kind of government agent, and certainly to use his chance to learn languages such as Spanish, Italian, French, and Dutch—all of which would prove useful in a diplomatic or political, as well as in a literary, career. He was certainly in Rome in 1645 or 1646, when he wrote the satirical verses about Richard Flecknoe (see p. 77); in December 1654 he met Lord Francis Villiers there.

Before his travels, Marvell had written a Latin and Greek poem addressed to the King in a Cambridge collection of 1637 (Marvell's first Horatian Ode). He must have been back in London by 1648, when he wrote the elegy for Lord Francis Villiers, killed in the second Civil War in July of that year (see p. 1).

Of Marvell's politics at this time little positive can be said; though Villiers was certainly a royalist, as was Lovelace, in whose book Marvell in 1649 published commendatory verses. Lord Hastings, for whom he wrote an elegy in the same year, was also royalist.

In 1650, a year of change both political and personal, Marvell went to Yorkshire and became tutor to Mary, daughter of the Parliamentary general Lord Fairfax, with whom he had a remote family connection. He remained there for more than two years. Fairfax, though in a position of great authority, had unsuccessfully opposed the execution of the king in 1649, and in 1650 the invasion of Scotland, which was undertaken in his place by Cromwell, fresh from his expedition to 'pacify' Ireland.

At this point Marvell, though not actively involved, could hardly avoid at least a contemplation of politics. The peculiar interest of Fairfax's position was that he had voluntarily and on principle withdrawn from eminence and settled into a studious country retirement. It is obvious from 'Upon Appleton House' that Marvell greatly admired Fairfax. Perhaps, as is sometimes suggested by critics examining the nuances of 'Upon Appleton House', he had mixed feelings about his employer's choice of life, though sympathizing with the motives for it. Later he was to say that men ought to have trusted the king, yet without making it plain at this stage that he was now, with whatever reservations, on the side of the new regime. The reason why there is so much uncertainty about his position is that at this period he expressed himself not in prose but in poetry, with a higher degree of obliquity and a concern for more than topical significance. 'An Horatian Ode on Cromwell's Return from Ireland' is a manifestly great poem, yet it is also baffling to anybody who wants simply to know where the poet stood with Cromwell in 1650. There has been a great deal of interpretation, some of it of high quality, but the poem remains resistant to decoding.

The palpable complexity of its political references is in itself a deterrent to over-simplification, and there are also certain rather intractable difficulties of a bibliographical kind. The Ode is, ostensibly, a poem of the sort that celebrates the arrival or departure of some great person on a notable occasion, the great man on this occasion being Cromwell. Yet at the very centre of the poem is a long and apparently sympathetic account of a different and even more solemn occasion—not the return of Cromwell, but the final departure of the king. There is an ambiguity of tone not unlike that of Horace's great Actium Ode (i.37), in which a somewhat brazen celebration of Octavian's victory modulates into a splendid, and as it were reluctant, tribute to the defeated Cleopatra.

The occasion dates the poem, which must have been written between Cromwell's return from Ireland in May 1650 and his preventive campaign against Scotland, begun in June 1650. The poet was at Nun Appleton, presumably in close contact with Fairfax and doubtless thinking hard about the great issues. Of course he had not been obliged, as Fairfax had, to make firm decisions and act on them, and it would be unsafe to assume that he fully shared his employer's disapproval of the Scottish campaign, or indeed of the regicide. That event he placed at the centre of the Ode, in an obvious tribute to its absolute importance: it is the point at which the horizontal and the vertical of history could be thought to cross. The echoes of Horace and Lucan reinforce this sense of historical crisis; Lucan is concerned with the crisis caused by Julius Caesar's violation of the ancient rights, his casting of the ancient republic into another mould, and his assumption of the role of dictator or protector. Horace celebrates Actium, the battle which assured the forced imperial power of Augustus. Charles was an English king and in terms of imperialist history not only the inheritor of Lucan's crisis but the negative image of Horace's—the ruin of the great work of time, not its beginnings.

The complexity of the entire poem—characteristic of the complexity of all Marvell's best poetry—can perhaps be briefly illustrated by a consideration of the central passage on the death of the king. It ostensibly develops the point about Cromwell's 'wiser art', his reputed skill in ensuring the capture of the king by the Parliament. Now the king is produced on a scaffold, such as might be used by strolling players, but fixed outside the Banqueting House built by the king's father for the masquing in which Charles had often taken part. Now he must perform the death of a great man, himself; he is an actor, and this is his last act. His performance on this 'scene' is admirable, and the verse enacts it in a soft cadence:

> Nor called the gods with vulgar spite
> To vindicate his helpless right,
> But bowed his comely head
> Down as upon a bed . . .

But this quiet solemnity is merely a preparation for the startling drum-roll of the next lines:

> This was that memorable hour
> Which first assured the forced power—

followed by the triumphal metaphor of the bleeding head, and the celebration of Cromwell's ruthless achievements as the servant of the new republic—a soldier who knows that rights hold or break not in so far as they are ancient, but as men are strong or weak; a man who hardly needs to have it explained that 'the same arts that did gain / A power must it maintain'.

The poem, whether by allusion or by significant silences, is extremely topical. As Michael Wilding among others has pointed out,[1] Cromwell was by this time rather more than an obedient servant of Parliament; if the poem hints at this it does no more than hint. Some think Marvell is endorsing a Machiavellian dictator, while others argue that his rhetorical tone is one of admiring caution. A Machiavellian (and Marvell had been called that) would say that in the real world what men do is more important than what they ought to do; things are as they are and not as we would have them. Perhaps the profession of admiration for the king, and the deliberative, concessive tone of some of the praise of Cromwell ('And if we would speak true, / Much to the man is due') are merely rhetorical feints, and Marvell really was, with only a few reservations, committed to the new order, to the crossing of this Rubicon. The means might appear wicked, but God sometimes works by such means, as when Samson married Delilah so as, in the process of a divine plot, to pull down the pillars on the Philistian nobles.

But the poem as a whole seems resistant to simple and even to sophisticated explanations, whether political or rhetorical. Perhaps it has always been so. Its first success seems to have been with royalists, but it was excluded from *1681* (though two extant copies include it) presumably because it was read, in the light of the poet's subsequent career, as being clearly on Cromwell's side. And despite the vast quantity of modern exegesis a certain obscurity remains. It is deepened by our failure fully to explain the relation between the Ode and the poem 'Tom May's Death', which speaks of 'ancient rights' and does not disparage them (see the notes on that poem, p. 309).

The chronology of Marvell's poems is far from certain, except of course when their occasion is obvious, as with 'An Horatian Ode' and such works as 'The Character of Holland' and the poem on the first anniversary of Cromwell's protectorate. It would be difficult to deny that it was during his Nun Appleton period that Marvell wrote the Bilbrough and Appleton House poems, and it is usually assumed, though there is no direct evidence, that in those years he also wrote 'The Garden' and the series of Mower poems. These are distinctive not only because mowers were unusual pastoral subjects, being considered lower than shepherds (a point contested by Marvell's mower), but also because there are so many differences between the poems themselves. All, however, are beautifully fashioned, and all extend the scope of the genres and sub-genres to which they can be assigned. They are works of extraordinary refinement, with the long tradition of ancient and Renaissance pastoral behind them, yet here transfigured. And it is true that Marvell's poems, even when they most resemble those of other poets, or are most in the fashions of the time, usually have a quite distinctive tone.

In 1653 he became tutor to William Dutton, a ward of Cromwell's and lived at Eton in the house of John Oxenbridge, whose knowledge of the Bermudas probably contributed to Marvell's poem on the subject. Now close to Cromwell, he wrote songs for the marriage of Cromwell's daughter Mary in 1657. In the same year he was appointed Latin Secretary to the Council of State, the post for which the blind Milton, now needing such assistance, had unsuccessfully recommended him four years earlier. (He may have repaid that debt by playing a part in saving Milton's life after the Restoration.) Together with Milton and Dryden, he walked in Cromwell's funeral procession in 1658 and wrote a poem on the Protector's death, which among other things pledged support to Richard Cromwell, the son who briefly succeeded Oliver in that quasi-regal office. In 1659 Marvell became Member of Parliament for Hull.

When Richard was dislodged and Parliament dissolved there followed a period of political confusion, but Marvell again represented Hull in the Convention Parliament of 1660, which recalled Charles II. Any difficulty his adherence to Cromwell may have caused him was ended by an Act of Oblivion of 1660, which excepted only a dozen names, Marvell's not being among them. Charles managed without elections, and Marvell continued to represent Hull with admirable assiduity, as his correspondence demonstrates, until his death, though for a time he quarrelled with the other Hull member, Colonel Gilby. John Kenyon remarks that Marvell was leading a double life, as pedestrian MP and 'a prominent though usually anonymous opposition spokesman', as in his 'Last Instructions to a Painter'.[2] He was habitually and savagely critical of the Court.

In the early 1660s Marvell went to Holland on an unknown mission and accompanied the Earl of Carlisle on a lengthy embassy to Russia, Sweden, and Denmark. In his remaining years he was primarily a politician, devotedly anti-Catholic and pro-Nonconformist, though the exact nature of his own religion is characteristically ambiguous and obscure. It was as defender of liberty against foreign and domestic threats—some of which took the forms of religion—and as the advocate of a monarchical state less absolute than the Stuarts had always wanted that he was remembered in the following century, and indeed in Wordsworth's time, though it was in that time also that poets began to look with interest at his lyric poetry.

Marvell's prose work *The Rehearsal Transprosed* of 1672, and its second part (1673)—which, unlike the first, was published under his name—were called by Gilbert Burnet 'the wittiest books that have appeared in this age'. His target was Samuel Parker, a successful parson, later a bishop, who had written in favour of legally enforced religious conformity. Affirming the country's need for more rather than less tolerance, Marvell in a famous passage looked back upon the Civil War. It was no longer worth asking whether it had been fought for religion or liberty; indeed, he said, 'I think the Cause was too good to have been fought for. Men ought to have trusted God; they ought and might have trusted the King with the whole matter.' It seems unlikely that he had felt the same way in 1650. Now, many years later, he appears to be saying that both sides had been wrong, that the king, being still the centre and source of power, ought himself to have prevented the conflict.

The interest of the remainder of Marvell's life is largely political. He was probably under government surveillance, for he was strong in opposition, and even invited, or stumbled into, a brawl with the Speaker of the House, who censured him for his unparliamentary conduct. In 1677 he wrote, anonymously, *An Account of the Growth of Popery and Arbitrary Government*, against the supposed plot of English Catholics, abetted by Louis XIV, to catholicize England—incidentally taking every chance of attacking ministerial and parliamentary corruption. (This was the year before the Popish Plot.) He seems to have had a hot temper, yet he was on occasion devious and often wrote anonymously. As remarked earlier, the government offered a reward for information about the authorship of *The Growth of Popery* and had in their hands an informer's report naming Marvell when he died in 1678 (poisoned—it was rumoured—by the Jesuits). He died suddenly after a trip to Hull, and was doubtless immersed in these and other dangerous troubles.

There is some mystery about the publication in 1681 of the *Miscel-*

*laneous Poems* 'by Andrew Marvell, Esq; late Member of the Honourable House of Commons'. The book is prefaced by a Note to the Reader: 'These are to certify every ingenious reader, that all these poems, as also the other things in this book contained, are printed according to the exact copies of my late dear husband, under his own hand-writing, being found since his death among his other papers, witness my hand this 15th day of October, 1680.' Despite the energetic advocacy of the late Sir William Empson, Mary Palmer's claim to have been married to Marvell is not generally believed. The poet was associated with a business founded in 1671 by Richard Thompson, a cousin, and Edward Nelthorpe, a more remote kinsman. Another cousin, Edward Thompson, was also involved. The firm was bankrupted in 1676, having lost, according to some accounts, the very large sum of £60,000. Marvell took lodgings in Great Russell Street for his associates to hide from the creditors. But he and Nelthorpe died within a month of each other, and the fiction, if such it was, that Mary Palmer was Marvell's wife seems to have been part of a trick to ensure that a bond of £500 entered into by the poet should not be forfeited to the creditors. Marvell's estate was apparently extremely small, and £500 would be useful. Mary Palmer's deposition maintained that Marvell had made her promise not to reveal their marriage, and she complied by behaving more like a servant than a wife. Empson's long, lively, detailed, and quixotic essay is an attempt to confute the standard account. He believed Mary was speaking the truth: the marriage, though clandestine and probably performed at a church specializing in such unions, was nevertheless valid, its object being, as in all such cases, 'to delay the announcement, not to hide the truth for ever'. However, the volume of the relevant church register needed to prove this theory is missing. Empson further speculates that Marvell was homosexual and that he must have been relieved to discover, somewhat belatedly, that he enjoyed sex with a woman.[3] However that may be, Thomas Cooke, who had talked to the poet's nieces, affirmed in the short biography included in his 1726 edition that Marvell 'never married', and Tupper's evidence, though damaged a little by Empson's genial ferocity, seems strong enough to confirm that view.

It would seem then that the ambiguities, the uncommittedness, of the poems reflected some of the same qualities in Marvell's life. John Aubrey described him as 'of middling stature, pretty strong set, roundish faced, cherry cheeked, hazel eye . . . he was in his conversation very modest, and of very few words; Though he loved wine he would never drink hard in company; and was wont to say, "that he would not play the good-fellow in any man's company in whose hands he would not trust his life." ' Aubrey also suggests that Marvell used drink as a stimulus to poetry. He sounds like a man often morose and always private. He had friends, including Harrington and Milton,[4] and he had a family and business associates; but on the whole we may think of him as a reserved, passionate man. Some of his political poetry deserves to be called odious, and no doubt there were equally reprehensible passages in his political dealings generally; but on the whole the respect in which he came to be held as a defender of liberty and rational government was justified.

Although a handful of his lyrics was republished in Tonson's various miscellanies early in the eighteeth century, only the satires achieved immediate posthumous celebrity. The two eighteenth-century editions were largely undertaken to ensure the survival of his reputation as a patriot. It was in the nineteenth century, in America as well as in England, that the emphasis altered, but full acceptance of the 'metaphysical' poetry came when the prejudice against 'conceited' or 'metaphysical' poetry was overthrown in the twentieth century. The foundation of much modern criticism is T. S. Eliot's tercentenary essay of 1921, still in many respects the finest single essay on Marvell. The first Oxford edition, Margoliouth's, followed in 1927 (2nd edn. 1952; 3rd edn. revised by Pierre Legouis and Elsie Duncan-Jones, 1971) and Pierre Legouis' vast doctoral study, *André Marvell: Poète, Puritain, Patriote*, in 1928 (there is a shorter version in English, published in 1965). The last two decades have seen many additions to the understanding of Marvell's politics as well as to that of his poetry, not least in the numerous and various contributions of E. E. Duncan-Jones, now the doyenne of Marvell scholarship. Some modern studies are included in our select list of Further Reading (p. 355): to have admitted more would have defeated our purpose, for the volume of criticism and scholarship on Marvell is now bewildering and continues to grow. It seems there is always something more to be said of Marvell.

[1] Michael Wilding, 'Marvell's "An Horatian Ode upon Cromwell's Return from Ireland", the Levellers, and the Junta', *Dragons Teeth: Literature of the English Revolution* (Oxford, 1987), 114–37.

[2] 'Andrew Marvell: Life and Times', *Andrew Marvell*, ed. R. L. Brett (Oxford, 1979), 1–35.

[3] 'The Marriage of Marvell', *Using Biography* (1984), 43–95. The standard view is based on F. S. Tupper, 'Mary Palmer, *alias* Mrs Andrew Marvell', *PMLA*, 53 (1938), 367–92.

[4] Christopher Hill, 'Milton and Marvell', *Approaches to Marvell*, ed. C. A. Patrides (1978), 1–30.

# CHRONOLOGY

| | |
|---|---|
| 1621 | Andrew Marvell born (31 March), at Winestead in Holderness, Yorkshire. |
| 1624 | Revd Andrew Marvell, the poet's father, appointed Master of the Hull Charterhouse. |
| 1629–33 | Marvell probably at Hull Grammar School. |
| 1633 | Sizar at Trinity College, Cambridge; later (1638) elected to a scholarship. |
| 1639 | Graduates BA. |
| 1641 | Revd Andrew Marvell drowned. |
| 1642 | Marvell moves to London. |
| 1642/3–47 | Marvell travels in Holland, France, Italy, and Spain. |
| 1647 | Marvell sells his property in Meldreth, Cambridgeshire. |
| 1648 | 'Elegy on Villiers' published. |
| 1649 | 'To his noble Friend Richard Lovelace' and 'Upon the death of Hastings' published. |
| 1650 | Writes 'An Horatian Ode' and 'Tom May's Death'. |
| 1651–2 | Tutor to Mary Fairfax at Nun Appleton, Yorkshire. |
| 1653 | Milton recommends Marvell for the post of Latin Secretary. Marvell tutor to Cromwell's ward William Dutton in Eton. |
| 1654 | 'The First Anniversary' published. |
| 1656 | Marvell at Saumur travelling with William Dutton. |
| 1657 | Writes 'On the Victory obtained by Blake over the Spaniards'. Latin Secretary. Celebrates wedding of Cromwell's daughter Mary. |
| 1658 | 'A Poem upon the Death of his late Highness the Lord Protector'. |
| 1659 | Elected MP for Hull. |
| 1660 | Re-elected MP for Hull (until 1678). |
| 1662–3 | In Holland on state business on behalf of the Earl of Carlisle. |
| 1663–5 | Secretary to the Earl of Carlisle in Russia, Sweden, and Denmark. |
| 1667 | Writes 'Clarendon's Housewarming' and 'The Last Instructions to a Painter'. |
| 1672–3 | In controversy with Samuel Parker. Publishes *The Rehearsal Transprosed* Part 1 (anonymously) and Part 2 (acknowledged). |

| | |
|---|---|
| 1674 | Commendatory verses on the second edition of *Paradise Lost*. |
| 1676 | Publishes *Mr Smirke*. |
| 1677 | *An Account of the Growth of Popery* published anonymously. |
| 1678 | Dies (16 August) in his house in Great Russell Street. |
| 1681 | *Miscellaneous Poems* published. |

# NOTE ON THE TEXT

THIS edition contains those English poems in *Miscellaneous Poems* (1681), which we are reasonably sure are by Marvell, a few poems separately published, those satires from after 1660 most generally accepted as Marvell's, the first part of *The Rehearsal Transprosed* (1672), and the mock *King's Speech*.

The texts of the poems are modernized versions of those given in H. M. Margoliouth's standard Oxford edition (3rd edn., revised by Pierre Legouis with the collaboration of E. E. Duncan-Jones, 1971), though the originals have also been consulted.

The degree sign (°) indicates a note at the end of the book. More general headnotes are not cued.

## *An Elegy upon the Death of my Lord Francis Villiers*

'Tis true that he is dead: but yet to choose,
Methinks thou, Fame, should not have brought the news;
Thou canst discourse at will and speak at large:
But wast not in the fight nor durst thou charge;
While he transported all with valiant rage
His name eternized, but cut short his age;
On the safe battlements of Richmond's bowers
Thou wast espied, and from the gilded towers
Thy silver trumpets sounded a retreat
Far from the dust and battle's sulphury heat. 10
Yet what couldst thou have done? 'Tis always late
To struggle with inevitable fate.
Much rather thou, I know, expect'st to tell
How heavy Cromwell gnashed the earth and fell.
Or how slow death far from the sight of day
The long-deceived Fairfax bore away.
But until then, let us young Francis praise:
And plant upon his hearse the bloody bays,°
Which we will water with our welling eyes.
Tears spring not still from spungy cowardice. 20
The purer fountains from the rocks more steep
Distil and stony valour best doth weep.
Besides revenge, if often quenched in tears,
Hardens like steel and daily keener wears.
Great Buckingham, whose death doth freshly strike
Our memories, because to this so like.
Ere that in the eternal court he shone,
And here a favourite, there found a throne,
The fatal night before he hence did bleed,
Left to his princess this immortal seed,° 30
As the wise Chinese in the fertile womb
Of earth doth a more precious clay entomb,°
Which dying, by his will he leaves consigned:
Till by mature delay of time refined
The crystal metal fit to be released
Is taken forth to crown each royal feast:

Such was the fate by which this posthume breathed,
Who scarcely seems begotten but bequeathed.
  Never was any human plant that grew
More fair than this and acceptably new.° 40
'Tis truth that beauty doth most men dispraise:°
Prudence and valour their esteem do raise.
But he that hath already these in store,
Can not be poorer sure for having more.
And his unimitable handsomeness
Made him indeed be more than man, not less.
We do but faintly God's resemblance bear
And like rough coins of careless mints appear:
But he of purpose made, did represent
In a rich medal every lineament. 50
  Lovely and admirable as he was,
Yet was his sword or armour all his glass.
Nor in his mistress' eyes that joy he took,
As in an enemy's himself to look.
I know how well he did, with what delight
Those serious imitations of fight.
Still in the trials of strong exercise
His was the first, and his the second prize.
  Bright lady, thou that rulest from above
The last and greatest monarchy of love: 60
Fair Richmond, hold thy brother or he goes.°
Try if the jasmine of thy hand or rose
Of thy red lip can keep him always here.
For he loves danger and doth never fear.
Or may thy tears prevail with him to stay?
  But he, resolved, breaks carelessly away.
Only one argument could now prolong
His stay and that most fair and so most strong:
The matchless Clora whose pure fires did warm°
His soul and only could his passions charm. 70
  You might with much more reason go reprove
The amorous magnet which the north doth love.
Or preach divorce, and say it is amiss
That with tall elms the twining vines should kiss,
Than chide two such so fit, so equal fair
That in the world they have no other pair,
Whom it might seem that heaven did create

To restore man unto his first estate.
Yet she for honour's tyrannous respect
Her own desires did, and his neglect. 80
And like the modest plant at every touch°
Shrunk in her leaves and feared it was too much.
  But who can paint the torments and that pain
Which he professed and now she could not feign?°
He like the sun but overcast and pale:
She like a rainbow, that ere long must fail,
Whose roseal cheek where heaven itself did view
Begins to separate and dissolve to dew.
  At last he leave obtains though sad and slow,
First of her and then of himself to go. 90
How comely and how terrible he sits
At once, and war as well as love befits!
Ride where thou wilt and bold adventures find:
But all the ladies are got up behind.
Guard them, though not thyself: for in thy death
The eleven thousand virgins lose their breath.
  So Hector issuing from the Trojan wall
The sad Iliads to the gods did call,
With hands displayed and with dishevelled hair,
That they the empire in his life would spare,° 100
While he secure through all the field doth spy
Achilles, for Achilles only cry.
Ah, ignorant that yet ere night he must
Be drawn by him inglorious through the dust.
  Such fell young Villiers in the cheerful heat
Of youth: his locks entangled all with sweat
And those eyes which the sentinel did keep
Of love, closed up in an eternal sleep.
While Venus of Adonis thinks no more
Slain by the harsh tusk of the savage boar. 110
Hither she runs and hath him hurried far
Out of the noise and blood, and killing war:
Where in her gardens of sweet myrtle laid
She kisses him in the immortal shade.
  Yet died he not revengeless: much he did
Ere he could suffer. A whole pyramid
Of vulgar bodies he erected high:
Scorning without a sepulchre to die.

And with his steel which did whole troops divide
He cut his epitaph on either side. 120
Till finding nothing to his courage fit
He rid up last to death and conquered it.
Such are the obsequies to Francis own:°
He best the pomp of his own death hath shown.
And we hereafter to his honour will
Not write so many, but so many kill.
Till the whole army by just vengeance come
To be at once his trophy and his tomb.

## *To his Noble Friend Mr Richard Lovelace, upon his Poems*

Sir,
Our times are much degenerate from those
Which your sweet muse with your fair fortune chose,°
And as complexions alter with the climes,
Our wits have drawn the infection of our times.
That candid age no other way could tell°
To be ingenious, but by speaking well.
Who best could praise had then the greatest praise,
'Twas more esteemed to give than wear the bays:
Modest ambition studied only then
To honour not herself but worthy men. 10
These virtues now are banished out of town,
Our Civil Wars have lost the civic crown.°
He highest builds, who with most art destroys,
And against others' fame his own employs.
I see the envious caterpillar sit
On the fair blossom of each growing wit.
The air's already tainted with the swarms
Of insects which against you rise in arms:
Word-peckers, paper-rats, book-scorpions,
Of wit corrupted, the unfashioned sons.° 20
The barbed censurers begin to look
Like the grim consistory on thy book;°
And on each line cast a reforming eye,

Severer than the young presbytery.°
Till when in vain they have thee all perused,
You shall, for being faultless, be accused.
Some reading your *Lucasta* will allege
You wronged in her the House's privilege.°
Some that you under sequestration are,°
Because you writ when going to the war,° 30
And one the book prohibits, because Kent
Their first petition by the author sent.°
But when the beauteous ladies came to know
That their dear Lovelace was endangered so:
Lovelace that thawed the most congealed breast—
He who loved best and them defended best,
Whose hand so rudely grasps the steely brand,
Whose hand so gently melts the lady's hand—
They all in mutiny though yet undressed
Sallied, and would in his defence contest. 40
And one, the loveliest that was yet e'er seen,
Thinking that I too of the rout had been,
Mine eyes invaded with a female spite,
(She knew what pain 'twould be to lose that sight.)
'O no, mistake not,' I replied, 'for I
In your defence, or in his cause, would die.'
But he, secure of glory and of time,
Above their envy, or mine aid, doth climb.
Him valiant'st men and fairest nymphs approve;
His book in them finds judgement, with you love. 50

## *Upon the death of Lord Hastings*

Go, intercept some fountain in the vein,
Whose virgin-source yet never steeped the plain.
Hastings is dead, and we must find a store
Of tears untouched, and never wept before.
Go, stand betwixt the morning and the flowers;
And, ere they fall, arrest the early showers.
Hastings is dead; and we, disconsolate,
With early tears must mourn his early fate.

Alas, his virtues did his death presage:
Needs must he die, that doth out-run his age. 10
The phlegmatic and slow prolongs his day,°
And on time's wheel sticks like a remora.°
What man is he that hath not heaven beguiled,
And is not thence mistaken for a child?
While those of growth more sudden, and more bold,
Are hurried hence, as if already old.°
For, there above, they number not as here,
But weigh to man the geometric year.°
Had he but at this measure still increased,
And on the Tree of Life once made a feast, 20
As that of Knowledge; what loves had he given
To earth, and then what jealousies to heaven!
But 'tis a maxim of the state, that none,°
Lest he become like them, taste more than one.°
Therefore the democratic stars did rise,
And all that worth from hence did ostracize.°
Yet as some prince, that, for state-jealousy,
Secures his nearest and most loved ally;
His thought with richest triumphs entertains,
And in the choicest pleasures charms his pains: 30
So he, not banished hence, but there confined,
There better recreates his active mind.
Before the crystal palace where he dwells,
The armed angels hold their carousels;°
And underneath, he views the tournaments
Of all these sublunary elements.
But most he doth the eternal book behold,
On which the happy names do stand enrolled;
And gladly there can all his kindred claim,
But most rejoices at his mother's name. 40
The gods themselves cannot their joy conceal,
But draw their veils, and their pure beams reveal:
Only they drooping Hymeneus note,°
Who, for sad purple, tears his saffron coat;
And trails his torches through the starry hall
Reversed at his darling's funeral.°
And Aesculapius, who, ashamed and stern,°
Himself at once condemneth, and Mayern°
Like some sad chemist, who, prepared to reap

The golden harvest, sees his glasses leap.° 50
For, how immortal must their race have stood,
Had Mayern once been mixed with Hastings' blood!
How sweet and verdant would these laurels be,
Had they been planted on that balsam tree!°
  But what could he, good man, although he bruised
All herbs, and them a thousand ways infused?
All he had tried, but all in vain, he saw,
And wept, as we, without redress or law.
For man (alas) is but the heaven's sport;
And art indeed is long, but life is short.° 60

# POEMS PUBLISHED IN *MISCELLANEOUS POEMS*, 1681

---

## *A Dialogue, between the Resolved Soul and Created Pleasure*

Courage, my soul, now learn to wield
The weight of thine immortal shield.
Close on thy head thy helmet bright.
Balance thy sword against the fight.°
See where an army, strong as fair,
With silken banners spreads the air.°
Now, if thou be'st that thing divine,
In this day's combat let it shine:
And show that Nature wants an art
To conquer one resolved heart. 10

PLEASURE

Welcome the creation's guest,
Lord of earth, and heaven's heir.
Lay aside that warlike crest,
And of Nature's banquet share:
Where the souls of fruits and flowers
Stand prepared to heighten yours.°

SOUL

I sup above, and cannot stay
To bait so long upon the way.°

PLEASURE

On these downy pillows lie,
Whose soft plumes will thither fly: 20
On these roses strewed so plain°
Lest one leaf thy side should strain°

SOUL

My gentler rest is on a thought,
Conscious of doing what I ought.

PLEASURE

If thou be'st with perfumes pleased,
Such as oft the gods appeased,
Thou in fragrant clouds shalt show
Like another god below.

SOUL

A soul that knows not to presume
Is heaven's and its own perfume. 30

PLEASURE

Everything does seem to vie
Which should first attract thine eye:
But since none deserves that grace,
In this crystal view *thy* face.°

SOUL

When the creator's skill is prized,
The rest is all but earth disguised.°

PLEASURE

Hark how music then prepares
For thy stay these charming airs;
Which the posting winds recall,°
And suspend the river's fall. 40

SOUL

Had I but any time to lose,
On this I would it all dispose.
Cease, tempter. None can chain a mind
Whom this sweet chordage cannot bind.°

CHORUS

Earth cannot show so brave a sight
As when a single soul does fence
The batteries of alluring sense,
And heaven views it with delight.°
  Then persevere: for still new charges sound:
  And if thou overcom'st, thou shalt be crowned. 50

PLEASURE

All this fair, and soft, and sweet,
  Which scatteringly doth shine,
Shall within one beauty meet,°
  And she be only thine.

SOUL

If things of sight such heavens be,
What heavens are those we cannot see?

PLEASURE

Wheresoe'er thy foot shall go
  The minted gold shall lie,
Till thou purchase all below,
  And want new worlds to buy. 60

SOUL

Were't not a price, who'd value gold?
And that's worth naught that can be sold.

PLEASURE

Wilt thou all the glory have
  That war or peace commend?
Half the world shall be thy slave
  The other half thy friend.

SOUL

What friends, if to my self untrue!
What slaves, unless I captive you!

PLEASURE

Thou shalt know each hidden cause;°
  And see the future time: 70
Try what depth the centre draws;°
  And then to heaven climb.

SOUL

None thither mounts by the degree°
Of knowledge, but humility.

CHORUS

Triumph, triumph, victorious soul;
The world has not one pleasure more:°
The rest does lie beyond the pole,
And is thine everlasting store.

## *On a Drop of Dew*

See how the orient dew,°
Shed from the bosom of the morn
  Into the blowing roses,°
Yet careless of its mansion new;
For the clear region where 'twas born°
  Round in itself encloses:°
  And in its little globe's extent,
Frames as it can its native element.°
  How it the purple flower does slight,
    Scarce touching where it lies, 10
  But gazing back upon the skies,
    Shines with a mournful light,
      Like its own tear,
Because so long divided from the sphere.
  Restless it rolls and unsecure,
    Trembling lest it grow impure,
  Till the warm sun pity its pain,
And to the skies exhale it back again.
    So the soul, that drop, that ray°
Of the clear fountain of eternal day, 20
Could it within the human flower be seen,
    Remembering still its former height,
    Shuns the swart leaves and blossoms green,°
    And recollecting its own light,°
Does, in its pure and circling thoughts, express
The greater heaven in a heaven less.°
    In how coy a figure wound,°
    Every way it turns away:
    So the world excluding round,°
    Yet receiving in the day, 30
    Dark beneath, but bright above,
    Here disdaining, there in love.
  How loose and easy hence to go,
  How girt and ready to ascend,°
  Moving but on a point below,
  It all about does upwards bend.
Such did the manna's sacred dew distil,

White and entire, though congealed and chill,
Congealed on earth: but does, dissolving, run°
Into the glories of the almighty sun. 40

## *The Coronet*

When for the thorns with which I long, too long,
With many a piercing wound,
My saviour's head have crowned,
I seek with garlands to redress that wrong:
Through every garden, every mead,
I gather flowers (my fruits are only flowers),
Dismantling all the fragrant towers°
That once adorned my shepherdess's head.
And now when I have summed up all my store,
Thinking (so I myself deceive) 10
So rich a chaplet thence to weave°
As never yet the king of glory wore:
Alas, I find the serpent old
That, twining in his speckled breast,°
About the flowers disguised does fold,
With wreaths of fame and interest.°
Ah, foolish man, that wouldst debase with them,
And mortal glory, heaven's diadem!
But thou who only couldst the serpent tame,°
Either his slippery knots at once untie; 20
And disentangle all his winding snare;
Or shatter too with him my curious frame,°
And let these wither, so that he may die,
Though set with skill and chosen out with care:
That they, while thou on both their spoils dost tread,
May crown thy feet, that could not crown thy head.°

(14)

## *Eyes and Tears*

1

How wisely Nature did decree,
With the same eyes to weep and see!
That, having viewed the object vain,°
They might be ready to complain.

2

And since the self-deluding sight,
In a false angle takes each height,°
These tears, which better measure all,
Like watery lines and plummets fall.

3

Two tears, which Sorrow long did weigh
Within the scales of either eye, 10
And then paid out in equal poise,
Are the true price of all my joys.

4

What in the world most fair appears,
Yea, even laughter, turns to tears:
And all the jewels which we prize,
Melt in these pendants of the eyes.

5

I have through every garden been,
Amongst the red, the white, the green,
And yet, from all the flowers I saw,
No honey but these tears, could draw. 20

6

So the all-seeing sun each day
Distils the world with chemic ray,
But finds the essence only showers
Which straight in pity back he pours.°

7

Yet happy they whom grief doth bless,
That weep the more, and see the less:
And, to preserve their sight more true,
Bathe still their eyes in their own dew.

8

So Magdalen, in tears more wise°
Dissolved those captivating eyes, 30
Whose liquid chains could flowing meet
To fetter her Redeemer's feet.

9

Not full sails hasting loaden home,
Nor the chaste lady's pregnant womb,
Nor Cynthia teeming shows so fair,°
As two eyes swollen with weeping are.

10

The sparkling glance that shoots desire,
Drenched in these waves does lose its fire.
Yea, oft the thunderer pity takes
And here the hissing lightning slakes. 40

11

The incense was to heaven dear,
Not as a perfume, but a tear.
And stars show lovely in the night,
But as they seem the tears of light.

12

Ope then, mine eyes, your double sluice,
And practise so your noblest use;
For others too can see, or sleep,
But only human eyes can weep.

13

Now. like two clouds dissolving, drop,
And at each tear in distance stop: 50
Now, like two fountains, trickle down;
Now, like two floods o'erturn and drown.

14

Thus let your streams o'erflow your springs,
Till eyes and tears be the same things:
And each the other's difference bears;°
These weeping eyes, those seeing tears.

## *Bermudas*

Where the remote Bermudas ride°
In the ocean's bosom unespied,
From a small boat, that rowed along,
The listening winds received this song.
'What should we do but sing his praise
That led us through the watery maze,
Unto an isle so long unknown,°
And yet far kinder than our own?
Where he the huge sea-monsters wracks,°
That lift the deep upon their backs, 10
He lands us on a grassy stage,
Safe from the storms, and prelate's rage.
He gave us this eternal spring,
Which here enamels everything,
And sends the fowls to us in care,°
On daily visits through the air.
He hangs in shades the orange bright,
Like golden lamps in a green night,°
And does in the pom'granates close
Jewels more rich than Ormus shows.° 20
He makes the figs our mouths to meet,
And throws the melons at our feet,°
But apples plants of such a price,
No tree could ever bear them twice.
With cedars, chosen by his hand,
From Lebanon, he stores the land,
And makes the hollow seas, that roar,
Proclaim the ambergris on shore.°
He cast (of which we rather boast)
The gospel's pearl upon our coast, 30

And in these rocks for us did frame
A temple, where to sound his name.
Oh let our voice his praise exalt,
Till it arrive at heaven's vault:
Which thence (perhaps) rebounding, may
Echo beyond the Mexique Bay."°
Thus sung they, in the English boat,
A holy and a cheerful note,
And all the way, to guide their chime,
With falling oars they kept the time.° 40

## *Clorinda and Damon*

C. Damon, come drive thy flocks this way.
D. No, 'tis too late; they went astray.
C. I have a grassy scutcheon spied,°
Where Flora blazons all her pride.°
The grass I aim to feast thy sheep:
The flowers I for thy temples keep.
D. Grass withers; and the flowers too fade.°
C. Seize the short joys then, ere they vade.°
Seest thou that unfrequented cave?°
D. That den?° 10
C. Love's shrine.
D. But virtue's grave.
C. In whose cool bosom we may lie
Safe from the sun.
D. Not heaven's eye.
C. Near this, a fountain's liquid bell
Tinkles within the concave shell.
D. Might a soul bathe there and be clean,
Or slake its drought?
C. What is't you mean?
D. These once had been enticing things,
Clorinda, pastures, caves, and springs.
C. And what late change?
D. The other day
Pan met me.
C. What did great Pan say?° 20

D. Words that transcend poor shepherds' skill,
But he e'er since my songs does fill:
And his name swells my slender oat.°
C. Sweet must Pan sound in Damon's note.
D. Clorinda's voice might make it sweet.
C. Who would not in Pan's praises meet?

CHORUS

Of Pan the flowery pastures sing,
Caves echo, and the fountains ring.
Sing then while he doth us inspire;
For all the world is our Pan's choir. 30

## *A Dialogue between the Soul and Body*

SOUL

O, who shall from this dungeon raise
A soul, enslaved so many ways,
With bolts of bones, that fettered stands
In feet, and manacled in hands.°
Here blinded with an eye; and there
Deaf with the drumming of an ear.
A soul hung up, as 'twere, in chains
Of nerves, and arteries, and veins.
Tortured, besides each other part,
In a vain head, and double heart?° 10

BODY

O, who shall me deliver whole,
From bonds of this tyrannic soul?°
Which, stretched upright, impales me so,°
That mine own precipice I go;°
And warms and moves this needless frame°
(A fever could but do the same),°
And, wanting where its spite to try,
Has made me live to let me die.
A body that could never rest,
Since this ill spirit it possessed. 20

SOUL

What magic could me thus confine°
Within another's grief to pine,
Where, whatsoever it complain,
I feel, that cannot feel, the pain.°
And all my care itself employs,
That to preserve, which me destroys:°
Constrained not only to endure
Diseases, but, what's worse, the cure:
And ready oft the port to gain,°
Am shipwrecked into health again?° 30

BODY

But physic yet could never reach
The maladies thou me dost teach:°
Whom first the cramp of hope does tear,
And then the palsy shakes of fear;
The pestilence of love does heat,
Or hatred's hidden ulcer eat;
Joy's cheerful madness does perplex,
Or sorrow's other madness vex;
Which knowledge forces me to know,
And memory will not forgo. 40
What but a soul could have the wit
To build me up for sin so fit?
So architects do square and hew,°
Green trees that in the forest grew.°

## *The Nymph complaining for the Death of her Fawn*

The wanton troopers riding by°
Have shot my fawn, and it will die.
Ungentle men! They cannot thrive
To kill thee. Thou ne'er didst alive
Them any harm: alas, nor could
Thy death yet do them any good.
I'm sure I never wished them ill;
Nor do I for all this; nor will:

But if my simple prayers may yet
Prevail with heaven to forget 10
Thy murder, I will join my tears
Rather than fail. But, O my fears!
It cannot die so. Heaven's King°
Keeps register of everything:
And nothing may we use in vain.
E'en beasts must be with justice slain,
Else men are made their deodands.°
Though they should wash their guilty hands
In this warm life-blood, which doth part
From thine, and wound me to the heart, 20
Yet could they not be clean: their stain
Is dyed in such a purple grain,
There is not such another in
The world, to offer for their sin.°
   Unconstant Sylvio, when yet
I had not found him counterfeit,
One morning (I remember well),
Tied in this silver chain and bell,
Gave it to me: nay, and I know
What he said then; I'm sure I do. 30
Said he, 'Look how your huntsman here
Hath taught a fawn to hunt his *dear*.'
But Sylvio soon had me beguiled.
This waxed tame, while he grew wild,
And quite regardless of my smart,
Left me his fawn, but took his heart.°
   Thenceforth I set myself to play
My solitary time away
With this: and very well content,
Could so mine idle life have spent. 40
For it was full of sport; and light
Of foot, and heart; and did invite
Me to its game; it seemed to bless
Itself in me. How could I less
Than love it? O I cannot be
Unkind, t'a beast that loveth me.
   Had it lived long, I do not know
Whether it too might have done so
As Sylvio did: his gifts might be

Perhaps as false or more than he. 50
But I am sure, for ought that I
Could in so short a time espy,
Thy love was far more better than
The love of false and cruel men.
With sweetest milk, and sugar, first
I it at mine own fingers nursed.
And as it grew, so every day
It waxed more white and sweet than they.
It had so sweet a breath! And oft
I blushed to see its foot more soft, 60
And white (shall I say than my hand?)
Nay, any lady's of the land.
It is a wondrous thing, how fleet
'Twas on those little silver feet.
With what a pretty skipping grace,
It oft would challenge me the race:
And when 't had left me far away,
'Twould stay, and run again, and stay.
For it was nimbler much than hinds;
And trod, as on the four winds.° 70
I have a garden of my own
But so with roses overgrown,
And lilies, that you would it guess
To be a little wilderness.
And all the springtime of the year
It only loved to be there.
Among the beds of lilies, I
Have sought it oft, where it should lie;
Yet could not, till itself would rise,
Find it, although before mine eyes. 80
For, in the flaxen lilies' shade,
It like a bank of lilies laid.
Upon the roses it would feed,
Until its lips e'en seemed to bleed:
And then to me 'twould boldly trip,
And print those roses on my lip.
But all its chief delight was still
On roses thus itself to fill:
And its pure virgin limbs to fold
In whitest sheets of lilies cold. 90

Had it lived long, it would have been
Lilies without, roses within.°
  O help! O help! I see it faint:
And die as calmly as a saint.
See how it weeps. The tears do come
Sad, slowly dropping like a gum.
So weeps the wounded balsam: so°
The holy frankincense doth flow.
The brotherless Heliades
Melt in such amber tears as these.° 100
  I in a golden vial will
Keep these two crystal tears; and fill
It till it do o'erflow with mine;
Then place it in Diana's shrine.°
  Now my sweet fawn is vanished to
Whither the swans and turtles go:°
In fair Elysium to endure,
With milk-white lambs, and ermines pure.
O do not run too fast: for I
Will but bespeak thy grave, and die. 110
  First my unhappy statue shall
Be cut in marble; and withal,
Let it be weeping too—but there
The engraver sure his art may spare,
For I so truly thee bemoan,
That I shall weep though I be stone:°
Until my tears, still dropping, wear
My breast, themselves engraving there.
There at my feet shalt thou be laid,°
Of purest alabaster made: 120
For I would have thine image be
White as I can, though not as thee.

(23)

## *Young Love*

1

Come, little infant, love me now,
While thine unsuspected years
Clear thine aged father's brow
From cold jealousy and fears.

2

Pretty, surely, 'twere to see
By young love old time beguiled,°
While our sportings are as free
As the nurse's with the child.

3

Common beauties stay fifteen;°
Such as yours should swifter move, 10
Whose fair blossoms are too green°
Yct for lust, but not for love.

4

Love as much the snowy lamb,
Or the wanton kid, does prize,
As the lusty bull or ram,
For his morning sacrifice.

5

Now then love me: time may take
Thee before thy time away:
Of this need we'll virtue make,
And learn love before we may. 20

6

So we win of doubtful fate;°
And if good she to us meant,
We that good shall antedate,°
Or, if ill, that ill prevent.

7

Thus as kingdoms, frustrating
  Other titles to their crown,
In the cradle crown their king,
  So all foreign claims to drown,

8

So, to make all rivals vain,
  Now I crown thee with my love: 30
Crown me with thy love again,
  And we both shall monarchs prove.

## *To his coy Mistress*

Had we but world enough, and time,
This coyness, lady, were no crime.
We would sit down, and think which way
To walk, and pass our long love's day.
Thou by the Indian Ganges' side
Shouldst rubies find: I by the tide
Of Humber would complain. I would
Love you ten years before the flood:°
And you should, if you please, refuse
Till the conversion of the Jews.° 10
My vegetable love should grow
Vaster than empires, and more slow.
A hundred years should go to praise
Thine eyes, and on thy forehead gaze.
Two hundred to adore each breast:
But thirty thousand to the rest.
An age at least to every part,
And the last age should show your heart:°
For, lady, you deserve this state;
Nor would I love at lower rate. 20
  But at my back I always hear
Time's winged chariot hurrying near:
And yonder all before us lie
Deserts of vast eternity.
Thy beauty shall no more be found;

Nor, in thy marble vault, shall sound
My echoing song: then worms shall try
That long-preserved virginity:
And your quaint honour turn to dust;°
And into ashes all my lust. 30
The grave's a fine and private place,
But none, I think, do there embrace.
  Now, therefore, while the youthful hue
Sits on thy skin like morning dew,°
And while thy willing soul transpires
At every pore with instant fires,°
Now let us sport us while we may;
And now, like amorous birds of prey,
Rather at once our time devour,
Than languish in his slow-chapped power.° 40
Let us roll all our strength, and all
Our sweetness, up into one ball:
And tear our pleasures with rough strife,
Thorough the iron gates of life.°
Thus, though we cannot make our sun
Stand still, yet we will make him run.°

## *The unfortunate Lover*

I

Alas, how pleasant are their days
With whom the infant Love yet plays!
Sorted by pairs, they still are seen
By fountains cool, and shadows green.
But soon these flames do lose their light,
Like meteors of a summer's night:
Nor can they to that region climb,
To make impression upon time.°

2

'Twas in a shipwreck, when the seas°
Ruled, and the winds did what they please, 10
That my poor lover floating lay,
And, ere brought forth, was cast away:

Till at the last the master-wave
Upon the rock his mother drave;
And there she split against the stone,
In a Caesarean section.°

3

The sea him lent these bitter tears
Which at his eyes he always bears:
And from the winds the sighs he bore,°
Which through his surging breast do roar. 20
No day he saw but that which breaks
Through frighted clouds in forked streaks,°
While round the rattling thunder hurled,
As at the funeral of the world.

4

While Nature to his birth presents
This masque of quarrelling elements,°
A numerous fleet of cormorants black,
That sailed insulting o'er the wrack,
Received into their cruel care
Th' unfortunate and abject heir: 30
Guardians most fit to entertain
The orphan of the hurricane.

5

They fed him up with hopes and air,
Which soon digested to despair,
And as one cormorant fed him, still
Another on his heart did bill.°
Thus while they famish him, and feast,
He both consumed, and increased:°
And languished with doubtful breath,°
The amphibium of life and death.° 40

6

And now, when angry heaven would
Behold a spectacle of blood,
Fortune and he are called to play
At sharp before it all the day:°

And tyrant Love his breast does ply
With all his winged artillery,
Whilst he, betwixt the flames and waves,
Like Ajax, the mad tempest braves.°

7

See how he nak'd and fierce does stand,
Cuffing the thunder with one hand, 50
While with the other he does lock,
And grapple, with the stubborn rock:
From which he with each wave rebounds,
Torn into flames, and ragg'd with wounds,
And all he says, a lover dressed°
In his own blood does relish best.

8

This is thc only banncrct°
That ever Love created yet:
Who though, by the malignant stars,
Forced to live in storms and wars;° 60
Yet dying leaves a perfume here,
And music within every ear:°
And he in story only rules,°
In a field sable a lover gules.°

## *The Gallery*

I

Clora, come view my soul, and tell
Whether I have contrived it well.
Now all its several lodgings lie
Composed into one gallery;
And the great arras-hangings, made
Of various faces, by are laid;
That, for all furniture, you'll find
Only your picture in my mind.°

2

Here thou are painted in the dress
Of an inhuman murderess; 10
Examining upon our hearts°
Thy fertile shop of cruel arts:°
Engines more keen than ever yet°
Adorned tyrant's cabinet;°
Of which the most tormenting are
Black eyes, red lips, and curled hair.

3

But, on the other side, th'art drawn
Like to Aurora in the dawn;
When in the East she slumbering lies,
And stretches out her milky thighs; 20
While all the morning choir does sing,
And manna falls, and roses spring;
And, at thy feet, the wooing doves
Sit perfecting their harmless loves.°

4

Like an enchantress here thou showst,
Vexing thy restless lover's ghost;
And, by a light obscure, dost rave°
Over his entrails, in the cave;
Divining thence, with horrid care,
How long thou shalt continue fair; 30
And (when informed) them throwst away,
To be the greedy vulture's prey.

5

But, against that, thou sitst afloat
Like Venus in her pearly boat.
The halcyons, calming all that's nigh,°
Betwixt the air and water fly;
Or, if some rolling wave appears,
A mass of ambergris it bears.°
Nor blows more wind than what may well
Convey the perfume to the smell.° 40

6

These pictures and a thousand more
Of thee my gallery does store°
In all the forms thou canst invent
Either to please me, or torment:
For thou alone to people me,
Art grown a numerous colony;
And a collection choicer far
Than or Whitehall's or Mantua's were.°

7

But, of these pictures and the rest,
That at the entrance likes me best: 50
Where the same posture, and the look
Remains, with which I first was took:
A tender shepherdess, whose hair
Hangs loosely playing in the air,
Transplanting flowers from the green hill,
To crown her head, and bosom fill.

## *The Fair Singer*

1

To make a final conquest of all me,
Love did compose so sweet an enemy,
In whom both beauties to my death agree,
Joining themselves in fatal harmony;
That while she with her eyes my heart does bind,
She with her voice might captivate my mind.

2

I could have fled from one but singly fair:
My disentangled soul itself might save,
Breaking the curled trammels of her hair;°
But how should I avoid to be her slave, 10
Whose subtle art invisibly can wreathe
My fetters of the very air I breathe?

3

It had been easy fighting in some plain,
Where victory might hang in equal choice.
But all resistance against her is vain,
Who has the advantage both of eyes and voice,
And all my forces needs must be undone,
She having gained both the wind and sun.°

## *Mourning*

I

You, that decipher out the fate°
Of human offsprings from the skies,
What mean these infants which of late°
Spring from the stars of Clora's eyes?

2

Her eyes confused, and doubled o'er,
With tears suspended ere they flow,
Seem bending upwards, to restore
To heaven, whence it came, their woe.

3

When, moulding of the watery spheres,°
Slow drops untie themselves away, 10
As if she, with those precious tears,
Would strow the ground where Strephon lay.

4

Yet some affirm, pretending art,
Her eyes have so her bosom drowned,
Only to soften near her heart
A place to fix another wound.

5

And, while vain pomp does her restrain
Within her solitary bower,
She courts herself in amorous rain;
Herself both Danae and the showr.° 20

6

Nay, others, bolder, hence esteem
Joy now so much her master grown,
That whatsoever does but seem
Like grief, is from her windows thrown.

7

Nor that she pays, while she survives,
To her dead love this tribute due,
But casts abroad these donatives,°
At the installing of a new.

8

How wide they dream! The Indian slaves°
That sink for pearl through seas profound 30
Would find her tears yet deeper waves
And not of one the bottom sound.

9

I yet my silent judgement keep,
Disputing not what they believe:
But sure as oft as women weep,
It is to be supposed they grieve.

## *Daphnis and Chloe*

1

Daphnis must from Chloe part:
Now is come the dismal hour
That must all his hopes devour,
All his labour, all his art.

2

Nature, her own sex's foe,
Long had taught her to be coy:
But she neither knew t'enjoy,
Nor yet let her lover go.

3

But with this sad news surprised,
Soon she let that niceness fall, 10
And would gladly yield to all,
So it had his stay comprised.°

4

Nature so herself does use°
To lay by her wonted state,
Lest the world should separate;°
Sudden parting closer glues.

5

He, well-read in all the ways
By which men their siege maintain,
Knew not that the fort to gain,
Better 'twas the siege to raise. 20

6

But he came so full possessed
With the grief of parting thence,
That he had not so much sense
As to see he might be blessed.

7

Till love in her language breathed
Words she never spake before,
But, than legacies no more,°
To a dying man bequeathed.

8

For, alas, the time was spent,
Now the latest minute's run 30
When poor Daphnis is undone,
Between joy and sorrow rent.

9

At that 'Why', that 'Stay, my dear',
His disordered locks he tare;
And with rolling eyes did glare,
And his cruel fate forswear.

10

As the soul of one scarce dead,
With the shrieks of friends aghast,
Looks distracted back in haste,
And then straight again is fled; 40

11

So did wretched Daphnis look,
Frighting her he loved most.°
At the last, this lover's ghost
Thus his leave resolved took.°

12

'Are my hell and heaven joined
More to torture him that dies?
Could departure not suffice,
But that you must then grow kind?

13

'Ah, my Chloe, how have I
Such a wretched minute found, 50
When thy favours should me wound
More than all thy cruelty?

14

'So to the condemned wight°
The delicious cup we fill;
And allow him all he will,
For his last and short delight.

15

'But I will not now begin
Such a debt unto my foe;
Nor to my departure owe
What my presence could not win. 60

16

'Absence is too much alone:°
Better 'tis to go in peace,
Than my losses to increase
By a late fruition.

17

'Why should I enrich my fate?°
'Tis a vanity to wear,
For my executioner,
Jewels of so high a rate.

18

'Rather I away will pine
In a manly stubbornness 70
Than be fatted up express
For the cannibal to dine.

19

'Whilst this grief does thee disarm,
All th' enjoyment of our love
But the ravishment would prove
Of a body dead while warm.

20

'And I parting should appear
Like the gourmand Hebrew dead,°
While with quails and manna fed,°
He does through the desert err.° 80

21

'Or the witch that midnight wakes
For the fern, whose magic weed
In one minute casts the seed,°
And invisible him makes.

22

'Gentler times for love are meant:
Who for parting pleasure strain
Gather roses in the rain,
Wet themselves, and spoil their scent.

23

'Farewell, therefore, all the fruit
Which I could from love receive: 90
Joy will not with sorrow weave,
Nor will I this grief pollute.

24

'Fate, I come, as dark, as sad,
As thy malice could desire;
Yet bring with me all the fire
That Love in his torches had.'

25

At these words away he broke;
As who long has praying lien,
To his headsman makes the sign,
And receives the parting stroke. 100

26

But hence, virgins, all beware:
Last night he with Phlogis slept;
This night for Dorinda kept;
And but rid to take the air.

27

Yet he does himself excuse;
Nor indeed without a cause:
For, according to the laws,
Why did Chloe once refuse?

## *The Definition of Love*

1

My love is of a birth as rare
As 'tis for object strange and high:
It was begotten by Despair
Upon Impossibility.

2

Magnanimous Despair alone
Could show me so divine a thing,
Where feeble Hope could ne'er have flown
But vainly flapped its tinsel wing.

3

And yet I quickly might arrive
Where my extended soul is fixed,° 10
But Fate does iron wedges drive,
And always crowds itself betwixt.

4

For Fate with jealous eye does see
Two perfect loves, nor lets them close:
Their union would her ruin be,
And her tyrannic power depose.

5

And therefore her decrees of steel
Us as the distant Poles have placed,
(Though Love's whole world on us doth wheel)
Not by themselves to be embraced, 20

6

Unless the giddy heaven fall,
And earth some new convulsion tear;
And, us to join, the world should all
Be cramped into a planisphere.°

7

As lines (so loves) oblique may well
Themselves in every angle greet:
But ours so truly parallel,
Though infinite, can never meet.

8

Therefore the love which us doth bind,
But Fate so enviously debars, 30
Is the conjunction of the mind,
And opposition of the stars.°

## *The Picture of Little T.C. in a Prospect of Flowers*

I

See with what simplicity
This nymph begins her golden days!
In the green grass she loves to lie,
And there with her fair aspect tames
The wilder flowers, and gives them names:°
But only with the roses plays;
And them does tell
What colour best becomes them, and what smell.

2

Who can foretell for what high cause
This darling of the gods was born!° 10
Yet this is she whose chaster laws
The wanton Love shall one day fear,
And, under her command severe,
See his bow broke and ensigns torn.
Happy, who can
Appease this virtuous enemy of man!

3

O, then let me in time compound,°
And parley with those conquering eyes;
Ere they have tried their force to wound,
Ere, with their glancing wheels, they drive 20
In triumph over hearts that strive,
And them that yield but more despise.°
Let me be laid,
Where I may see thy glories from some shade.

4

Meantime, whilst every verdant thing
Itself does at thy beauty charm,
Reform the errors of the spring;
Make that the tulips may have share
Of sweetness, seeing they are fair;
And roses of their thorns disarm: 30
But most procure
That violets may a longer age endure.

5

But, O young beauty of the woods,
Whom Nature courts with fruits and flowers,
Gather the flowers, but spare the buds;
Lest Flora angry at thy crime,
To kill her infants in their prime,
Do quickly make the example yours;
And, ere we see,
Nip in the blossom all our hopes and thee.

## *The Match*

1

Nature had long a treasure made
Of all her choicest store;
Fearing, when she should be decayed,
To beg in vain for more.

2

Her orientest colours there,
And essences most pure,
With sweetest perfumes hoarded were,
All, as she thought, secure.

3

She seldom them unlocked, or used,
But with the nicest care; 10
For, with one grain of them diffused,
She could the world repair.

4

But likeness soon together drew
What she did separate lay;
Of which one perfect beauty grew,
And that was Celia.

5

Love wisely had of long foreseen
  That he must once grow old;
And therefore stored a magazine,°
  To save him from the cold. 20

6

He kept the several cells replete
  With nitre thrice refined;
The naphtha's and the sulphur's heat,
  And all that burns the mind.

7

He fortified the double gate,
  And rarely thither came;
For, with one spark of these, he straight
  All Nature could inflame.

8

Till, by vicinity so long,°
  A nearer way they sought; 30
And, grown magnetically strong,
  Into each other wrought.

9

Thus all his fuel did unite
  To make one fire high:
None ever burned so hot, so bright:
  And, Celia, that am I.

10

So we alone the happy rest,
  Whilst all the world is poor,
And have within ourselves possessed
  All Love's and Nature's store. 40

(40)

## *The Mower against Gardens*

Luxurious man, to bring his vice in use,°
  Did after him the world seduce,
And from the fields the flowers and plants allure,
  Where nature was most plain and pure.
He first enclosed within the garden's square
  A dead and standing pool of air,°
And a more luscious earth for them did knead,°
  Which stupefied them while it fed.°
The pink grew then as double as his mind;
  The nutriment did change the kind. 10
With strange perfumes he did the roses taint,
  And flowers themselves were taught to paint.
The tulip, white, did for complexion seek,
  And learned to interline its cheek:
Its onion root they then so high did hold,°
  That one was for a meadow sold.°
Another world was searched, through oceans new,
  To find the *marvel of Peru*.°
And yet these rarities might be allowed
  To man, that sovereign thing and proud, 20
Had he not dealt between the bark and tree,°
  Forbidden mixtures there to see.°
No plant now knew the stock from which it came;
  He grafts upon the wild the tame:°
That the uncertain and adulterate fruit°
  Might put the palate in dispute.
His green seraglio has its eunuchs too,
  Lest any tyrant him outdo.
And in the cherry he does nature vex,
  To procreate without a sex.° 30
'Tis all enforced, the fountain and the grot,
  While the sweet fields do lie forgot:
Where willing nature does to all dispense
  A wild and fragrant innocence:
And fauns and fairies do the meadows till,
  More by their presence than their skill.
Their statues, polished by some ancient hand,

May to adorn the gardens stand:
But howsoe'er the figures do excel,
The gods themselves with us do dwell. 40

## *Damon the Mower*

1

Hark how the Mower Damon sung,
With love of Juliana stung!°
While everything did seem to paint
The scene more fit for his complaint.°
Like her fair eyes the day was fair,
But scorching like his amorous care.
Sharp like his scythe his sorrow was,
And withered like his hopes the grass.

2

'Oh what unusual heats are here,
Which thus our sunburned meadows sear! 10
The grasshopper its pipe gives o'er;
And hamstringed frogs can dance no more.°
But in the brook the green frog wades;
And grasshoppers seek out the shades.
Only the snake, that kept within,
Now glitters in its second skin.

3

'This heat the sun could never raise,
Nor Dog Star so inflame the days.
It from a higher beauty groweth,
Which burns the fields and mower both: 20
Which mads the dog, and makes the sun°
Hotter than his own Phaeton.°
Not July causeth these extremes,
But Juliana's scorching beams.

4

'Tell me where I may pass the fires
Of the hot day, or hot desires.
To what cool cave shall I descend,
Or to what gelid fountain bend?°

Alas! I look for ease in vain,
When remedies themselves complain. 30
No moisture but my tears do rest,
Nor cold but in her icy breast.

5

'How long wilt thou, fair shepherdess,
Esteem me, and my presents less?
To thee the harmless snake I bring,
Disarmed of its teeth and sting;
To thee chameleons, changing hue,
And oak leaves tipped with honey dew.
Yet thou, ungrateful, hast not sought
Nor what they are, nor who them brought.° 40

6

'I am the Mower Damon, known
Through all the meadows I have mown.
On me the morn her dew distills
Before her darling daffodils.
And, if at noon my toil me heat,
The sun himself licks off my sweat.
While, going home, the evening sweet
In cowslip-water bathes my feet.°

7

'What, though the piping shepherd stock
The plains with an unnumbered flock, 50
This scythe of mine discovers wide
More ground than all his sheep do hide.
With this the golden fleece I shear°
Of all these closes every year.°
And though in wool more poor than they,
Yet am I richer far in hay.

8

'Nor am I so deformed to sight,°
If in my scythe I looked right;
In which I see my picture done,
As in a crescent moon the sun. 60

The deathless fairies take me oft
To lead them in their dances soft:
And, when I tune myself to sing,
About me they contract their ring.°

9

'How happy might I still have mowed,
Had not Love here his thistles sowed!
But now I all the day complain,
Joining my labour to my pain;
And with my scythe cut down the grass,
Yet still my grief is where it was: 70
But, when the iron blunter grows,
Sighing, I whet my scythe and woes.'

10

While thus he threw his elbow round,
Depopulating all the ground,
And, with his whistling scythc, docs cut
Each stroke between the earth and root,
The edged steel by careless chance
Did into his own ankle glance;
And there among the grass fell down,
By his own scythe, the Mower mown. 80

11

'Alas!' said he, 'these hurts are slight
To those that die by love's despite.
With shepherd's-purse, and clown's-all-heal,°
The blood I staunch, and wound I seal.
Only for him no cure is found,
Whom Juliana's eyes do wound.
'Tis death alone that this must do:
For Death thou art a Mower too.'

## *The Mower to the Glow-worms*

1

Ye living lamps, by whose dear light°
The nightingale does sit so late,

And studying all the summer night,
Her matchless songs does meditate;

2

Ye country comets, that portend
No war, nor prince's funeral,
Shining unto no higher end
Than to presage the grass's fall;°

3

Ye glow-worms, whose officious flame°
To wandering mowers shows the way, 10
That in the night have lost their aim,
And after foolish fires do stray;°

4

Your courteous lights in vain you waste,
Since Juliana here is come,
For she my mind hath so displaced
That I shall never find my home.

## *The Mower's Song*

1

My mind was once the true survey°
Of all these meadows fresh and gay,
And in the greenness of the grass
Did see its hopes as in a glass;°
When Juliana came, and she
What I do to the grass, does to my thoughts and me.

2

But these, while I with sorrow pine,
Grew more luxuriant still and fine,
That not one blade of grass you spied,
But had a flower on either side; 10
When Juliana came, and she
What I do to the grass, does to my thoughts and me.

3

Unthankful meadows, could you so
A fellowship so true forgo,
And in your gaudy May-games meet,
While I lay trodden under feet?
When Juliana came, and she
What I do to the grass, does to my thoughts and me.

4

But what you in compassion ought,
Shall now by my revenge be wrought: 20
And flowers, and grass, and I and all,
Will in one common ruin fall.
For Juliana comes, and she
What I do to the grass, does to my thoughts and me.

5

And thus, ye meadows, which have been
Companions of my thoughts more green,
Shall now the heraldry become
With which I shall adorn my tomb;
For Juliana comes, and she
What I do to the grass, does to my thoughts and me. 30

## *Ametas and Thestylis making Hay-ropes*

I

AMETAS

Thinkst thou that this love can stand,
Whilst thou still dost say me nay?
Love unpaid does soon disband:
Love binds love as hay binds hay.

2

THESTYLIS

Thinkst thou that this rope would twine
If we both should turn one way?
Where both parties so combine,
Neither love will twist nor hay.

3

AMETAS

Thus you vain excuses find,
Which yourselves and us delay:° 10
And love ties a woman's mind
Looser than with ropes of hay.

4

THESTYLIS

What you cannot constant hope
Must be taken as you may.

5

AMETAS

Then let's both lay by our rope,
And go kiss within the hay.

## *Music's Empire*

1

First was the world as one great cymbal made,
Where jarring winds to infant Nature played.
All music was a solitary sound,
To hollow rocks and murmuring fountains bound.

2

Jubal first made the wilder notes agree;
And Jubal tuned music's jubilee:°
He called the echoes from their sullen cell,°
And built the organ's city where they dwell.

3

Each sought a consort in that lovely place;°
And virgin trebles wed the manly base. 10
From whence the progeny of numbers new
Into harmonious colonies withdrew.°

4

Some to the lute, some to the viol went,
And others chose the cornet eloquent,
These practising the wind, and those the wire,
To sing men's triumphs, or in heaven's choir.°

5

Then music, the mosaic of the air,°
Did of all these a solemn noise prepare:
With which she gained the empire of the ear,
Including all between the earth and sphere. 20

6

Victorious sounds! Yet here your homage do
Unto a gentler conqueror than you:°
Who though he flies the music of his praise,
Would with you heaven's hallelujahs raise.

## *The Garden*

1

How vainly men themselves amaze°
To win the palm, the oak, or bays,°
And their uncessant labours see
Crowned from some single herb or tree,
Whose short and narrow verged shade
Does prudently their toils upbraid,°
While all flowers and all trees do close°
To weave the garlands of repose.

2

Fair Quiet, have I found thee here,
And Innocence, thy sister dear! 10
Mistaken long, I sought you then
In busy companies of men.
Your sacred plants, if here below,
Only among the plants will grow.°
Society is all but rude,
To this delicious solitude.°

3

No white nor red was ever seen°
So amorous as this lovely green.°
Fond lovers, cruel as their flame,
Cut in these trees their mistress' name.° 20
Little, alas, they know, or heed,
How far these beauties hers exceed!
Fair trees! wheres'e'er your barks I wound,
No name shall but your own be found.°

4

When we have run our passion's heat,°
Love hither makes his best retreat.°
The gods, that mortal beauty chase,
Still in a tree did end their race.°
Apollo hunted Daphne so,
Only that she might laurel grow. 30
And Pan did after Syrinx speed,
Not as a nymph, but for a reed.°

5

What wondrous life in this I lead!
Ripe apples drop about my head;
The luscious clusters of the vine
Upon my mouth do crush their wine;
The nectarine, and curious peach,
Into my hands themselves do reach;
Stumbling on melons, as I pass,
Ensnared with flowers, I fall on grass.° 40

6

Meanwhile the mind, from pleasure less,°
Withdraws into its happiness:
The mind, that ocean where each kind°
Does straight its own resemblance find,°
Yet it creates, transcending these,
Far other worlds, and other seas,°
Annihilating all that's made
To a green thought in a green shade.°

7

Here at the fountain's sliding foot,°
Or at some fruit-tree's mossy root, 50
Casting the body's vest aside,
My soul into the boughs does glide:
There like a bird it sits, and sings,
Then whets, and combs its silver wings;°
And, till prepared for longer flight,°
Waves in its plumes the various light.°

8

Such was that happy garden-state,
While man there walked without a mate:°
After a place so pure, and sweet,
What other help could yet be meet!° 60
But 'twas beyond a mortal's share
To wander solitary there:
Two paradises 'twere in one
To live in paradise alone.

9

How well the skilful gardener drew
Of flowers and herbs this dial new,°
Where from above the milder sun
Does through a fragrant zodiac run;°
And, as it works, the industrious bee
Computes its time as well as we.° 70
How could such sweet and wholesome hours
Be reckoned but with herbs and flowers!

## *The second Chorus from Seneca's Tragedy 'Thyestes'* °

Climb at court for me that will
Tottering favour's pinnacle;°
All I seek is to lie still.
Settled in some secret nest,
In calm leisure let me rest,
And far off the public stage
Pass away my silent age.
Thus when without noise, unknown,

I have lived out all my span,
I shall die, without a groan, 10
An old honest country man.
Who exposed to others' eyes,
Into his own heart ne'er pries,
Death to him's a strange surprise.

## *An Epitaph upon Frances Jones*

Enough: and leave the rest to Fame.
'Tis to commend her but to name.
Courtship, which living she declined,
When dead to offer were unkind.
Where never any could speak ill,
Who would officious praises spill?
Nor can the truest wit or friend,
Without detracting, her commend.
To say she lived a virgin chaste,
In this age loose and all unlaced; 10
Nor was, when vice is so allowed,
Of virtue or ashamed, or proud;
That her soul was on heaven so bent;
No minute but it came and went;°
That ready her last debt to pay
She summed her life up every day;
Modest as morn; as midday bright;
Gentle as evening; cool as night;
'Tis true: but all so weakly said,
'Twere more significant, *she's dead.* 20

## *Upon the Hill and Grove at Bilbrough*

### TO THE LORD FAIRFAX

I

See how the arched earth does here
Rise in a perfect hemisphere!

The stiffest compass could not strike
A line more circular and like;
Nor softest pencil draw a brow°
So equal as this hill does bow.
It seems as for a model laid,°
And that the world by it was made.

2

Here learn, ye mountains more unjust,
Which to abrupter greatness thrust, 10
That do with your hook-shouldered height
The earth deform and heaven fright,
For whose excrescence ill-designed,°
Nature must a new centre find,°
Learn here those humble steps to tread,
Which to securer glory lead.

3

See what a soft access and wide
Lies open to its grassy side;
Nor with the rugged path deters
The feet of breathless travellers. 20
See then how courteous it ascends,
And all the way it rises bends;
Nor for itself the height does gain,
But only strives to raise the plain.

4

Yet thus it all the field commands,
And in unenvied greatness stands,
Discerning further than the cliff
Of heaven-daring Tenerife.°
How glad the weary seamen haste°
When they salute it from the mast! 30
By night the Northern Star their way
Directs, and this no less by day.

5

Upon its crest this mountain grave
A plume of aged trees does wave.°
No hostile hand durst ere invade
With impious steel the sacred shade.°

For something always did appear
Of the great master's terror there:°
And men could hear his armour still
Rattling through all the grove and hill. 40

6

Fear of the master, and respect
Of the great nymph, did it protect,
Vera the nymph that him inspired,°
To whom he often here retired,
And on these oaks engraved her name;
Such wounds alone these woods became:
But ere he well the barks could part
'Twas writ already in their heart.

7

For they ('tis credible) have sense,
As we, of love and reverence, 50
And underneath the coarser rind
The genius of the house do bind.
Hence they successes seem to know,
And in their lord's advancement grow;
But in no memory were seen,
As under this so straight and green;°

8

Yet now no further strive to shoot,
Contented if they fix their root.
Nor to the wind's uncertain gust,
Their prudent heads too far entrust. 60
Only sometimes a fluttering breeze
Discourses with the breathing trees,
Which in their modest whispers name
Those acts that swelled the cheek of fame.

9

'Much other groves', say they, 'than these
And other hills him once did please.
Through groves of pikes he thundered then,
And mountains raised of dying men.

For all the civic garlands due
To him our branches are but few. 70
Nor are our trunks enow to bear
The trophies of one fertile year.'

10

'Tis true, ye trees, nor ever spoke°
More certain oracles in oak.°
But peace, (if you his favour prize):
That courage its own praises flies.
Therefore to your obscurer seats
From his own brightness he retreats:
Nor he the hills without the groves,
Nor height, but with retirement, loves. 80

## *Upon Appleton House*

### TO THE LORD FAIRFAX°

1

Within this sober frame expect
Work of no foreign architect,
That unto caves the quarries drew,
And forests did to pastures hew,
Who of his great design in pain
Did for a model vault his brain,°
Whose columns should so high be raised
To arch the brows that on them gazed.°

2

Why should of all things man unruled
Such unproportioned dwellings build? 10
The beasts are by their dens expressed:
And birds contrive an equal nest;°
The low-roofed tortoises do dwell
In cases fit of tortoise shell:
No creature loves an empty space;
Their bodies measure out their place.

3

But he, superfluously spread,
Demands more room alive than dead;
And in his hollow palace goes
Where winds (as he) themselves may lose; 20
What need of all this marble crust
T'impark the wanton mote of dust,°
That thinks by breadth the world t'unite
Though the first builders failed in height?°

4

But all things are composed here
Like Nature, orderly and near:
In which we the dimensions find
Of that more sober age and mind,
When larger-sized men did stoop
To enter at a narrow loop;° 30
As practising, in doors so strait,°
To strain themselves through heaven's gate.

5

And surely when the after age
Shall hither come in pilgrimage,
These sacred places to adore,
By Vere and Fairfax trod before,°
Men will dispute how their extent
Within such dwarfish confines went:
And some will smile at this, as well
As Romulus his bee-like cell.° 40

6

Humility alone designs
Those short but admirable lines,
By which, ungirt and unconstrained,
Things greater are in less contained.
Let others vainly strive t'immure°
The circle in the quadrature!°
These holy mathematics can
In every figure equal man.

7

Yet thus the laden house does sweat,
And scarce endures the master great: 50
But where he comes the swelling hall
Stirs, and the square grows spherical,
More by his magnitude distressed,
Then he is by its straitness pressed:
And too officiously it slights
That in itself which him delights.°

8

So honour better lowness bears,
Than that unwonted greatness wears:
Height with a certain grace does bend,
But low things clownishly ascend. 60
And yet what needs there here excuse,
Where everything does answer use?
Where neatness nothing can condemn,
Nor pride invent what to contemn?°

9

A stately frontispiece of poor°
Adorns without the open door:
Nor less the rooms within commends
Daily new furniture of friends.
The house was built upon the place
Only as for a mark of grace; 70
And for an inn to entertain°
Its lord a while, but not remain.

10

Him Bishop's Hill or Denton may,°
Or Bilbrough, better hold than they:
But Nature here hath been so free
As if she said, 'Leave this to me.'
Art would more neatly have defaced
What she had laid so sweetly waste,
In fragrant gardens, shady woods,
Deep meadows, and transparent floods. 80

11

While with slow eyes we these survey,
And on each pleasant footstep stay,
We opportunely may relate
The progress of this house's fate.
A nunnery first gave it birth
(For virgin buildings oft brought forth);
And all that neighbour-ruin shows
The quarries whence this dwelling rose.

12

Near to this gloomy cloister's gates
There dwelt the blooming virgin Thwaites,° 90
Fair beyond measure, and an heir
Which might deformity make fair.
And oft she spent the summer suns
Discoursing with the subtle nuns.
Whence in these words one to her weaved,
(As 'twere by chance) thoughts long conceived.

13

'Within this holy leisure we
Live innocently, as you see.
These walls restrain the world without,
But hedge our liberty about. 100
These bars enclose that wider den
Of those wild creatures called men.
The cloister outward shuts its gates,
And, from us, locks on them the grates.

14

'Here we, in shining armour white,°
Like virgin Amazons do fight.
And our chaste lamps we hourly trim,
Lest the great bridegroom find them dim.°
Our orient breaths perfumed are
With incense of incessant prayer. 110
And holy-water of our tears
Most strangely our complexion clears.

15

'Not tears of grief; but such as those
With which calm pleasure overflows;
Or pity, when we look on you
That live without this happy vow.
How should we grieve that must be seen
Each one a spouse, and each a queen,
And can in heaven hence behold
Our brighter robes and crowns of gold? 120

16

'When we have prayed all our beads,°
Some one the holy legend reads;°
While all the rest with needles paint
The face and graces of the saint.
But what the linen can't receive
They in their lives do interweave.
This work the saints best represents;
That serves for altar's ornaments.

17

'But much it to our work would add
If here your hand, your face we had: 130
By it we would Our Lady touch;
Yet thus she you resembles much.
Some of your features, as we sewed,
Through every shrine should be bestowed.
And in one beauty we would take
Enough a thousand saints to make.

18

'And (for I dare not quench the fire
That me does for your good inspire)
'Twere sacrilege a man t'admit
To holy things, for heaven fit. 140
I see the angels in a crown°
On you the lilies showering down:
And around about you glory breaks,
That something more than human speaks.

19

'All beauty, when at such a height,
Is so already consecrate.
Fairfax I know; and long ere this
Have marked the youth, and what he is.
But can he such a rival seem
For whom you heaven should disesteem? 150
Ah, no! and 'twould more honour prove
He your *devoto* were, than love.°

20

'Here live beloved, and obeyed:
Each one your sister, each your maid.
And, if our rule seem strictly penned,
The rule itself to you shall bend.
Our abbess too, now far in age,
Doth your succession near presage.
How soft the yoke on us would lie
Might such fair hands as yours it tie! 160

21

'Your voice, the sweetest of the choir,
Shall draw heaven nearer, raise us higher.
And your example, if our head,
Will soon us to perfection lead.
Those virtues to us all so dear,
Will straight grow sanctity when here:
And that, once sprung, increase so fast
Till miracles it work at last.

22

'Nor is our order yet so nice,°
Delight to banish as a vice. 170
Here pleasure piety doth meet;
One perfecting the other sweet.°
So through the mortal fruit we boil
The sugar's uncorrupting oil:
And that which perished while we pull,
Is thus preserved clear and full.

23

'For such indeed are all our arts,
Still handling Nature's finest parts.
Flowers dress the altars; for the clothes,
The sea-born amber we compose;° 180
Balms for the grieved we draw; and pastes°
We mold, as baits for curious tastes.°
What need is here of man? unless
These as sweet sins we should confess.°

24

'Each night among us to your side
Appoint a fresh and virgin bride;
Whom if our Lord at midnight find,
Yet neither should be left behind.
Where you may lie as chaste in bed,
As pearls together billeted, 190
All night embracing arm in arm
Like crystal pure with cotton warm.

25

'But what is this to all the store
Of joys you see, and may make more!
Try but a while, if you be wise:
The trial neither costs, nor ties.'
Now, Fairfax, seek her promised faith:
Religion that dispensed hath,
Which she henceforward does begin;°
The nun's smooth tongue has sucked her in. 200

26

Oft, though he knew it was in vain,
Yet would he valiantly complain.
'Is this that sanctity so great,
An art by which you finelier cheat?
Hypocrite witches, hence avaunt,
Who though in prison yet enchant!
Death only can such thieves make fast,
As rob though in the dungeon cast.

27

'Were there but, when this house was made,
One stone that a just hand had laid, 210
It must have fall'n upon her head
Who first thee from thy faith misled.
And yet, how well soever meant,
With them 'twould soon grow fraudulent;
For like themselves they alter all,
And vice infects the very wall.°

28

'But sure those buildings last not long,
Founded by folly, kept by wrong.
I know what fruit their gardens yield,
When they it think by night concealed. 220
Fly from their vices. 'Tis thy state,°
Not thee, that they would consecrate.
Fly from their ruin. How I fear,
Though guiltless, lest thou perish there.'

29

What should he do? He would respect
Religion, but not right neglect:
For first religion taught him right,
And dazzled not but cleared his sight.
Sometimes resolved, his sword he draws,
But reverenceth then the laws: 230
For justice still that courage led;
First from a judge, then soldier bred.°

30

Small honour would be in the storm.°
The court him grants the lawful form;
Which licensed either peace or force,
To hinder the unjust divorce.
Yet still the nuns his right debarred,
Standing upon their holy guard.
Ill-counselled women, do you know
Whom you resist, or what you do? 240

31

Is not this he whose offspring fierce
Shall fight through all the universe;
And with successive valour try
France, Poland, either Germany;°
Till one, as long since prophesied,°
His horse through conquered Britain ride?
Yet, against fate, his spouse they kept,
And the great race would intercept.°

32

Some to the breach against their foes
Their wooden saints in vain oppose. 250
Another bolder stands at push
With their old holy-water brush.
While the disjointed abbess threads°
The jingling chain-shot of her beads.
But their loudest cannon were their lungs;
And sharpest weapons were their tongues.

33

But waving these aside like flies,
Young Fairfax through the wall does rise.
Then th' unfrequented vault appeared,
And superstitions vainly feared. 260
The relics false were set to view;
Only the jewels there were true—
But truly bright and holy Thwaites
That weeping at the altar waits.

34

But the glad youth away her bears,
And to the nuns bequeaths her tears;
Who guiltily their prize bemoan,
Like gypsies that a child had stolen.°
Thenceforth (as when the enchantment ends,
The castle vanishes or rends) 270
The wasting cloister with the rest
Was in one instant dispossessed.

35

At the demolishing, this seat
To Fairfax fell as by escheat.°
And what both nuns and founders willed
'Tis likely better thus fulfilled.
For if the virgin proved not theirs,
The cloister yet remained hers.
Though many a nun there made her vow,
'Twas no religious house till now. 280

36

From that blest bed the hero came,
Whom France and Poland yet does fame:°
Who, when retired here to peace,
His warlike studies could not cease;
But laid these gardens out in sport
In the just figure of a fort;
And with five bastions it did fence,
As aiming one for every sense.°

37

When in the east the morning ray
Hangs out the colours of the day, 290
The bee through these known alleys hums,
Beating the *dian* with its drums.°
Then flowers their drowsy eyelids raise,
Their silken ensigns each displays,
And dries its pan yet dank with dew,°
And fills its flask with odours new.°

38

These, as their governor goes by,
In fragrant volleys they let fly;
And to salute their governess
Again as great a charge they press: 300
None for the virgin nymph; for she°
Seems with the flowers a flower to be.
And think so still! though not compare°
With breath so sweet, or cheek so fair.

39

Well shot, ye firemen! Oh how sweet,°
And round your equal fires do meet,
Whose shrill report no ear can tell,
But echoes to the eye and smell.
See how the flowers, as at parade,
Under their colours stand displayed: 310
Each regiment in order grows,
That of the tulip, pink, and rose.

40

But when the vigilant patrol
Of stars walks round about the Pole,
Their leaves, that to the stalks are curled,
Seem to their staves the ensigns furled.
Then in some flower's beloved hut
Each bee as sentinel is shut,
And sleeps so too: but, if once stirred,
She runs you through, or asks the word.° 320

41

Oh thou, that dear and happy isle
The garden of the world ere while,°
Thou paradise of four seas,°
Which heaven planted us to please,
But, to exclude the world, did guard
With watery if not flaming sword;°
What luckless apple did we taste,
To make us mortal, and thee waste?°

42

Unhappy! shall we never more
That sweet militia restore,° 330
When gardens only had their towers,
And all the garrisons were flowers,
When roses only arms might bear,
And men did rosy garlands wear?
Tulips, in several colours barred,
Were then the Switzers of our Guard.°

43

The gardener had the soldier's place,
And his more gentle forts did trace.
The nursery of all things green
Was then the only magazine. 340
The winter quarters were the stoves,°
Where he the tender plants removes.
But war all this doth overgrow;
We ordnance plant and powder sow.

44

And yet there walks one on the sod°
Who, had it pleased him and God,
Might once have made our gardens spring
Fresh as his own and flourishing.
But he preferred to the Cinque Ports°
These five imaginary forts, 350
And, in those half-dry trenches, spanned°
Power which the ocean might command.

45

For he did, with his utmost skill,
Ambition weed, but conscience till—
Conscience, that heaven-nursed plant,
Which most our earthy gardens want.°
A prickling leaf it bears, and such
As that which shrinks at every touch;°
But flowers eternal, and divine,
That in the crowns of saints do shine. 360

46

The sight does from these bastions ply,
The invisible artillery;
And at proud Cawood Castle seems°
To point the battery of its beams.
As if it quarrelled in the seat°
The ambition of its prelate great.
But o'er the meads below it plays,
Or innocently seems to gaze.°

47

And now to the abyss I pass
Of that unfathomable grass, 370
Where men like grasshoppers appear,
But grasshoppers are giants there:°
They, in their squeaking laugh, contemn
Us as we walk more low than them:
And, from the precipices tall
Of the green spires, to us do call.

48

To see men through this meadow dive,
We wonder how they rise alive,
As, under water, none does know
Whether he fall through it or go.° 380
But, as the mariners that sound,
And show upon their lead the ground,°
They bring up flowers so to be seen,
And prove they've at the bottom been.

49

No scene that turns with engines strange°
Does oftener than these meadows change.
For when the sun the grass hath vexed,
The tawny mowers enter next;
Who seem like Israelites to be,
Walking on foot through a green sea.° 390
To them the grassy deeps divide,
And crowd a lane to either side.°

50

With whistling scythe, and elbow strong,
These massacre the grass along:
While one, unknowing, carves the rail,°
Whose yet unfeathered quills her fail.
The edge all bloody from its breast
He draws, and does his stroke detest,
Fearing the flesh untimely mowed°
To him a fate as black forebode. 400

51

But bloody Thestylis, that waits°
To bring the mowing camp their cates,°
Greedy as kites, has trussed it up,
And forthwith means on it to sup:
When on another quick she lights,
And cries, 'He called us Israelites;°
But now, to make his saying true,
Rails rain for quails, for manna, dew.'°

52

Unhappy birds! what does it boot
To build below the grass's root;
When lowness is unsafe as height, 410
And chance o'ertakes, what 'scapeth spite?
And now your orphan parents' call
Sounds your untimely funeral.
Death-trumpets creak in such a note,
And 'tis the sourdine in their throat.°

53

Or sooner hatch or higher build:°
The mower now commands the field,
In whose new traverse seemeth wrought°
A camp of battle newly fought: 420
Where, as the meads with hay, the plain
Lies quilted o'er with bodies slain:
The women that with forks it fling,
Do represent the pillaging.

54

And now the careless victors play,
Dancing the triumphs of the hay;°
Where every mower's wholesome heat
Smells like an Alexander's sweat.°
Their females fragrant as the mead
Which they in fairy circles tread:° 430
When at their dance's end they kiss,
Their new-made hay not sweeter is.

55

When after this 'tis piled in cocks,
Like a calm sea it shows the rocks,
We wondering in the river near
How boats among them safely steer.
Or, like the desert Memphis sand,°
Short pyramids of hay do stand.
And such the Roman camps do rise°
In hills for soldiers' obsequies. 440

56

This scene again withdrawing brings°
A new and empty face of things,°
A levelled space, as smooth and plain
As cloths for Lely stretched to stain.°
The world when first created sure
Was such a table rase and pure.°
Or rather such is the *toril*°
Ere the bulls enter at Madril.°

57

For to this naked equal flat,
Which Levellers take pattern at,° 450
The villagers in common chase°
Their cattle, which it closer rase;
And what below the scythe increased°
Is pinched yet nearer by the beast.°
Such, in the painted world, appeared
Davenant with the universal herd.°

58

They seem within the polished grass
A landskip drawn in looking-glass,°
And shrunk in the huge pasture show
As spots, so shaped, on faces do— 460
Such fleas, ere they approach the eye,
In multiplying glasses lie.°
They feed so wide, so slowly move,
As constellations do above.

59

Then, to conclude these pleasant acts,
Denton sets ope its cataracts,°
And makes the meadow truly be
(What it but seemed before) a sea.
For, jealous of its lord's long stay,
It tries t'invite him thus away. 470
The river in itself is drowned,
And isles the astonished cattle round.°

60

Let others tell the paradox,
How eels now bellow in the ox;
How horses at their tails do kick,
Turned as they hang to leeches quick;°
How boats can over bridges sail;
And fishes do the stables scale.
How salmons trespassing are found;
And pikes are taken in the pound.° 480

61

But I, retiring from the flood,
Take sanctuary in the wood,
And, while it lasts, myself embark
In this yet green, yet growing ark,
Where the first carpenter might best°
Fit timber for his keel have pressed.°
And where all creatures might have shares,
Although in armies, not in pairs.

62

The double wood of ancient stocks,
Linked in so thick, a union locks,° 490
It like two pedigrees appears,°
On th' one hand Fairfax, th' other Vere's:
Of whom though many fell in war,°
Yet more to heaven shooting are:
And, as they Nature's cradle decked,
Will in green age her hearse expect.°

63

When first the eye this forest sees
It seems indeed as wood not trees:
As if their neighbourhood so old°
To one great trunk them all did mould. 500
There the huge bulk takes place, as meant
To thrust up a fifth element,°
And stretches still so closely wedged
As if the night within were hedged.

64

Dark all without its knits; within
It opens passable and thin;
And in as loose an order grows,
As the Corinthian porticoes.°
The arching boughs unite between
The columns of the temple green; 510
And underneath the winged choirs
Echo about their tuned fires.°

65

The nightingale does here make choice
To sing the trials of her voice.
Low shrubs she sits in, and adorns
With music high the squatted thorns.
But highest oaks stoop down to hear,
And listening elders prick the ear.
The thorn, lest it should hurt her, draws
Within the skin its shrunken claws. 520

66

But I have for my music found
A sadder, yet more pleasing sound:
The stockdoves, whose fair necks are graced
With nuptial rings, their ensigns chaste;
Yet always, for some cause unknown,
Sad pair unto the elms they moan.°
O why should such a couple mourn,
That in so equal flames do burn!

67

Then as I careless on the bed
Of gelid strawberries do tread, 530
And through the hazels thick espy
The hatching throstle's shining eye,
The heron from the ash's top,
The eldest of its young lets drop,
As if it stork-like did pretend°
That tribute to its lord to send.

68

But most the hewel's wonders are,°
Who here has the holt-felster's care.°
He walks still upright from the root,
Measuring the timber with his foot, 540
And all the way, to keep it clean,
Doth from the bark the woodmoths glean.
He, with his beak, examines well
Which fit to stand and which to fell.

69

The good he numbers up, and hacks,
As if he marked them with the axe.
But where he, tinkling with his beak,
Does find the hollow oak to speak,
That for his building he designs,
And through the tainted side he mines. 550
Who could have thought the tallest oak
Should fall by such a feeble stroke!

70

Nor would it, had the tree not fed
A traitor-worm, within it bred,
(As first our flesh corrupt within
Tempts impotent and bashful sin).
And yet that worm triumphs not long,
But serves to feed the hewel's young,
While the oak seems to fall content,
Viewing the treason's punishment. 560

71

Thus I, easy philosopher,
Among the birds and trees confer.
And little now to make me wants
Or of the fowls, or of the plants:
Give me but wings as they, and I
Straight floating on the air shall fly:
Or turn me but, and you shall see
I was but an inverted tree.°

72

Already I begin to call
In their most learned original: 570
And where I language want, my signs
The bird upon the bough divines;
And more attentive there doth sit
Than if she were with lime-twigs knit.
No leaf does tremble in the wind
Which I, returning, cannot find.

73

Out of these scattered Sibyl's leaves°
Strange prophecies my fancy weaves:
And in one history consumes,
Like Mexique paintings, all the plumes.° 580
What Rome, Greece, Palestine, ere said
I in this light mosaic read.°
Thrice happy he who, not mistook,
Hath read in Nature's mystic book.

74

And see how chance's better wit
Could with a mask my studies hit!°
The oak leaves me embroider all,
Between which caterpillars crawl:
And ivy, with familiar trails,
Me licks, and clasps, and curls, and hales. 590
Under this antic cope I move°
Like some great prelate of the grove.°

75

Then, languishing with ease, I toss
On pallets swoll'n of velvet moss,
While the wind, cooling through the boughs,
Flatters with air my panting brows.
Thanks for my rest, ye mossy banks;
And unto you, cool zephyrs, thanks,
Who, as my hair, my thoughts too shed,°
And winnow from the chaff my head. 600

76

How safe, methinks, and strong, behind
These trees have I encamped my mind:
Where beauty, aiming at the heart,
Bends in some tree its useless dart;
And where the world no certain shot
Can make, or me it toucheth not.
But I on it securely play,
And gall its horsemen all the day.

77

Bind me, ye woodbines, in your twines,
Curl me about, ye gadding vines,° 610
And, oh, so close your circles lace,
That I may never leave this place:
But lest your fetters prove too weak,
Ere I your silken bondage break,
Do you, O brambles, chain me too,
And, courteous briars, nail me through.

78

Here in the morning tie my chain,
Where the two woods have made a lane,
While, like a guard on either side,
The trees before their lord divide; 620
This, like a long and equal thread,
Betwixt two labyrinths does lead.
But where the floods did lately drown,
There at the evening stake me down.

79

For now the waves are fall'n and dried,
And now the meadows fresher dyed,
Whose grass, with moister colour dashed,
Seems as green silks but newly washed.
No serpent new nor crocodile
Remains behind our little Nile,° 630
Unless itself you will mistake,°
Among these meads the only snake.

80

See in what wanton harmless folds
It everywhere the meadow holds;
And its yet muddy back doth lick,
Till as a crystal mirror slick,°
Where all things gaze themselves, and doubt
If they be in it or without.
And for his shade which therein shines,°
Narcissus-like, the sun too pines.° 640

81

Oh what a pleasure 'tis to hedge
My temples here with heavy sedge,
Abandoning my lazy side,
Stretched as a bank unto the tide,
Or to suspend my sliding foot
On th' osier's undermined root,
And in its branches tough to hang,
While at my lines the fishes twang!

82

But now away my hooks, my quills,°
And angles—idle utensils.° 650
The young Maria walks tonight:°
Hide, trifling youth, thy pleasures slight.
'Twere shame that such judicious eyes
Should with such toys a man surprise;
She, that already is the law
Of all her sex, her age's awe.

83

See how loose Nature, in respect
To her, itself doth recollect;
And everything so whisht and fine,°
Starts forthwith to its *bonne mine.*° 660
The sun himself, of her aware,
Seems to descend with greater care;
And lest she see him go to bed,
In blushing clouds conceals his head.

84

So when the shadows laid asleep
From underneath these banks do creep,
And on the river as it flows
With eben shuts begin to close;°
The modest halcyon comes in sight,°
Flying betwixt the day and night; 670
And such a horror calm and dumb,°
Admiring Nature does benumb.

85

The viscous air, wheres'e'er she fly,°
Follows and sucks her azure dye;
The jellying stream compacts below,
If it might fix her shadow so;
The stupid fishes hang, as plain°
As flies in crystal overta'en;
And men the silent scene assist,°
Charmed with the sapphire-winged mist.° 680

86

Maria such, and so doth hush
The world, and through the evening rush.°
No new-born comet such a train
Draws through the sky, nor star new-slain.°
For straight those giddy rockets fail,
Which from the putrid earth exhale,
But by her flames, in heaven tried,
Nature is wholly vitrified.°

87

'Tis she that to these gardens gave
That wondrous beauty which they have; 690
She straightness on the woods bestows;
To her the meadow sweetness owes;
Nothing could make the river be
So crystal pure but only she;
She yet more pure, sweet, straight, and fair,
Than gardens, woods, meads, rivers are.

88

Therefore what first she on them spent,
They gratefully again present:
The meadow, carpets where to tread;
The garden, flow'rs to crown her head; 700
And for a glass, the limpid brook,
Where she may all her beauties look;
But, since she would not have them seen,
The wood about her draws a screen.

89

For she, to higher beauties raised,
Disdains to be for lesser praised.
She counts her beauty to converse
In all the languages as hers;°
Nor yet in those herself employs
But for the wisdom, not the noise; 710
Nor yet that wisdom would affect,
But as 'tis heaven's dialect.

90

Blest nymph! that couldst so soon prevent
Those trains by youth against thee meant:°
Tears (watery shot that pierce the mind);
And signs (love's cannon charged with wind);
True praise (that breaks through all defence);
And feigned complying innocence;
But knowing where this ambush lay,
She 'scaped the safe, but roughest way. 720

91

This 'tis to have been from the first
In a domestic heaven nursed,
Under the discipline severe
Of Fairfax, and the starry Vere;°
Where not one object can come nigh
But pure, and spotless as the eye;
And goodness doth itself entail
On females, if there want a male.

92

Go now, fond sex, that on your face
Do all your useless study place, 730
Nor once at vice your brows dare knit
Lest the smooth forehead wrinkled sit:
Yet your own face shall at you grin,°
Thorough the black-bag of your skin,°
When knowledge only could have filled
And virtue all those furrows tilled.

93

Hence she with graces more divine
Supplies beyond her sex the line;°
And like a sprig of mistletoe
On the Fairfacian oak does grow; 740
Whence, for some universal good,
The priest shall cut the sacred bud,
While her glad parents most rejoice,
And make their destiny their choice.°

94

Meantime, ye fields, springs, bushes, flowers,
Where yet she leads her studious hours,
(Till fate her worthily translates,
And find a Fairfax for our Thwaites),
Employ the means you have by her,
And in your kind yourselves prefer; 750
That, as all virgins she precedes,
So you all woods, streams, gardens, meads.

95

For you, Thessalian Tempe's seat°
Shall now be scorned as obsolete;
Aranjuez, as less, disdained;°
The Bel-Retiro as constrained;°
But name not the Idalian grove—°
For 'twas the seat of wanton love—
Much less the dead's Elysian Fields,
Yet nor to them your beauty yields. 760

96

'Tis not, what once it was, the world,°
But a rude heap together hurled,
All negligently overthrown,
Gulfs, deserts, precipices, stone.
Your lesser world contains the same,
But in more decent order tame;
You, heaven's centre, Nature's lap,
And paradise's only map.

97

But now the salmon-fishers moist
Their leathern boats begin to hoist, 770
And like Antipodes in shoes,°
Have shod their heads in their canoes.°
How tortoise-like, but not so slow,
These rational amphibii go!°
Let's in: for the dark hemisphere
Does now like one of them appear.°

## *Flecknoe, an English Priest at Rome*

Obliged by frequent visits of this man,
Whom as priest, poet, and musician,
I for some branch of Melchizedek took°
(Though he derives himself from 'my Lord Brooke');°
I sought his lodging, which is at the sign
Of The Sad Pelican—subject divine°

For poetry. There, three staircases high—
Which signifies his triple property—°
I found at last a chamber, as 'twas said,
But seemed a coffin set on the stairs' head 10
Not higher than seven, nor larger than three feet;
Only there was nor ceiling, nor a sheet,°
Save that the ingenious door did, as you come,
Turn in, and show to wainscot half the room.°
Yet of his state no man could have complained,
There being no bed where he entertained:
And though within one cell so narrow pent,
He'd stanzas for a whole *appartement*.°
   Straight without further information,°
In hideous verse, he, and a dismal tone,° 20
Begins to exorcise, as if I were
Possessed; and sure the Devil brought me there.
But I, who now imagined myself brought
To my last trial, in a serious thought
Calmed the disorders of my youthful breast,
And to my martyrdom prepared rest.°
Only this frail ambition did remain,
The last distemper of the sober brain,°
That there had been some present to assure
The future ages how I did endure: 30
And how I, silent, turned my burning ear
Towards the verse; and when that could not hear,
Held him the other; and unchanged yet,
Asked still for more, and prayed him to repeat:
Till the tyrant, weary to persecute,
Left off, and tried to allure me with his lute.
   Now as two instruments, to the same key
Being tuned by art, if the one touched be
The other opposite as soon replies,
Moved by the air and hidden sympathies; 40
So while he with his gouty fingers crawls
Over the lute, his murmuring belly calls,
Whose hungry guts to the same straitness twined°
In echo to the trembling strings repined.
   I, that perceived now what his music meant,
Asked civilly if he had eat this Lent.
He answered yes, with such and such an one.

For he has this of generous, that alone
He never feeds, save only when he tries
With gristly tongue to dart the passing flies. 50
I asked if he eat flesh. And he, that was
So hungry that, though ready to say mass,
Would break his fast before, said he was sick,°
And the ordinance was only politic.°
Nor was I longer to invite him scant,°
Happy at once to make him Protestant,
And silent. Nothing now our dinner stayed°
But till he had himself a body made—
I mean till he were dressed: for else so thin
He stands, as if he only fed had been 60
With consecrated wafers: and the host
Hath sure more flesh and blood than he can boast.
This *basso relievo* of a man,°
Who as a camel tall, yet easily can
The needle's eye thread without any stitch,°
(His only impossible is to be rich),°
Lest his too subtle body, growing rare,
Should leave his soul to wander in the air,
He therefore circumscribes himself in rhymes;°
And swaddled in's own papers seven times, 70
Wears a close jacket of poetic buff,
With which he doth his third dimension stuff.
Thus armed underneath, he over all
Does make a primitive *sottana* fall;°
And above that yet casts an antic cloak,°
Worn at the first Council of Antioch,°
Which by the Jews long hid, and disesteemed,
He heard of by tradition, and redeemed.°
But were he not in this black habit decked,
This half-transparent man would soon reflect 80
Each colour that he passed by, and be seen,
As the chameleon, yellow, blue, or green.
He dressed, and ready to disfurnish now°
His chamber, whose compactness did allow
No empty place for complimenting doubt,
But who came last is forced first to go out;
I meet one on the stairs who made me stand,
Stopping the passage, and did him demand.

I answered, 'He is here, Sir; but you see
You cannot pass to him but thorough me.'° 90
He thought himself affronted, and replied,
'I whom the palace never has denied°
Will make the way here;' I said, 'Sir, you'll do
Me a great favour, for I seek to go.'
He gathering fury still made sign to draw;
But himself there closed in a scabbard saw
As narrow as his sword's; and I, that was
Delightful, said, 'There can no body pass°
Except by penetration hither, where°
Two make a crowd; nor can three persons here° 100
Consist but in one substance.' Then, to fit°
Our peace, the priest said I too had some wit.
To prov't, I said, 'The place doth us invite
By its own narrowness, Sir, to unite.'
He asked me pardon; and to make me way
Went down, as I him followed to obey.
But the propitiatory priest had straight
Obliged us, when below, to celebrate
Together our atonement: so increased
Betwixt us two the dinner to a feast. 110

Let it suffice that we could eat in peace;
And that both poems did and quarrels cease
During the table; though my new-made friend
Did, as he threatened, ere 'twere long intend
To be both witty and valiant: I, loath,
Said 'twas too late, he was already both.

But now, alas, my first tormentor came,
Who satisfied with eating, but not tame,
Turns to recite; though judges most severe
After the assize's dinner mild appear, 120
And on full stomach do condemn but few,
Yet he more strict my sentence doth renew,
And draws out of the black box of his breast
Ten quire of paper in which he was dressed.
Yet that which was a greater cruelty
Than Nero's poem, he calls charity:°
And so the pelican at his door hung
Picks out the tender bosom to its young.

Of all his poems there he stands ungirt

Save only two foul copies for his shirt:° 130
Yet these he promises as soon as clean.
But how I loathed to see my neighbour glean
Those papers which he peeled from within
Like white flakes rising from a leper's skin!
More odious than those rags which the French youth
At ordinaries after dinner show'th°
When they compare their chancres and poulains.°
Yet he first kissed them, and after takes pains
To read; and then, because he understood
Not one word, thought and swore that they were good. 140
But all his praises could not now appease
The provoked author, whom it did displease
To hear his verses, by so just a curse,
That were ill made, condemned to be read worse;
And how (impossible) he made yet more
Absurdities in them than were before.
For he his untuned voice did fall or raisc
As a deaf man upon a viol plays,
Making the half points and the periods run
Confuseder than the atoms in the sun. 150
Thereat the poet swelled, with anger full,
And roared out, like Perillus in's own bull:°
'Sir, you read false.' 'That, any one but you,
Should know the contrary.' Whereat, I, now
Made mediator, in my room, said, 'Why,
To say that you read false, Sir, is no lie.'°
Thereat the waxen youth relented straight;
But saw with sad despair that 'twas too late.
For the disdainful poet was retired
Home, his most furious satire to have fired 160
Against the rebel, who, at this struck dead,
Wept bitterly as disinherited.
Who should commend his mistress now? Or who
Praise him? Both difficult indeed to do
With truth. I counselled him to go in time,
Ere the fierce poet's anger turned to rhyme.
He hasted; and I, finding myself free,
As one scaped strangely from captivity,
Have made the chance be painted; and go now
To hang it in Saint Peter's for a vow.° 170

## *An Horatian Ode upon Cromwell's Return from Ireland*

The forward youth that would appear
Must now forsake his muses dear,°
  Nor in the shadows sing
  His numbers languishing.
'Tis time to leave the books in dust,
And oil the unused armour's rust:
  Removing from the wall
  The corslet of the hall.
So restless Cromwell could not cease°
In the inglorious arts of peace, 10
  But through adventurous war
  Urged his active star.
And, like the three-forked lightning, first
Breaking the clouds where it was nursed,
  Did thorough his own side°
  His fiery way divide.
(For 'tis all one to courage high
The emulous or enemy:
  And with such to enclose
  Is more than to oppose.)° 20
Then burning through the air he went,
And palaces and temples rent:
  And Caesar's head at last°
  Did through his laurels blast.°
'Tis madness to resist or blame
The force of angry heaven's flame:
  And, if we would speak true,
  Much to the man is due,
Who, from his private gardens, where
He lived reserved and austere, 30
  As if his highest plot
  To plant the bergamot,°
Could by industrious valour climb
To ruin the great work of time,
  And cast the kingdom old°
  Into another mould.

Though justice against fate complain,
And plead the ancient rights in vain:°
  But those do hold or break
  As men are strong or weak. 40
Nature, that hateth emptiness,
Allows of penetration less:°
  And therefore must make room
  Where greater spirits come.
What field of all the Civil Wars,
Where his were not the deepest scars?°
  And Hampton shows what part
  He had of wiser art,
Where, twining subtile fears with hope,°
He wove a net of such a scope, 50
  That Charles himself might chase
  To Carisbrooke's narrow case:°
That thence the royal actor borne°
The tragic scaffold might adorn:
  While round the armed bands
  Did clap their bloody hands.°
*He* nothing common did or mean
Upon that memorable scene:
  But with his keener eye°
  The axe's edge did try:° 60
Nor called the gods with vulgar spite
To vindicate his helpless right,
  But bowed his comely head
  Down as upon a bed.°
This was that memorable hour
Which first assured the forced power.°
  So when they did design
  The Capitol's first line,
A bleeding head where they begun,
Did fright the architects to run; 70
  And yet in that the State
  Foresaw its happy fate.°
And now the Irish are ashamed
To see themselves in one year tamed:°
  So much one man can do,
  That does both act and know.°

They can affirm his praises best,°
And have, though overcome, confessed
  How good he is, how just,
  And fit for highest trust: 80
Nor yet grown stiffer with command,°
But still in the Republic's hand:°
  How fit he is to sway
  That can so well obey.
He to the Commons' feet presents°
A kingdom, for his first year's rents:
  And, what he may, forbears°
  His fame, to make it theirs:
And has his sword and spoils ungirt,
To lay them at the public's skirt. 90
  So when the falcon high
  Falls heavy from the sky,
She, having killed, no more does search
But on the next green bough to perch,
  Where, when he first does lure,°
  The falconer has her sure.
What may not then our isle presume
While Victory his crest does plume?
  What may not others fear
  If thus he crown each year?° 100
A Caesar, he, ere long to Gaul,
To Italy a Hannibal,
  And to all states not free
  Shall climacteric be.°
The Pict no shelter now shall find°
Within his parti-coloured mind,°
  But from this valour sad°
  Shrink underneath the plaid:
Happy, if in the tufted brake
The English hunter him mistake,° 110
  Nor lay his hounds in near
  The Caledonian deer.
But thou, the Wars' and Fortune's son,
March indefatigably on,
  And for the last effect
  Still keep thy sword erect:°

Besides the force it has to fright
The spirits of the shady night,°
The same arts that did gain
A power, must it maintain.° 120

## *Tom May's Death*

As one put drunk into the packet-boat,°
Tom May was hurried hence and did not know't.
But was amazed on the Elysian side,
And with an eye uncertain, gazing wide,
Could not determine in what place he was,
(For whence, in Stephen's Alley, trees or grass?)°
Nor where The Pope's Head, nor The Mitre lay,°
Signs by which still he found and lost his way.°
At last while doubtfully he all compares,
He saw near hand, as he imagined, Ares.° 10
Such did he seem for corpulence and port,
But 'twas a man much of another sort;
'Twas Ben that in the dusky laurel shade°
Amongst the chorus of old poets layed,
Sounding of ancient heroes, such as were
The subjects' safety, and the rebels' fear,
But how a double-headed vulture eats
Brutus and Cassius, the people's cheats.°
But seeing May, he varied straight his song,
Gently to signify that he was wrong. 20
'Cups more than civil of Emathian wine,°
I sing' (said he) 'and the Pharsalian Sign,
Where the historian of the commonwealth
In his own bowels sheathed the conquering health.'
By this, May to himself and them was come,
He found he was translated, and by whom,°
Yet then with foot as stumbling as his tongue°
Pressed for his place among the learned throng.
But Ben, who knew not neither foe nor friend,°
Sworn enemy to all that do pretend, 30
Rose; more than ever he was seen severe,
Shook his gray locks, and his own bays did tear

At this intrusion. Then with laurel wand—
The awful sign of his supreme command,
At whose dread whisk Virgil himself does quake,
And Horace patiently its stroke does take—
As he crowds in, he whipped him o'er the pate
Like Pembroke at the masque, and then did rate:°
'Far from these blessed shades tread back again
Most servile wit, and mercenary pen, 40
Polydore, Lucan, Alan, Vandal, Goth,°
Malignant poet and historian both,
Go seek the novice statesmen, and obtrude
On them some Roman-cast similitude,
Tell them of liberty, the stories fine,
Until you all grow consuls in your wine;
Or thou, Dictator of the glass, bestow
On him the Cato, this the Cicero,°
Transferring old Rome hither in your talk,
As Bethlem's House did to Loreto walk.° 50
Foul architect, that hadst not eye to see
How ill the measures of these states agree,
And who by Rome's example England lay,
Those but to Lucan do continue May.°
But thee nor ignorance nor seeming good
Misled, but malice fixed and understood.
Because some one than thee more worthy wears°
The sacred laurel, hence are all these tears?
Must therefore all the world be set on flame,
Because a gazette-writer missed his aim?° 60
And for a tankard-bearing muse must we
As for the basket, Guelphs and Ghibellines be?°
When the sword glitters o'er the judge's head,
And fear has coward churchmen silenced,
Then is the poet's time, 'tis then he draws,
And single fights forsaken virtue's cause.
He, when the wheel of empire whirleth back,
And though the world's disjointed axle crack,°
Sings still of ancient rights and better times,°
Seeks wretched good, arraigns successful crimes. 70
But thou, base man, first prostituted hast
Our spotless knowledge and the studies chaste,
Apostatizing from our arts and us,

To turn the chronicler to Spartacus.°
Yet wast thou taken hence with equal fate,°
Before thou couldst great Charles's death relate.
But what will deeper wound thy little mind,
Hast left surviving Davenant still behind,
Who laughs to see in this thy death renewed,
Right Roman poverty and gratitude. 80
Poor poet thou, and grateful senate they,
Who thy last reckoning did so largely pay,°
And with the public gravity would come,
When thou hadst drunk thy last to lead thee home,
If that can be thy home where Spenser lies,
And reverend Chaucer, but their dust does rise
Against thee, and expels thee from their side,
As the eagle's plumes from other birds divide.°
Nor here thy shade must dwell. Return, return,
Where sulphury Phlegethon does ever burn.° 90
Thee Cerberus with all his jaws shall gnash,°
Megaera thee with all her serpents lash.°
Thou riveted unto Ixion's wheel°
Shalt break, and the perpetual vulture feel.°
'Tis just, what torments poets e'er did feign,
Thou first historically shouldst sustain.'
Thus, by irrevocable sentence cast,
May, only Master of these Revels, passed.
And straight he vanished in a cloud of pitch,
Such as unto the Sabbath bears the witch. 100

## *To his worthy Friend Doctor Witty upon his Translation of the 'Popular Errors'*

Sit further, and make room for thine own fame,
Where just desert enrolls thy honoured name—
The good interpreter. Some in this task
Take off the cypress veil, but leave a mask,°
Changing the Latin, but do more obscure
That sense in English which was bright and pure.
So of translators they are authors grown,

For ill translators make the book their own.
Others do strive with words and forced phrase
To add such lustre, and so many rays, 10
That but to make the vessel shining, they
Much of the precious metal rub away.
He is translation's thief that addeth more,
As much as he that taketh from the store
Of the first author. Here he maketh blots
That mends; and added beauties are but spots.°
Celia whose English doth more richly flow°
Than Tagus, purer than dissolved snow,
And sweet as are her lips that speak it, she
Now learns the tongues of France and Italy; 20
But she is Celia still: no other grace
But her own smiles commend that lovely face;
Her native beauty's not Italianated,
Nor her chaste mind into the French translated:
Her thoughts are English, though her sparkling wit
With other language doth them fitly fit.
Translators learn of her: but stay, I slide
Down into error with the vulgar tide;
Women must not teach here: the Doctor doth
Stint them to caudles, almond-milk, and broth.° 30
Now I reform, and surely so will all
Whose happy eyes on thy translation fall.
I see the people hastening to thy book,
Liking themselves the worse the more they look,
And so disliking, that they nothing see
Now worth the liking, but thy book and thee.
And (if I judgement have) I censure right;
For something guides my hand that I must write.
You have translation's statutes best fulfilled,
That handling neither sully nor would gild. 40

## *The Character of Holland*

Holland, that scarce deserves the name of land,
As but the off-scouring of the British sand;
And so much earth as was contributed

By English pilots when they heaved the lead;
Or what by th' ocean's slow alluvion fell°
Of shipwrecked cockle and the mussel shell;
This indigested vomit of the sea
Fell to the Dutch by just propriety.
  Glad then, as miners that have found the ore,
They with mad labour fished the land to shore, 10
And dived as desperately for each piece
Of earth, as if't had been of ambergris,
Collecting anxiously small loads of clay,
Less than what building swallows bear away,
Or than those pills which sordid beetles roll,
Transfusing into them their dunghill soul.
  How did they rivet, with gigantic piles,
Thorough the centre their new-catched miles,°
And to the stake a struggling country bound,
Where barking waves still bait the forced ground,° 20
Building their watery Babel far more high
To reach the sea, than those to scale the sky.
  Yet still his claim the injured ocean laid,
And oft at leap-frog o'er their steeples played:°
As if on purpose it on land had come
To show them what's their *Mare Liberum*.°
A daily deluge over them does boil;
The earth and water play at level-coil;°
The fish ofttimes the burger dispossessed,
And sat not as a meat but as a guest. 30
And oft the tritons and the sea nymphs saw
Whole shoals of Dutch served up for cabillau;°
Or as they over the new level ranged
For pickled herring, pickled *Heeren* changed.°
Nature, it seemed, ashamed of her mistake,
Would throw their land away at duck and drake.°
  Therefore necessity, that first made kings,
Something like government among them brings.
For as with pygmies; who best kills the crane,
Among the hungry, he that treasures grain, 40
Among the blind, the one-eyed blinkard reigns,
So rules among the drowned, he that drains.
Not who first see the rising sun commands,
But who could first discern the rising lands.

Who best could know to pump an earth so leak°
Him they their Lord and country's Father speak.
To make a bank was a great plot of state;
Invent a shovel and be a magistrate.
Hence some small dyke-grave unperceived invades°
The power, and grows, as 'twere, a King of Spades. 50
But for less envy some joint states endures,
Who look like a Commission of the Sewers.
For these Half-anders, half wet, and half dry,°
Nor bear strict service, nor pure liberty.
'Tis probable religion after this
Came next in order, which they could not miss.
How could the Dutch but be converted, when
The Apostles were so many fishermen?
Besides, the waters of themselves did rise,
And, as their land, so them did re-baptize, 60
Though herring for their god few voices missed,
And Poor-John to have been the Evangelist.°
Faith, that could never twins conceive before,
Never so fertile, spawned upon this shore,
More pregnant than their Margaret, that laid down°
For *Hans-in-Kelder* of a whole Hans-town.°
Sure when religion did itself embark,
And from the East would Westward steer its ark,
It struck, and splitting on this unknown ground,
Each one thence pillaged the first piece he found: 70
Hence Amsterdam, Turk-Christian-Pagan-Jew,
Staple of sects and mint of schism grew,
That bank of conscience, where not one so strange
Opinion but finds credit, and exchange.
In vain for catholics ourselves we bear;
The Universal Church is only there.
Nor can civility there want for tillage,
Where wisely for their court they chose a village.°
How fit a title clothes their governors,
Themselves the *Hogs*, as all their subjects *Bores*!° 80
Let it suffice to give their country fame
That it had one Civilis called by name,°
Some fifteen hundred and more years ago;
But surely never any that was so.

See but their mermaids with their tails of fish,
Reeking at church over the chafing dish:°
A vestal turf enshrined in earthen ware
Fumes through the loopholes of a wooden square.°
Each to the temple with these altars tend
But still does place it at her western end,° 90
While the fat steam of female sacrifice
Fills the priest's nostrils and puts out his eyes.
Or what a spectacle the skipper gross,
A water-Hercules butter-coloss,°
Tunned up with all their several towns of *Beer*;°
When staggering upon some land, snick and sneer,°
They try, like statuaries, if they can
Cut out each other's Athos to a man:°
And carve in their large bodies, where they please,
The arms of the United Provinces. 100
But when such amity at home is showed,
What then are their confederacies abroad?
Let this one court'sy witness all the rest:
When their whole navy they together pressed—
Not Christian captives to redeem from bands,
Or intercept the western golden sands—
No, but all ancient rights and leagues must vail,°
Rather than to the English strike their sail;
To whom their weather-beaten province owes
Itself—when as some greater vessel tows 110
A cockboat tossed with the same wind and fate—
We buoyed so often up their sinking state.
Was this *Jus Belli et Pacis?* Could this be°
Cause why their burgomaster of the sea°
Rammed with gun powder, flaming with brand wine,°
Should raging hold his linstock to the mine,°
While, with feigned treaties, they invade by stealth
Our sore new circumcised Commonwealth?°
Yet of his vain attempt no more he sees
Than of case-butter shot and bullet-cheese.° 120
And the torn navy staggered with him home,
While the sea laughed itself into a foam.
'Tis true since that (as fortune kindly sports),°
A wholesome danger drove us to our ports,°
While half their banished keels the tempest tossed,

Half bound at home in prison to the frost:
That ours meantime at leisure might careen,°
In a calm winter, under skies serene,
As the obsequious air and waters rest,
Till the dear halcyon hatch out all its nest.° 130
The Commonwealth doth by its losses grow;
And, like its own seas, only ebbs to flow.
Besides, that very agitation laves,
And purges out the corruptible waves.°
  And now again our armed *Bucentore*°
Doth yearly their sea nuptials restore.
And now the hydra of seven provinces°
Is strangled by our infant Hercules.°
Their tortoise wants its vainly stretched neck;°
Their navy all our conquest or our wreck; 140
Or, what is left, their Carthage overcome
Would render fain unto our better Rome,
Unless our Senate, lest their youth disuse
The war, (but who would?) peace, if begged, refuse.
  For now of nothing may our state despair,
Darling of heaven, and of men the care;
Provided that they be what they have been,
Watchful abroad, and honest still within.
For while our Neptune doth a trident shake,
Steeled with those piercing heads, Deane, Monck, and
  Blake° 150
And while Jove governs in the highest sphere,
Vainly in hell let Pluto domineer.

## *The First Anniversary of the Government under His Highness the Lord Protector.*°

Like the vain curlings of the watery maze,°
Which in smooth streams a sinking weight does raise,
So man, declining always, disappears
In the weak circles of increasing years;
And his short tumults of themselves compose,
While flowing time above his head does close.

Cromwell alone with greater vigour runs,
(Sun-like) the stages of succeeding suns:
And still the day which he doth next restore,
Is the just wonder of the day before. 10
Cromwell alone doth with new lustre spring,
And shines the jewel of the yearly ring.°
'Tis he the force of scattered time contracts,
And in one year the work of ages acts:°
While heavy monarchs make a wide return,
Longer, and more malignant than Saturn:°
And though they all Platonic years should reign,°
In the same posture would be found again.
Their earthy projects under ground they lay,
More slow and brittle than the China clay:° 20
Well may they strive to leave them to their son,
For one thing never was by one king done.
Yet some more active for a frontier town,°
Took in by proxy, begs a false renown;
Another triumphs at the public cost,
And will have won, if he no more have lost;°
They fight by others, but in person wrong,°
And only are against their subjects strong;
Their other wars seem but a feigned contest,
This common enemy is still oppressed;° 30
If conquerors, on them they turn their might;
If conquered, on them they wreak their spite:
They neither build the temple in their days,°
Nor matter for succeeding founders raise;
Nor sacred prophecies consult within,
Much less themselves to perfect them begin;°
No other care they bear of things above,
But with astrologers divine and Jove
To know how long their planet yet reprieves
From the deserved fate their guilty lives: 40
Thus (image-like) a useless time they tell,°
And with vain sceptre strike the hourly bell,
Nor more contribute to the state of things,
Than wooden heads unto the viol's strings.
While indefatigable Cromwell hies,
And cuts his way still nearer to the skies,
Learning a music in the region clear,

To tune this lower to that higher sphere.°
So when Amphion did the lute command,°
Which the god gave him, with his gentle hand,° 50
The rougher stones, unto his measures hewed,
Danced up in order from the quarries rude;
This took a lower, that a higher place,
As he the treble altered, or the bass:
No note he struck, but a new storey was laid,
And the great work ascended while he played.
The listening structures he with wonder eyed,
And still new stops to various time applied:
Now through the strings a martial rage he throws,
And joining straight the Theban tower arose;° 60
Then as he strokes them with a touch more sweet,
The flocking marbles in a palace meet;
But for he most the graver notes did try,
Therefore the temples reared their columns high:
Thus, ere he ceased, his sacred lute creates
Th' harmonious city of the seven gates.°
Such was that wondrous order and consent,
When Cromwell tuned the ruling Instrument,°
While tedious statesmen many years did hack,°
Framing a liberty that still went back,° 70
Whose numerous gorge could swallow in an hour
That island, which the sea cannot devour:
Then our Amphion issues out and sings,
And once he struck, and twice, the powerful strings.
The Commonwealth then first together came,
And each one entered in the willing frame;
All other matter yields, and may be ruled;
But who the minds of stubborn men can build?
No quarry bears a stone so hardly wrought,
Nor with such labour from its centre brought; 80
None to be sunk in the foundation bends,°
Each in the house the highest place contends,
And each the hand that lays him will direct,
And some fall back upon the architect;
Yet all composed by his attractive song,°
Into the animated city throng.
The Commonwealth does through their centres all
Draw the circumference of the public wall;

The crossest spirits here do take their part,
Fastening the contignation which they thwart;° 90
And they, whose nature leads them to divide,
Uphold this one, and that the other side;
But the most equal still sustain the height,
And they as pillars keep the work upright,
While the resistance of opposed minds,°
The fabric (as with arches) stronger binds,
Which on the basis of a senate free,
Knit by the roof's protecting weight, agree.
  When for his foot he thus a place had found,°
He hurls e'er since the world about him round, 100
And in his several aspects, like a star,°
Here shines in peace, and thither shoots a war,
While by his beams observing princes steer,
And wisely court the influence they fear.°
O would they rather by his pattern won
Kiss the approaching, nor yet angry Son;
And in their numbered footsteps humbly tread
The path where holy oracles do lead;°
How might they under such a captain raise
The great designs kept for the latter days!° 110
But mad with reason (so miscalled) of state
They know them not, and what they know not, hate.
Hence still they sing hosanna to the whore,
And her, whom they should massacre, adore:
But Indians, whom they should convert, subdue;°
Nor teach, but traffic with, or burn the Jew.°
  Unhappy princes, ignorantly bred,
By malice some, by error more misled,
If gracious heaven to my life give length,
Leisure to time, and to my weakness strength, 120
Then shall I once with graver accents shake
Your regal sloth, and your long slumbers wake:
Like the shrill huntsman that prevents the east,°
Winding his horn to kings that chase the beast.
  Till then my muse shall hollo far behind°
Angelic Cromwell who outwings the wind,
And in dark nights, and in cold days alone
Pursues the monster thorough every throne:°
Which shrinking to her Roman den impure,

Gnashes her gory teeth; nor there secure. 130
  Hence oft I think if in some happy hour
High grace should meet in one with highest power,
And then a seasonable people still
Should bend to his, as he to heaven's will,
What we might hope, what wonderful effect
From such a wished conjuncture might reflect.
Sure, the mysterious work, where none withstand,
Would forthwith finish under such a hand:
Foreshortened time its useless course would stay,
And soon precipitate the latest day.° 140
But a thick cloud about that morning lies,
And intercepts the beams of mortal eyes,
That 'tis the most which we determine can,
If these the times, then this must be the man.
And well he therefore does, and well has guessed,
Who in his age has always forward pressed:
And knowing not where heaven's choice may light,
Girds yet his sword, and ready stands to fight;
But men, alas, as if they nothing cared,
Look on, all unconcerned, or unprepared; 150
And stars still fall, and still the dragon's tail
Swinges the volumes of its horrid flail.°
For the great justice that did first suspend°
The world by sin, does by the same extend.
Hence that blest day still counterpoised wastes,
The ill delaying what the elected hastes;
Hence landing nature to new seas is tossed,°
And good designs still with their authors lost.
  And thou, great Cromwell, for whose happy birth
A mould was chosen out of better earth; 160
Whose saint-like mother we did lately see°
Live out an age, long as a pedigree;
That she might seem (could we the Fall dispute),
T'have smelled the blossom, and not eat the fruit;
Though none does of more lasting parents grow,
But never any did them honour so,
Though thou thine heart from evil still unstained,
And always hast thy tongue from fraud refrained;
Thou, who so oft through storms of thundering lead
Hast born securely thine undaunted head, 170

Thy breast through poniarding conspiracies,
Drawn from the sheath of lying prophecies;
Thee proof beyond all other force or skill,
Our sins endanger, and shall one day kill.
How near they failed, and in thy sudden fall°
At once assayed to overturn us all.
Our brutish fury struggling to be free,
Hurried thy horses, while they hurried thee,°
When thou hadst almost quit thy mortal cares,
And soiled in dust thy crown of silver hairs. 180
Let this one sorrow interweave among
The other glories of our yearly song.°
Like skilful looms, which through the costly thread
Of purling ore, a shining wave do shed:°
So shall the tears we on past grief employ,
Still as they trickle, glitter in our joy.
So with more modesty we may be true,
And speak, as of the dead, the praises due:
While impious men deceived with pleasure short,
On their own hopes shall find the fall retort. 190
But the poor beasts, wanting their noble guide,
(What could they more?) shrunk guiltily aside.
First winged fear transports them far away,
And leaden sorrow then their flight did stay.
See how they each his towering crest abate,
And the green grass, and their known mangers hate,
Nor through wide nostrils snuff the wanton air,°
Nor their round hooves, or curled manes compare;
With wandering eyes, and restless ears they stood,
And with shrill neighings asked him of the wood. 200
Thou, Cromwell, falling, not a stupid tree,
Or rock so savage, but it mourned for thee:
And all about was heard a panic groan,°
As if that Nature's self were overthrown.
It seemed the earth did from the centre tear;
It seemed the sun was fallen out of the sphere:°
Justice obstructed lay, and reason fooled;
Courage disheartened, and religion cooled.
A dismal silence through the palace went,
And then loud shrieks the vaulted marbles rent, 210
Such as the dying chorus sings by turns,

And to deaf seas, the ruthless tempests mourns,
When now they sink, and now the plundering streams
Break up each deck, and rip the oaken seams.
  But thee triumphant hence the fiery car,
And fiery steeds had borne out of the war,
From the low world, and thankless men above,
Unto the kingdom blest of peace and love:
We only mourned ourselves, in thine ascent,
Whom thou hadst left beneath with mantle rent.° 220
  For all delight of life thou then didst lose,
When to command, thou didst thyself depose;
Resigning up thy privacy so dear,
To turn the headstrong people's charioteer;
For to be Cromwell was a greater thing,
Than aught below, or yet above a king:
Therefore thou rather didst thyself depress,
Yielding to rule, because it made thee less.
  For neither didst thou from the first apply
Thy sober spirit unto things too high, 230
But in thine own fields exercised'st long,
A healthful mind within a body strong;°
Till at the seventh time thou in the skies,
As a small cloud, like a man's hand, didst rise;
Then did thick mists and winds the air deform,
And down at last thou poured'st the fertile storm,
Which to the thirsty land did plenty bring,
But, though forewarned, o'ertook and wet the king.°
  What since he did, a higher force him pushed°
Still from behind, and it before him rushed, 240
Though undiscerned among the tumult blind,
Who think those high decrees by man designed.
'Twas heaven would not that his power should cease,
But walk still middle betwixt war and peace:
Choosing each stone, and poising every weight,
Trying the measures of the breadth and height;
Here pulling down, and there erecting new,
Founding a firm state by proportions true.
  When Gideon so did from the war retreat,
Yet by the conquest of two kings grown great, 250
He on the peace extends a warlike power,
And Israel silent saw him raze the tower;

And how he Succoth's Elders durst suppress,
With thorns and briars of the wilderness.
No king might ever such a force have done;
Yet would not he be Lord, nor yet his son.°
  Thou with the same strength, and a heart as plain,
Didst (like thine olive) still refuse to reign,
Though why should others all thy labour spoil,
And brambles be anointed with thine oil, 260
Whose climbing flame, without a timely stop,
Had quickly levelled every cedar's top?°
Therefore first growing to thyself a law,
Th' ambitious shrubs thou in just time didst awe.°
  So have I seen at sea, when whirling winds,
Hurry the bark, but more the seamen's minds,
Who with mistaken course salute the sand,
And threatening rocks misapprehend for land,
While baleful Tritons to the shipwreck guide,°
And corposants along the tacklings slide,° 270
The passengers all wearied out before,
Giddy, and wishing for the fatal shore,
Some lusty mate, who with more careful eye
Counted the hours, and every star did spy,
The helm does from the artless steersman strain,
And doubles back unto the safer main.
What though a while they grumble discontent,
Saving himself, he does their loss prevent.°
  'Tis not a freedom, that where all command;
Nor tyranny, where one does them withstand: 280
But who of both the bounders knows to lay°
Him as their father must the state obey.
  Thou, and thine house (like Noah's eight) did rest,
Left by the wars' flood on the mountains' crest:
And the large vale lay subject to thy will,
Which thou but as a husbandman wouldst till:
And only didst for others plant the vine
Of liberty, not drunken with its wine.
  That sober liberty which men may have,
That they enjoy, but more they vainly crave: 290
And such as to their parents' tents do press,
May show their own, not see his nakedness.
  Yet such a Chammish issue still does rage,°

The shame and plague both of the land and age,
Who watched thy halting, and thy fall deride,
Rejoicing when thy foot had slipped aside,
That their new king might the fifth sceptre shake,
And make the world, by his example, quake:
Whose frantic army should they want for men
Might muster heresies, so one were ten.° 300
What thy misfortune, they the spirit call,
And their religion only is to fall.°
Oh Mahomet! now couldst thou rise again,°
Thy falling-sickness should have made thee reign,
While Feake and Simpson would in many a tome,°
Have writ the comments of thy sacred foam:
For soon thou mightst have passed among their rant°
Were't but for thine unmoved tulipant;°
As thou must needs have owned them of thy band
For prophecies fit to be *Alcoraned*.° 310

Accursed locusts, whom your king does spit°
Out of the centre of the unbottomed pit;
Wanderers, adulterers, liars, Munser's rest,°
Sorcerers, atheists, Jesuits possessed;
You who the scriptures and the laws deface°
With the same liberty as points and lace;
Oh race most hypocritically strict!
Bent to reduce us to the ancient Pict;
Well may you act the Adam and the Eve;°
Ay, and the serpent too that did deceive. 320

But the great captain, now the danger's o'er,
Makes you for his sake tremble one fit more;
And, to your spite, returning yet alive
Does with himself all that is good revive.

So when first man did through the morning new
See the bright sun his shining race pursue,
All day he followed with unwearied sight,
Pleased with that other world of moving light;
But thought him when he missed his setting beams,
Sunk in the hills, or plunged below the streams. 330
While dismal blacks hung round the universe,
And stars (like tapers) burned upon his hearse:
And owls and ravens with their screeching noise
Did make the funerals sadder by their joys.

His weeping eyes the doleful vigils keep,
Not knowing yet the night was made for sleep:
Still to the west, where he him lost, he turned,
And with such accents as despairing mourned:
'Why did mine eyes once see so bright a ray;
Or why day last no longer than a day?' 340
When straight the sun behind him he descried,
Smiling serenely from the further side.°
  So while our star that gives us light and heat,
Seemed now a long and gloomy night to threat,
Up from the other world his flame he darts,
And princes (shining through their windows) starts,
Who their suspected counsellors refuse,
And credulous ambassadors accuse.
  'Is this', saith one, 'the nation that we read
Spent with both wars, under a captain dead,° 350
Yet rig a navy while we dress us late,
And ere we dine, raze and rebuild their state?°
What oaken forests, and what golden mines!
What mints of men, what union of designs!
(Unless their ships, do, as their fowl proceed
Of shedding leaves, that with their ocean breed).°
Theirs are not ships, but rather arks of war
And beaked promontories sailed from far;
Of floating islands a new hatched nest;
A fleet of worlds, of other worlds in quest; 360
A hideous shoal of wood-leviathans,
Armed with three tier of brazen hurricanes,°
That through the centre shoot their thundering side°
And sink the earth that does at anchor ride.
What refuge to escape them can be found,
Whose watery leaguers all the world surround?°
Needs must we all their tributaries be,
Whose navies hold the sluices of the sea.
The ocean is the fountain of command,
But that once took, we captives are on land. 370
And those that have the waters for their share,
Can quickly leave us neither earth nor air.
Yet if through these our fears could find a pass,
Through double oak, and lined with treble brass,
That one man still, although but named, alarms

More than all men, all navies, and all arms.
Him, all the day, him, in late nights I dread,
And still his sword seems hanging o'er my head.
The nation had been ours, but his one soul
Moves the great bulk, and animates the whole. 380
He secrecy with number hath enchased,°
Courage with age, maturity with haste:
The valiant's terror, riddle of the wise,
And still his falchion all our knots unties.°
Where did he learn those arts that cost us dear?
Where below earth, or where above the sphere?
He seems a king by long succession born,
And yet the same to be a king does scorn.
Abroad a king he seems, and something more,
At home a subject on the equal floor. 390
O could I once him with our title see,
So should I hope that he might die as we.
But let them write his praise that love him best,
It grieves me sore to have thus much confessed.'
Pardon, great Prince, if thus their fear or spite
More than our love and duty do thee right.
I yield, nor further will the prize contend,
So that we both alike may miss our end:
While thou thy venerable head dost raise
As far above their malice as my praise, 400
And as the angel of our commonweal,
Troubling the waters, yearly mak'st them heal.°

## *On the Victory obtained by Blake over the Spaniards in the Bay of Santa Cruz, in the Island of Tenerife, 1657°*

Now does Spain's fleet her spacious wings unfold,
Leaves the new world and hastens for the old:
But though the wind was fair, they slowly swum
Freighted with acted guilt, and guilt to come:°
For this rich load, of which so proud they are,
Was raised by tyranny, and raised for war;

Every capacious galleon's womb was filled,
With what the womb of wealthy kingdoms yield,
The new world's wounded entrails they had tore,
For wealth wherewith to wound the old once more: 10
Wealth which all others' avarice might cloy,
But yet in them caused as much fear as joy.
For now upon the main, themselves they saw—
That boundless empire, where you give the law—
Of winds' and waters' rage, they fearful be,
But much more fearful are your flags to see.
Day, that to those who sail upon the deep,
More wished for, and more welcome is than sleep,
They dreaded to behold, lest the sun's light,
With English streamers, should salute their sight:° 20
In thickest darkness they would choose to steer,
So that such darkness might suppress their fear;
At length theirs vanishes, and fortune smiles;
For they behold the sweet Canary Isles;
One of which doubtless is by nature blessed°
Above both worlds, since 'tis above the rest.
For lest some gloominess might strain her sky,
Trees there the duty of the clouds supply;°
O noble trust which heaven on this isle pours,
Fertile to be, yet never need her showers. 30
A happy people, which at once do gain
The benefits without the ills of rain.
Both health and profit fate cannot deny;
Where still the earth is moist, the air still dry;
The jarring elements no discord know,
Fuel and rain together kindly grow;
And coolness there, with heat doth never fight,
This only rules by day, and that by night.
  Your worth to all these isles, a just right brings,°
The best of lands should have the best of kings.° 40
And these want nothing heaven can afford,
Unless it be—the having you their lord;
But this great want will not a long one prove,
Your conquering sword will soon that want remove.
For Spain had better—she'll ere long confess—
Have broken all her swords, than this one peace,°
Casting that league off, which she held so long,

She cast off that which only made her strong.
Forces and art, she soon will feel, are vain,
Peace, against you, was the sole strength of Spain. 50
By that alone those islands she secures,
Peace made them hers, but war will make them yours.
There the indulgent soil that rich grape breeds,
Which of the gods the fancied drink exceeds;°
They still do yield, such is their precious mould,
All that is good, and are not cursed with gold—
With fatal gold, for still where that does grow,
Neither the soil, not people, quiet know.
Which troubles men to raise it when 'tis ore,
And when 'tis raised, does trouble them much more. 60
Ah, why was thither brought that cause of war,
Kind Nature had from thence removed so far?
In vain doth she those islands free from ill,
If fortune can make guilty what she will.
But whilst I draw that scene, where you ere long,
Shall conquests act, your present are unsung.°
For Santa Cruz the glad fleet takes her way,
And safely there casts anchor in the bay.
Never so many with one joyful cry,
That place saluted, where they all must die. 70
Deluded men! Fate with you did but sport,
You 'scaped the sea, to perish in your port.
'Twas more for England's fame you should die there,
Where you had most of strength, and least of fear.
The peak's proud height the Spaniards all admire,
Yet in their breasts carry a pride much higher.
Only to this vast hill a power is given,
At once both to inhabit earth and heaven.
But this stupendous prospect did not near,
Make them admire, so much as they did fear. 80
For here they met with news, which did produce,
A grief, above the cure of grapes' best juice.
They learned with terror that nor summer's heat,
Nor winter's storms, had made your fleet retreat.
To fight against such foes was vain, they knew,
Which did the rage of elements subdue,
Who on the ocean that does horror give,
To all besides, triumphantly do live.

With haste they therefore all their galleons moor,
And flank with cannon from the neighbouring shore. 90
Forts, lines, and sconces all the bay along,°
They build and act all that can make them strong.
Fond men who know not whilst such works they raise,
They only labour to exalt your praise.
Yet they by restless toil became at length,
So proud and confident of their made strength,
That they with joy their boasting general heard,
Wish then for that assault he lately feared.
His wish he has, for now undaunted Blake,
With winged speed, for Santa Cruz does make. 100
For your renown, his conquering fleet does ride,
O'er seas as vast as is the Spaniards' pride.
Whose fleet and trenches viewed, he soon did say,
'We to their strength are more obliged than they.
Were't not for that, they from their fate would run,
And a third world seek out, our arms to shun.
Those forts, which there so high and strong appear,
Do not so much suppress, as show their fear.
Of speedy victory let no man doubt,
Our worst work's past, now we have found them out. 110
Behold their navy does at anchor lie,
And they are ours, for now they cannot fly.'
This said, the whole fleet gave it their applause,
And all assumes your courage, in your cause.
That bay they enter, which unto them owes,
The noblest wreaths, that victory bestows.
Bold Stayner leads: this fleet's designed by fate,°
To give him laurel, as the last did plate.°
The thundering cannon now begins the fight,
And though it be at noon creates a night. 120
The air was soon after the fight begun,
Far more enflamed by it than by the sun.
Never so burning was that climate known,
War turned the temperate to the torrid zone.
Fate these two fleets between both worlds had brought,
Who fight, as if for both those worlds they fought.
Thousands of ways thousands of men there die,
Some ships are sunk, some blown up in the sky.
Nature ne'er made cedars so high aspire,

As oaks did then, urged by the active fire, 130
Which by quick powder's force, so high was sent,
That it returned to its own element.°
Torn limbs some leagues into the island fly,
Whilst others lower in the sea do lie.
Scarce souls from bodies severed are so far
By death, as bodies there were by the war.
The all-seeing sun, ne'er gazed on such a sight,
Two dreadful navies there at anchor fight.
And neither have or power or will to fly,°
There one must conquer, or there both must die. 140
Far different motives yet engaged them thus,
Necessity did them, but choice did us.
  A choice which did the highest worth express,
And was attended by as high success.
For your resistless genius there did reign,
By which we laurels reaped e'en on the main.
So prosperous stars, though absent to the sense,
Bless those they shine for, by their influence.
  Our cannon now tears every ship and sconce,
And o'er two elements triumphs at once. 150
Their galleons sunk, their wealth the sea does fill—
The only place where it can cause no ill.
  Ah, would those treasures which both Indies have,
Were buried in as large, and deep a grave,
Wars' chief support with them would buried be,
And the land owe her peace unto the sea.
Ages to come your conquering arms will bless,
There they destroy what had destroyed their peace.
And in one war the present age may boast
The certain seeds of many wars are lost. 160
  All the foe's ships destroyed, by sea or fire,
Victorious Blake, does from the bay retire,
His siege of Spain he then again pursues,
And there first brings of his success the news:
The saddest news that e'er to Spain was brought,
Their rich fleet sunk, and ours with laurel fraught,
Whilst fame in every place her trumpet blows,
And tells the world how much to you it owes.

# *Two Songs at the Marriage of the Lord Fauconberg and the Lady Mary Cromwell*

## FIRST SONG

*Chorus Endymion Luna*

CHORUS

The astrologer's own eyes are set,°
And even wolves the sheep forget;
Only this shepherd, late and soon,
Upon this hill outwakes the moon.
Hark how he sings, with sad delight,
Thorough the clear and silent night.

ENDYMION

Cynthia, O Cynthia, turn thine ear,
Nor scorn Endymion's plaints to hear.
As we our flocks, so you command
The fleecy clouds with silver wand. 10

CYNTHIA

If thou a mortal, rather sleep;
Or if a shepherd, watch thy sheep.

ENDYMION

The shepherd, since he saw thine eyes,
And sheep are both thy sacrifice.
Nor merits he a mortal's name,
That burns with an immortal flame.

CYNTHIA

I have enough for me to do,
Ruling the waves that ebb and flow.

ENDYMION

Since thou disdain'st not then to share
On sublunary things thy care; 20
Rather restrain these double seas,
Mine eyes' uncessant deluges.

CYNTHIA

My wakeful lamp all night must move,
Securing their repose above.

ENDYMION

If therefore thy resplendent ray
Can make a night more bright than day,
Shine thorough this obscurer breast,°
With shades of deep despair oppressed.

CHORUS

Courage, Endymion, boldly woo;
Anchises was a shepherd too:° 30
Yet is her younger sister laid
Sporting with him in Ida's shade:
And Cynthia, though the strongest,
Seeks but the honour to have held out longest.

ENDYMION

Here unto Latmos' top I climb:°
How far below thine orb sublime?
O why, as well as eyes to see,
Have I not arms that reach to thee?

CYNTHIA

'Tis needless then that I refuse,
Would you but your own reason use. 40

ENDYMION

Though I so high may not pretend,
It is the same so you descend.

CYNTHIA

These stars would say I do them wrong,
Rivals each one for thee too strong.

ENDYMION

The stars are fixed unto their sphere,
And cannot, though they would, come near.
Less loves set off each other's praise,
While stars eclipse by mixing rays.

CYNTHIA

That cave is dark.

ENDYMION

Then none can spy: 50
Or shine thou there and 'tis the sky.

CHORUS

Joy to Endymion,
For he has Cynthia's favour won.
And Jove himself approves
With his serenest influence their loves.
For he did never love to pair
His progeny above the air;
But to be honest, valiant, wise,
Makes mortals matches fit for deities.

SECOND SONG

*Hobbinol Phyllis Tomalin*

HOBBINOL

Phyllis, Tomalin, away:
Never such a merry day.
For the northern shepherd's son°
Has Menalca's daughter won.°

PHYLLIS

Stay till I some flowers ha' tied
In a garland for the bride.

TOMALIN

If thou wouldst a garland bring,
Phyllis, you may wait the spring:
They ha' chosen such an hour
When *she* is the only flower. 10

PHYLLIS

Let's not then at least be seen
Without each a sprig of green.

HOBBINOL

Fear not; at Menalca's hall
There is bays enough for all.
He, when young, as we did graze,
But when old, he planted bays.

TOMALIN

Here *she* comes; but with a look
Far more catching than my hook.
'Twas those eyes, I now darc swear,
Led our lambs we knew not where. 20

HOBBINOL

Not our lambs' own fleeces are
Curled so lovely as her hair:
Nor our sheep new washed can be
Half so white or sweet as *she*.

PHYLLIS

*He* so looks as fit to keep
Somewhat else than silly sheep.°

HOBBINOL

Come, let's in some carol new
Pay to love and them their due.

ALL

Joy to that happy pair,
Whose hopes united banish our despair. 30
What shepherd could for love pretend,
Whilst all the nymphs on Damon's choice attend?
What shepherdess could hope to wed
Before Marina's turn were sped?°
Now lesser beauties may take place,
And meaner virtues come in play;
While they,
Looking from high,
Shall grace
Our flocks and us with a propitious eye. 40
But what is most, the gentle swain
No more shall need of love complain;
But virtue shall be beauty's hire,°
And those be equal that have equal fire.
Marina yields. Who dares be coy?
Or who despair, now Damon does enjoy?
Joy to that happy pair,
Whose hopes united banish our despair.

## *A Poem upon the Death of his late Highness the Lord Protector*

That Providence which had so long the care
Of Cromwell's head, and numbered every hair,°

Now in itself (the glass where all appears)
Had seen the period of his golden years:°
And thenceforth only did attend to trace
What death might least so fair a life deface.
The people, which what most they fear esteem,
Death when more horrid, so more noble deem,
And blame the last act, like spectators vain,
Unless the prince whom they applaud be slain. 10
Nor fate indeed can well refuse that right
To those that lived in war, to die in fight.
But long his valour none had left that could
Endanger him, or clemency that would.°
And he whom Nature all for peace had made,
But angry heaven unto war had swayed,°
And so less useful where he most desired,
For what he least affected was admired,
Deserved yet an end whose every part,
Should speak the wondrous softness of his heart. 20
To Love and Grief the fatal writ was signed;°
(Those nobler weaknesses of human mind,°
From which those powers that issued the decree,
Although immortal, found they were not free),
That they, to whom his breast still open lies,
In gentle passions should his death disguise:
And leave succeeding ages cause to mourn,
As long as Grief shall weep, or Love shall burn.
Straight does a slow and languishing disease
Eliza, Nature's and his darling, seize.° 30
Her when an infant, taken with her charms,
He oft would flourish in his mighty arms,
And, lest their force the tender burden wrong,
Slacken the vigour of his muscles strong;
Then to the mother's breast her softly move,
Which while she drained of milk, she filled with love.
But as with riper years her virtue grew,
And every minute adds a lustre new,
When with meridian height her beauty shined,
And thorough that sparkled her fairer mind,° 40
When she with smiles serene and words discreet
His hidden soul at every turn could meet;
Then might y'ha' daily his affection spied,

Doubling that knot which destiny had tied,
While they by sense, not knowing, comprehend°
How on each other both their fates depend.
With her each day the pleasing hours he shares,
And at her aspect calms his growing cares;
Or with a grandsire's joy her children sees
Hanging about her neck or at his knees. 50
Hold fast, dear infants, hold them both or none;
This will not stay when once the other's gone.
  A silent fire now wastes those limbs of wax,
And him within his tortured image racks.°
So the flower withering which the garden crowned,
The sad root pines in secret under ground.
Each groan he doubled and each sigh he sighed,
Repeated over to the restless night.
No trembling string composed to numbers new,
Answers the touch in notes more sad, more true. 60
She, lest he grieve, hides what she can her pains,
And he to lessen hers his sorrow feigns:°
Yet both perceived, yet both concealed their skills,
And so diminishing increased their ills:
That whether by each other's grief they fell,
Or on their own redoubled, none can tell.
  And now Eliza's purple locks were shorn,
Where she so long her father's fate had worn:°
And frequent lightning to her soul that flies,
Divides the air, and opens all the skies: 70
And now his life, suspended by her breath,
Ran out impetuously to hasting death.
Like polished mirrors, so his steely breast
Had every figure of her woes expressed,
And with the damp of her last gasp obscured,
Had drawn such stains as were not to be cured.
Fate could not either reach with single stroke,
But the dear image fled, the mirror broke.
  Who now shall tell us more of mournful swans,
Of halcyons kind, or bleeding pelicans? 80
No downy breast did e'er so gently beat,
Or fan with airy plumes so soft a heat.
For he no duty by his height excused,
Nor, though a prince, to be a man refused:

But rather than in his Eliza's pain
Not love, not grieve, would neither live nor reign:
And in himself so oft immortal tried,°
Yet in compassion of another died.
So have I seen a vine, whose lasting age
Of many a winter hath survived the rage, 90
Under whose shady tent men every year
At its rich blood's expense their sorrow cheer,
If some dear branch where it extends its life
Chance to be pruned by an untimely knife,
The parent tree unto the grief succeeds,
And through the wound its vital humour bleeds,
Trickling in watery drops, whose flowing shape
Weeps that it falls ere fixed into a grape.
So the dry stock, no more that spreading vine,
Frustrates the autumn and the hopes of wine. 100
A secret cause does sure those signs ordain
Foreboding princes' falls, and seldom vain.
Whether some kinder powers that wish us well,
What they above cannot prevent foretell;
Or the great world do by consent presage,
As hollow seas with future tempests rage;
Or rather heaven, which us so long foresees,
Their funerals celebrates while it decrees.°
But never yet was any human fate
By Nature solemnized with so much state. 110
He unconcerned the dreadful passage crossed;
But, oh, what pangs that death did Nature cost!
First the great thunder was shot off, and sent°
The signal from the starry battlement.
The winds receive it, and its force outdo,
As practising how they could thunder too;
Out of the binder's hand the sheaves they tore,
And thrashed the harvest in the airy floor;
Or of huge trees, whose growth with his did rise,
The deep foundations opened to the skies. 120
Then heavy showers the winged tempests lead,°
And pour the deluge o'er the chaos' head.
The race of warlike horses at his tomb
Offer themselves in many a hecatomb;
With pensive head towards the ground they fall,

And helpless languish at the tainted stall.
Numbers of men decrease with pains unknown,
And hasten, not to see his death, their own.°
Such tortures all the elements unfixed,
Troubled to part where so exactly mixed. 130
And as through air his wasting spirits flowed,°
The universe laboured beneath their load.
  Nature, it seemed with him would Nature vie;
He with Eliza, it with him would die.
  He without noise still travelled to his end,
As silent suns to meet the night descend.
The stars that for him fought had only power°
Left to determine now his fatal hour,
Which, since they might not hinder, yet they cast°
To choose it worthy of his glories past. 140
  No part of time but bare his mark away
Of honour; all the year was Cromwell's day:
But this, of all the most auspicious found,
Twice had in open field him victor crowned:°
When up the armed mountains of Dunbar
He marched, and through deep Severn ending war.°
What day should him eternize but the same
That had before immortalized his name?
That so who ere would at his death have joyed,
In their own griefs might find themselves employed; 150
But those that sadly his departure grieved,
Yet joyed, remembering what he once achieved.
And the last minute his victorious ghost
Gave chase to Ligny on the Belgic coast.°
Here ended all his mortal toils: he laid
And slept in peace under the laurel shade.
  O Cromwell, heaven's favourite! To none
Have such high honours from above been shown:
For whom the elements we mourners see,
And heaven itself would the great herald be, 160
Which with more care set forth his obsequies
Than those of Moses hid from human eyes,°
As jealous only here lest all be less,
That we could to his memory express.
  Then let us to our course of mourning keep:
Where heaven leads, 'tis piety to weep.

Stand back, ye seas, and shrunk beneath the veil
Of your abyss, with covered head bewail
Your monarch: we demand not your supplies
To compass in our isle; our tears suffice: 170
Since him away the dismal tempest rent,
Who once more joined us to the continent;
Who planted England on the Flandric shore,
And stretched our frontier to the Indian ore;°
Whose greater truths obscure the fables old,
Whether of British saints or worthies told;°
And in a valour lessening Arthur's deeds,
For holiness the Confessor exceeds.°
He first put arms into Religion's hand,
And timorous Conscience unto Courage manned:° 180
The soldier taught that inward mail to wear,°
And fearing God how they should nothing fear.
'Those strokes', he said, 'will pierce through all below
Where those that strike from heaven fetch their blow.'
Astonished armies did their flight prepare,
And cities strong were stormed by his prayer;
Of that, forever Preston's field shall tell°
The story, and impregnable Clonmel.°
And where the sandy mountain Fenwick scaled,°
The sea between, yet hence his prayer prevailed.° 190
What man was ever so in heaven obeyed
Since the commanded sun o'er Gibeon stayed?°
In all his wars needs must he triumph when
He conquered God still ere he fought with men.
Hence, though in battle none so brave or fierce,
Yet him the adverse steel could never pierce.
Pity it seemed to hurt him more that felt
Each wound himself which he to others dealt;
Danger itself refusing to offend
So loose an enemy, so fast a friend. 200
Friendship, that sacred virtue, long does claim
The first foundation of his house and name:°
But within one its narrow limits fall,
His tenderness extended unto all.°
And that deep soul through every channel flows,
Where kindly nature loves itself to lose.
More strong affections never reason served,

Yet still affected most what best deserved.
If he Eliza loved to that degree,
(Though who more worthy to be loved than she?) 210
If so indulgent to his own, how dear
To him the children of the highest were?
For her he once did nature's tribute pay:
For these his life adventured every day:
And 'twould be found, could we his thoughts have cast,°
Their griefs struck deepest, if Eliza's last.
What prudence more than human did he need
To keep so dear, so differing minds agreed?
The worser sort, as conscious of their ill,
Lie weak and easy to the ruler's will; 220
But to the good (too many or too few)
All law is useless, all reward is due.
Oh ill-advised, if not for love, for shame,
Spare yet your own, if you neglect his fame;
Lest others dare to think your zeal a mask,
And you to govern, only heaven's task.
Valour, religion, friendship, prudence died
At once with him, and all that's good beside;
And we death's refuse, nature's dregs, confined
To loathsome life, alas! are left behind. 230
Where we (so once we used) shall now no more
To fetch day, press about his chamber door—
From which he issued with that awful state,
It seemed Mars broke through Janus' double gate,°
Yet always tempered with an air so mild,
No April suns that e'er so gently smiled—
No more shall hear that powerful language charm,
Whose force oft spared the labour of his arm:
No more shall follow where he spent the days
In war, in counsel, or in prayer and praise, 240
Whose meanest acts he would himself advance,
As ungirt David to the ark did dance.°
All, all is gone of ours or his delight
In horses fierce, wild deer, or armour bright;
Francisca fair can nothing now but weep,°
Nor with soft notes shall sing his cares asleep.
  I saw him dead. A leaden slumber lies
And mortal sleep over those wakeful eyes:

Those gentle rays under the lids were fled,
Which through his looks that piercing sweetness shed; 250
That port which so majestic was and strong,
Loose and deprived of vigour, stretched along:
All withered, all discoloured, pale and wan—
How much another thing, nor more that man?
Oh human glory vain, oh death, oh wings,
Oh worthless world, oh transitory things!
Yet dwelt that greatness in his shape decayed,
That still though dead, greater than death he laid;
And in his altered face you something feign°
That threatens death he yet will live again. 260
Not much unlike the sacred oak which shoots
To heaven its branches and through earth its roots,
Whose spacious boughs are hung with trophies round,
And honoured wreaths have oft the victor crowned.°
When angry Jove darts lightning through the air,
At mortals' sins, nor his own plant will spare,
(It groans, and bruises all below, that stood
So many years the shelter of the wood.)
The tree erewhile foreshortened to our view,
When fallen shows taller yet than as it grew:° 270
So shall his praise to after times increase,
When truth shall be allowed, and faction cease,
And his own shadows with him fall. The eye
Detracts from objects than itself more high:
But when death takes them from that envied seat,°
Seeing how little, we confess how great.°
Thee, many ages hence in martial verse
Shall the English soldier, ere he charge, rehearse,
Singing of thee, inflame themselves to fight,
And with the name of Cromwell, armies fright. 280
As long as rivers to the seas shall run,
As long as Cynthia shall relieve the sun,°
While stags shall fly unto the forests thick,
While sheep delight the grassy downs to pick,
As long as future time succeeds the past,
Always thy honour, praise, and name shall last.
Thou in a pitch how far beyond the sphere°
Of human glory towerst, and reigning there
Despoiled of mortal robes, in seas of bliss,

Plunging dost bathe, and tread the bright abyss: 290
There thy great soul at once a world does see,°
Spacious enough, and pure enough for thee.
How soon thou Moses hast, and Joshua found,
And David for the sword and harp renowned?
How straight canst to each happy mansion go?
(Far better known above than here below)
And in those joys dost spend the endless day,
Which in expressing we ourselves betray.
For we, since thou art gone, with heavy doom,
Wander like ghosts about thy loved tomb; 300
And lost in tears, have neither sight nor mind
To guide us upward through this region blind.
Since thou art gone, who best that way couldst teach,
Only our sighs, perhaps, may thither reach.
And Richard yet, where his great parent led,°
Beats on the rugged track: he, virtue dead,
Revives, and by his milder beams assures;
And yet how much of them his grief obscures?
He, as his father, long was kept from sight
In private, to be viewed by better light; 310
But opened once, what splendour does he throw?
A Cromwell in an hour a prince will grow.
How he becomes that seat, how strongly strains,
How gently winds at once the ruling reins?
Heaven to this choice prepared a diadem,
Richer than any Eastern silk or gem;
A pearly rainbow, where the sun enchased°
His brows, like an imperial jewel graced.
We find already what those omens mean,
Earth ne'er more glad, nor heaven more serene. 320
Cease now our griefs, calm peace succeeds a war,
Rainbows to storms, Richard to Oliver.
Tempt not his clemency to try his power,
He threats no deluge, yet foretells a shower.

## *On Mr Milton's 'Paradise Lost'*

When I beheld the poet blind, yet bold,
In slender book his vast design unfold,
Messiah crowned, God's reconciled decree,
Rebelling angels, the Forbidden Tree,
Heaven, hell, earth, chaos, all; the argument
Held me a while, misdoubting his intent
That he would ruin (for I saw him strong)
The sacred truths to fable and old song,
(So Samson groped the temple's posts in spite)
The world o'erwhelming to revenge his sight. 10
Yet as I read, soon growing less severe,
I liked his project, the success did fear;
Through that wide field how he his way should find
O'er which lame faith leads understanding blind;
Lest he perplexed the things he would explain,
And what was easy he should render vain.
Or if a work so infinite he spanned,
Jealous I was that some less skilful hand
(Such as disquiet always what is well,
And by ill imitating would excel) 20
Might hence presume the whole creation's day
To change in scenes, and show it in a play.°
Pardon me, mighty poet, nor despise
My causeless, yet not impious, surmise.
But I am now convinced that none will dare°
Within thy labours to pretend a share.
Thou hast not missed one thought that could be fit,
And all that was improper dost omit:
So that no room is here for writers left,
But to detect their ignorance or theft.° 30
That majesty which through thy work doth reign
Draws the devout, deterring the profane.
And things divine thou treat'st of in such state°
As them preserves, and thee, inviolate.
At once delight and horror on us seize,
Thou sing'st with so much gravity and ease;
And above human flight dost soar aloft,

With plume so strong, so equal, and so soft.
The bird named from that paradise you sing°
So never flags, but always keeps on wing. 40
  Where couldst thou words of such a compass find?
Whence furnish such a vast expense of mind?
Just heaven thee, like Tiresias, to requite,°
Rewards with prophecy thy loss of sight.
  Well mightst thou scorn thy readers to allure°
With tinkling rhyme, of thy own sense secure;°
While the *Town-Bays* writes all the while and spells,°
And like a pack-horse tires without his bells.
Their fancies like our bushy points appear,°
The poets tag them; we for fashion wear. 50
I too, transported by the mode, offend,
And while I meant to *praise* thee must commend.°
Thy verse created like thy theme sublime,
In number, weight, and measure, needs not rhyme.°

# SATIRES OF THE REIGN OF CHARLES II

## *Clarendon's Housewarming*

When Clarendon had discerned beforehand
  (As the cause can easily foretell the effect)
At once three deluges threatening our land,°
  'Twas the season, he thought, to turn architect.°

Us Mars, and Apollo, and Vulcan consume
  While he, the betrayer of England and Flander,
Like the kingfisher chooseth to build in the brume,°
  And nestle in flames like the salamander.

But observing that mortals run often behind,
  (So unreasonable are the rates that they buy at) 10
His omnipotence therefore much rather designed,
  How he might create a house with a fiat.

He had read of Rhodopis, a lady of Thrace,°
  Who was digged up so often ere she did marry,
And wished that his daughter had had as much grace,
  To erect him a pyramid out of her quarry.°

But then recollecting how the harper Amphion°
  Made Thebes dance aloft while he fiddled and sung,
He thought (as an instrument he was most free on)
  To build with the Jew's-trump of his own tongue.° 20

Yet a precedent fitter in Virgil he found,°
  Of African Poultney and Tyrian Did';°
That he begged for a palace so much of his ground
  As might carry the measure and name of a Hyde.°

Thus daily his gouty inventions he pained,
And all for to save the expenses of brickbat;°
That engine so fatal, which Denham had brained,°
And too much resembled his wife's chocolate.°

But while these devices he all doth compare,
None solid enough seemed for his thong-caster;° 30
He himself would not dwell in a castle of air,
Though he had built full many a one for his master.

Already he had got all our money and cattle,
To buy us for slaves, and purchase our lands;
What Joseph by famine, he wrought by sea battle;
Nay, scarce the priest's portion could scape from his hands.°

And hence like Pharaoh, that Israel pressed
To make mortar and brick, yet allowed them no straw,
He cared not though Egypt's ten plagues us distressed,
So he could to build but make policy law. 40

The Scotch forts and Dunkirk, but that they were sold,°
He would have demolished to raise up his walls;
Nay e'en from Tangier have sent back for the mold,°
But that he had nearer the stones of St Paul's.°

His wood would come in at the easier rate,
So long as the yards had a deal or a spar:
His friends in the navy would not be ingrate,
To grudge him some timber who framed them the war.

To proceed in the model, he called in his Allens,
The two Allens when jovial, who ply him with gallons; 50
The two Allens who served his blind justice for balance
The two Allens who served his injustice for talons.°

They approve it thus far, and said it was fine;
Yet his lordship to finish it would be unable,
Unless all abroad he divulged the design,
For his house then would grow like a vegetable.

His rent would no more in arrear run to Worcester;°
  He should dwell more noble and cheaper too at home;
While into a fabric the presents would muster,
  As by hook and by crook the world clustered of atom.° 60

He liked the advice and then soon it assayed,
  And presents crowd headlong to give good example;
So the bribes overlaid her that Rome once betrayed:°
  The tribes ne'er contributed so to the temple.

Straight judges, priests, bishops, true sons of the seal,°
  Sumners, governors, farmers, bankers, patentees,°
Bring in the whole milk of a year at a meal,
  As all Cheddar dairy's club to the incorporate cheese.

Bulteale's, Beaken's, Morley's, Wren's finger with telling
  Were shrivelled, and Clutterbuck's, Eager's, and Kipp's;° 70
Since the Act of Oblivion was never such selling,°
  As at this benevolence out of the snips.°

'Twas then that the chimney contractors he smoked,°
  Nor would take his beloved canary in kind:
But he swore that the patent should ne'er be revoked,
  No, would the whole parliament kiss him behind.°

Like Jove under Etna o'erwhelming the giant,
  For foundation the Bristol sank in the earth's bowel;°
And St John must now for the leads be compliant,°
  Or his right hand shall be cut off with the trowel. 80

For surveying the building, Pratt did the feat;
  But for the expense he relied upon Worstenholm,°
Who sat heretofore at the king's receipt,
  But received now and paid the Chancellor's custom.

By subsidies thus both cleric and laic,
  And with matter profane, cemented with holy;
He finished at last his palace mosaic,
  By a model more excellent than Leslie's folly.°

And upon the terrace, to consummate all,
A lantern like Fawkes' surveys the burnt town,° 90
And shows on the top, by the regal gilt ball,
Where you are to expect the sceptre and crown.

Fond city, its rubbish and ruins that builds,
Like vain chemists, a flower from its ashes returning,
Your metropolis house is in St James's fields,
And till there you remove, you shall never leave burning.

This temple of war and of peace is the shrine,
Where this idol of state sits adorned and accursed;
And to handsel his altar and nostrils divine,°
Great Buckingham's sacrifice must be the first.° 100

Now some (as all buildings must censure abide)
Throw dust in its front, and blame situation:
And other as much reprehend his backside,°
As too narrow by far for his expatiation;

But do not consider how in process of times,
That for namesake he may with Hyde Park it enlarge,
And with that convenience he soon for his crimes,
At Tyburn may land, and spare the Tower barge.°

Or rather how wisely his stall was built near,
Lest with driving too far his tallow impair; 110
When like the good ox, for public good cheer,
He comes to be roasted next St James's fair.°

## *The Last Instructions to a Painter*

*London, 4 September 1667*°

After two sittings, now our Lady State
To end her picture does the third time wait.°
But ere thou fall'st to work, first, painter, see
It ben't too slight grown or too hard for thee.
Canst thou paint without colors? Then 'tis right:
For so we too without a fleet can fight.°
Or canst thou daub a signpost, and that ill?°

'Twill suit our great debauch and little skill.
Or hast thou marked how antique masters limn°
The alley-roof with snuff of candle dim,° 10
Sketching in shady smoke prodigious tools?°
'Twill serve this race of drunkards, pimps, and fools.
But if to match our crimes thy skill presumes,
As the Indians, draw our luxury in plumes.°
Or if to score out our compendious fame,°
With Hooke, then, through the microscope take aim,
Where, like the new Comptroller, all men laugh
To see a tall louse brandish the white staff.°
Else shalt thou oft thy guiltless pencil curse,°
Stamp on thy palette, nor perhaps the worse. 20
The painter so, long having vexed his cloth—
Of his hound's mouth to feign the raging froth—
His desperate pencil at the work did dart:
His anger reached that rage which passed his art;
Chance finished that which art could but begin,
And he sat smiling how his dog did grin.°
So mayst thou perfect by a lucky blow°
What all thy softest touches cannot do.
  Paint then St Albans full of soup and gold,°
The new court's pattern, stallion of the old. 30
Him neither wit nor courage did exalt,
But Fortune chose him for her pleasure salt.°
Paint him with drayman's shoulders, butcher's mien,
Membered like mules, with elephantine chine.
Well he the title of St Albans bore,
For never Bacon studied nature more.°
But age, allaying now that youthful heat,
Fits him in France to play at cards and treat.°
Draw no commission lest the court should lie,
That, disavowing treaty, asks supply.° 40
He needs no seal but to St James's lease,°
Whose breeches were the instrument of peace;
Who, if the French dispute his power, from thence
Can straight produce them a plenipotence.°
Nor fears he the most Christian should trepan°
Two saints at once, St Germain, St Alban,°
But thought the Golden Age was now restored,
When men and women took each other's word.

Paint then again her Highness to the life,°
Philosopher beyond Newcastle's wife.° 50
She, nak'd, can Archimedes' self put down,°
For an experiment upon the crown.°
She perfected that engine, oft assayed,°
How after childbirth to renew a maid,
And found how royal heirs might be matured
In fewer months than mothers once endured.
Hence Crowther made the rare inventress free°
Of's Highness's Royal Society—°
Happiest of women, if she were but able
To make her glassen d---- once malleable!° 60
Paint her with oyster lip and breath of fame,
Wide mouth that 'sparagus may well proclaim;°
With Chancellor's belly and so large a rump,
There—not behind the coach—her pages jump.
Express her study now if China clay
Can, without breaking, venomed juice convey,
Or how a mortal poison she may draw
Out of the cordial meal of the cacao.°
Witness, ye stars of night, and thou the pale
Moon, that o'ercome with the sick steam didst fail; 70
Ye neighboring elms, that your green leaves did shed,
And fawns that from the womb abortive fled;°
Not unprovoked, she tries forbidden arts,
But in her soft breast love's hid cancer smarts,
While she revolves, at once, Sidney's disgrace°
And her self scorned for emulous Denham's face,°
And nightly hears the hated guards, away
Galloping with the Duke to other prey.
Paint Castlemaine in colours that will hold°
(Her, not her picture, for she now grows old): 80
She through her lackey's drawers, as he ran,
Discerned love's cause and a new flame began.
Her wonted joys thenceforth and court she shuns,
And still within her mind the footman runs:
His brazen calves, his brawny thighs—the face
She slights—his feet shaped for a smoother race.
Poring within her glass she readjusts
Her looks, and oft-tried beauty now distrusts,
Fears lest he scorn a woman once assayed,

And now first wished she e'er had been a maid. 90
Great Love, how dost thou triumph and how reign,
That to a groom couldst humble her disdain!
Stripped to her skin, see how she stooping stands,
Nor scorns to rub him down with those fair hands,
And washing (lest the scent her crime disclose)
His sweaty hooves, tickles him 'twixt the toes.
But envious Fame, too soon, began to note
More gold in's fob, more lace upon his coat;
And he, unwary, and of tongue too fleet,
No longer could conceal his fortune sweet. 100
Justly the rogue was whipped in porter's den,
And Jermyn straight has leave to come again.°
Ah, painter, now could Alexander live,
And this Campaspe thee, Apelles, give!°
Draw next a pair of tables opening, then°
The House of Commons clattering like the men.°
Describe the court and country, both set right
On opposite points, the black against the white.
Those having lost the nation at trick-track,°
These now adventuring how to win it back. 110
The dice betwixt them must the fate divide
(As chance doth still in multitudes decide).
But here the Court does its advantage know,
For the cheat Turner for them both must throw.°
As some from boxes, he so from the chair
Can strike the die and still with them goes share.°
Here, painter, rest a little, and survey
With what small arts the public game they play.
For so too Rubens, with affairs of state,
His labouring pencil oft would recreate.° 120
The close Cabal marked how the navy eats,°
And thought all lost that goes not to the cheats,
So therefore secretly for peace decrees,
Yet as for war the Parliament should squeeze,
And fix to the revenue such a sum°
Should Goodrick silence and strike Paston dumb,°
Should pay land armies, should dissolve the vain
Commons, and ever such a court maintain;
Hyde's avarice, Bennet's luxury should suffice,°
And what can these defray but the excise? 130

Excise a monster worse than e'er before
Frighted the midwife and the mother tore.
A thousand hands she has and thousand eyes,
Breaks into shops and into cellars pries,
With hundred rows of teeth the shark exceeds,
And on all trade like cassowar she feeds:°
Chops off the piece where e'er she close the jaw,
Else swallows all down her indented maw.°
She stalks all day in streets concealed from sight
And flies, like bats with leathern wings, by night; 140
She wastes the country and on cities preys.
Her, of a female harpy, in dog days,
Black Birch, of all the earth-born race most hot°
And most rapacious, like himself, begot,
And, of his brat enamoured, as't increased,
Buggered in incest with the mongrel beast.
  Say, Muse, for nothing can escape thy sight
(And, painter, wanting other, draw this fight),
Who, in an English senate, fierce debate
Could raise so long for this new whore of state. 150
  Of early wittols first the troop marched in—°
For diligence renowned and discipline—
In loyal haste they left young wives in bed,
And Denham these by one consent did head.
Of the old courtiers, next a squadron came,
That sold their master, led by Ashburnham.°
To them succeeds a despicable rout,
But knew the word and well could face about;°
Expectants pale, with hopes of spoil allured,
Though yet but pioneers, and led by Steward.° 160
Then damning cowards ranged the vocal plain,
Wood these commands, Knight of the Horn and Cane.°
Still his hook-shoulder seems the blow to dread,
And under's ampit he defends his head.
The posture strange men laughed at of his poll,
Hid with his elbow like the spice he stole.
Headless St Denis so his head does bear,°
And both of them alike French martyrs were.°
Court officers, as used, the next place took,°
And followed Fox, but with disdainful look.° 170
His birth, his youth, his brokage all dispraise

In vain, for always he commands that pays.
Then the procurers under Progers filed—°
Gentlest of men—and his lieutenant mild,
Brouncker—Love's squire—through all the field arrayed,°
No troop was better clad, nor so well paid.
Then marched the troop of Clarendon, all full
Haters of fowl, to teal preferring bull:°
Gross bodies, grosser minds, and grossest cheats,
And bloated Wren conducts them to their seats.° 180
Charlton advances next, whose coif does awe.°
The Mitre troop, and with his looks gives law.°
He marched with beaver cocked of bishop's brim,
And hid much fraud under an aspect grim.
Next the lawyers' mercenary band appear:
Finch in the front, and Thurland in the rear.°
The troop of privilege, a rabble bare°
Of debtors deep, fell to Trelawney's care.°
Their fortune's error they supplied in rage,
Nor any further would than these engage. 190
Then marched the troop, whose valiant acts before
(Their public acts) obliged them still to more.
For chimney's sake they all Sir Pool obeyed,°
Or in his absence him that first it laid.
Then comes the thrifty troop of privateers,°
Whose horses each with other interferes.
Before them Higgons rides with brow compact,
Mourning his Countess, anxious for his Act.°
Sir Frederick and Sir Solomon draw lots°
For the command of politics or sots,° 200
Thence fell to words, but quarrel to adjourn;
Their friends agreed they should command by turn.
Carteret the rich did the accountants guide°
And in ill English all the world defied.
The Papists—but of these the House had none
Else Talbot offered to have led them on.°
Bold Duncombe next, of the projectors chief,°
And old Fitz-Hardinge of the Eaters Beef.°
Late and disordered out the drinkers drew,
Scarce them their leaders, they their leaders knew. 210
Before them entered, equal in command,
Apsley and Broderick marching hand in hand.°

Last then but one, Powell that could not ride,°
Led the French standard, weltering in his stride.°
He, to excuse his slowness, truth confessed
That 'twas so long before he could be dressed.
The Lord's sons, last, all these did reinforce:
Cornbury before them managed hobby-horse.°
  Never before nor since, a host so steeled
Trooped on to muster in the Tothill Field:° 220
Not the first cock-horse that with cork were shod
To rescue Albermarle from the sea-cod,
Nor the late feather-men, whom Tomkins fierce
Shall with one breath, like thistledown disperse.°
All the two Coventrys their generals chose°
For one had much, the other nought to lose;°
Nor better choice all accidents could hit.
While hector Harry steers by Will the wit.
They both accept the charge with merry glee,
To fight a battle, from all gunshot free.° 230
Pleased with their numbers, yet in valour wise,
They feign a parley, better to surprise;
They that ere long shall the rude Dutch upbraid,
Who in a time of treaty durst invade.°
  Thick was the morning, and the House was thin,
The Speaker early, when they all fell in.
Propitious heavens, had not you them crossed,
Excise had got the day, and all been lost.
For the other side all in loose quarters lay,°
Without intelligence, command, or pay: 240
A scattered body, which the foe ne'er tried,
But oftener did among themselves divide.
And some ran o'er each night, while others sleep,
And undescried returned ere morning peep.
But Strangeways, that all night still walked the round°
(For vigilance and courage both renowned)
First spied the enemy and gave the alarm,
Fighting it single till the rest might arm.
Such Roman Cocles strode before the foe,°
The falling bridge behind, the stream below. 250
  Each ran, as chance him guides to several post,
And all to pattern his example boast.
Their former trophies they recall to mind

And to new edge their angry courage grind.
First entered forward Temple conqueror
Of Irish cattle and Solicitor;°
Then daring Seymour that with spear and shield,
Had stretched the monster Patent on the field;°
Keen Whorwood next, in aid of damsel frail,°
That pierced the giant Mordaunt through his mail;° 260
And surly Williams, the accountants' bane;°
And Lovelace young, of chimney-men the cane.°
Old Waller, trumpet-general, swore he'd write
This combat truer than the naval fight.°
Of birth, state, wit, strength, courage, Howard presumes
And in his breast wears many Montezumes.°
These and some more with single valour stay
The adverse troops, and hold them all at bay.
Each thinks his person represents the whole,
And with that thought does multiply his soul, 270
Believes himself an army, theirs, one man
As easily conquered, and believing can,°
With heart of bees so full, and head of mites,
That each, though duelling, a battle fights.
Such once Orlando, famous in romance,
Broached whole brigades like larks upon his lance.°
But strength at last still under number bows,
And the faint sweat trickled down Temple's brows.
E'en iron Strangeways, chafing, yet gave back,
Spent with fatigue, to breathe a while toback.° 280
When marching in, a seasonable recruit°
Of citizens and merchants held dispute;
And, charging all their pikes, a sullen band
Of Presbyterian Switzers made a stand.
Nor could all these the field have long maintained
But for the unknown reserve that still remained:
A gross of English gentry, nobly born,°
Of clear estates, and to no faction sworn,
Dear lovers of their king, and death to meet
For country's cause, that glorious think and sweet; 290
To speak not forward, but in action brave,
In giving generous, but in counsel grave;
Candidly credulous for once, nay twice,
But sure the devil cannot cheat them thrice.

The van and battle, though retiring, falls°
Without disorder in their intervals.°
Then, closing all in equal front, fall on,
Led by great Garway and great Lyttleton.
Lee, equal to obey or to command,°
Adjutant-general, was still at hand. 300
The martial standard, Sandys displaying, shows°
St Dunstan in it, tweaking Satan's nose.°
See sudden chance of war! To paint or write
Is longer work and harder than to fight.
At the first charge the enemy give out,
And the excise receives a total rout.
Broken in courage, yet the men the same
Resolve henceforth upon their other game:
Where force had failed, with stratagem to play,
And what haste lost, recover by delay. 310
St Albans straight is sent to, to forbear,
Lest the sure peace, forsooth, too soon appear.
The seamen's clamour to three ends they use:°
To cheat their pay, feign want, the House accuse.
Each day they bring the tale, and that too true,
How strong the Dutch their equipage renew.
Meantime through all the yards their orders run
To lay the ships up, cease the keels begun.
The timber rots, and useless axe does rust,
The unpractised saw lies buried in its dust, 320
The busy hammer sleeps, the ropes untwine,
The stores and wages all are mine and thine.
Along the coast and harbours they take care
That money lack, nor forts be in repair.
Long thus they could against the House conspire,
Load them with envy, and with sitting tire.
And the loved king, and never yet denied,
Is brought to beg in public and to chide;°
But when this failed, and months enow were spent,
They with the first day's proffer seem content, 330
And to land tax from the excise turn round,
Bought off with eighteen-hundred-thousand pound.
Thus like fair thieves, the Commons' purse they share,
But all the members' lives, consulting, spare.
Blither than hare that hath escaped the hounds,
The House prorogued, the Chancellor rebounds.°

Not so decrepit Aeson, hashed and stewed,°
With magic herbs, rose from the pot renewed,
And with fresh age felt his glad limbs unite;
His gout (yet still he cursed) had left him quite. 340
What frosts to fruit, what arsenic to the rat,
What to fair Denham, mortal chocolate,
What an account to Carteret, that, and more,
A Parliament is to the Chancellor.
So the sad tree shrinks from the morning's eye,°
But blooms all night and shoots its branches high.
So, at the sun's recess, again returns
The comet dread, and earth and heaven burns.
Now Mordaunt may, within his castle tower,
Imprison parents, and the child deflower.° 350
The Irish herd is now let loose and comes°
By millions over, not by hecatombs;°
And now, now the Canary Patent may°
Be broached again for the great holiday.
See how he reigns in his new palace culminant,
And sits in state divine like Jove the fulminant!°
First Buckingham, that durst to him rebel,°
Blasted with lightning, struck with thunder, fell.
Next the twelve Commons are condemned to groan°
And roll in vain at Sisyphus's stone.° 360
But still he cared, while in revenge he braved
That peace secured and money might be saved:
Gain and revenge, revenge and gain are sweet
United most, else when by turns they meet.
France had St Albans promised (so they sing),
St Albans promised him, and he the King:
The Count forthwith is ordered all to close,
To play for Flanders and the stake to lose,°
While, chained together, two ambassadors
Like slaves shall beg for peace at Holland's doors.° 370
This done, among his Cyclops he retires°
To forge new thunder and inspect their fires.
The court as once of war, now fond of peace,
All to new sports their wanton fears release.°
From Greenwich (where intelligence they hold)
Comes news of pastime martial and old,
A punishment invented first to awe

Masculine wives transgressing Nature's law,
Where, when the brawny female disobeys
And beats the husband till for peace he prays, 380
No concerned jury for him damage finds,
Nor partial justice her behaviour binds,
But the just street does the next house invade,
Mounting the neighbour couple on lean jade,
The distaff knocks, the grains from kettle fly,
And boys and girls in troops run hooting by:
Prudent antiquity, that knew by shame,
Better than law, domestic crimes to tame,
And taught youth by spectacle innocent!
So thou and I, dear painter, represent 390
In quick effigy, others' faults, and feign°
By making them ridiculous, to restrain.
With homely sight they chose thus to relax
The joys of state, for the new peace and tax.
So Holland with us had the mastery tried,
And our next neighbours, France and Flanders, ride.
  But a fresh news the great designment nips,
Of, at the Isle of Candy, Dutch and ships!°
Bab May and Arlington did wisely scoff°
And thought all safe, if they were so far off.° 400
Modern geographers, 'twas there, they thought,
Where Venice twenty years the Turk had fought,
While the first year our navy is but shown,
The next divided, and the third we've none.
They, by the name, mistook it for that isle
Where pilgrim Palmer travelled in exile°
With the bull's horn to measure his own head
And on Pasiphae's tomb to drop a bead.°
But Morice learned demonstrates, by the post,°
This Isle of Candy was on Essex coast. 410
  Fresh messengers still the sad news assure;
More timorous now we are than first secure.
False terrors our believing fears devise,
And the French army one from Calais spies.
Bennet and May and those of shorter reach°
Change all for guineas, and a crown for each,
But wiser men and well foreseen in chance
In Holland theirs had lodged before, and France.

Whitehall's unsafe; the court all meditates
To fly to Windsor and mure up the gates. 420
Each does the other blame, and all distrust;
But Mordaunt, new obliged, would sure be just.
Not such a fatal stupefaction reigned
At London's flame, nor so the court complained.
The Bloodworth-Chancellor gives, then does recall°
Orders; amazed, at last gives none at all.
St Albans's writ to, that he may bewail
To Master Louis, and tell coward tale
How yet the Hollanders do make a noise,
Threaten to beat us, and are naughty boys. 430
Now Dolman's disobedient, and they still°
Uncivil; his unkindness would us kill.
Tell him our ships unrigged, our forts unmanned,
Our money spent; else 'twere at his command.
Summon him therefore of his word and prove°
To move him out of pity, if not love;
Pray him to make De Witt and Ruyter cease,°
And whip the Dutch unless they'll hold their peace.
But Louis was of memory but dull
And to St Albans too undutiful,° 440
Nor word nor near relation did revere,
But asked him bluntly for his character.°
The gravelled count did with the answer feint—°
His character was that which thou didst paint—°
And so enforced, like enemy or spy,
Trusses his baggage and the camp does fly.
Yet Louis writes and, lest our heart should break,
Consoles us morally out of Seneque.°
Two letters next unto Breda are sent:
In cipher one to Harry excellent;° 450
The first instructs our (verse the name abhors)
Plenipotentiary ambassadors
To prove by Scripture treaty does imply
Cessation, as the look adultery,°
And that, by law of arms, in martial strife,
Who yields his sword has title to his life.
Presbyter Holles the first point should clear,°
The second Coventry the Cavalier;
But, would they not be argued back from sea,

Then to return home straight, *infecta re.*° 460
But Harry's ordered, if they won't recall
Their fleet, to threaten—we will grant them all.
The Dutch are then in proclamation shent°
For sin against the eleventh commandment.°
Hyde's flippant style there pleasantly curvets,
Still his sharp wit on states and princes whets
(So Spain could not escape his laughter's spleen:
None but himself must choose the King a Queen),
But when he came the odious clause to pen
That summons up the Parliament again, 470
His writing master many a time he banned°
And wished himself the gout to seize his hand.
Never old lecher more repugnance felt,
Consenting, for his rupture, to be gelt;
But still in hope he solaced, ere they come,
To work the peace and so to send them home,
Or in their hasty call to find a flaw,
Their acts to vitiate, and them overawe;
But most relied upon this Dutch pretence
To raise a two-edged army for's defence.° 480
First then he marched our whole militia's force
(As if, alas, we ships or Dutch had horse);
Then from the usual commonplace, he blames
These, and in standing army's praise declaims;
And the wise court that always loved it dear,
Now thinks all but too little for their fear.
Hyde stamps, and straight upon the ground the swarms
Of current Myrmidons appear in arms,°
And for their pay he writes, as from the King—
With that cursed quill plucked from a vulture's wing— 490
Of the whole nation now to ask a loan
(The eighteen-hundred-thousand pound was gone).
This done, he pens a proclamation stout,°
In rescue of the *banquiers banquerout*,°
His minion imps that, in his secret part,
Lie nuzzling at the sacramental wart,
Horse-leeches circling at the haemorrhoid vein:°
He sucks the King, they him, he them again.
The kingdom's farm he lets to them bid least
(Greater the bribe, and that's at interest). 500

Here men, induced by safety, gain, and ease,
Their money lodge; confiscate when he please.
These can at need, at instant, with a scrip°
(This liked him best) his cash beyond sea whip.
When Dutch invade, when Parliament prepare,
How can he engines so convenient spare?
Let no man touch them or demand his own,
Pain of displeasure of great Clarendon.
  The state affairs thus marshalled, for the rest
Monck in his shirt against the Dutch is pressed.° 510
Often, dear painter, have I sat and mused
Why he should still be 'n all adventures used,
If they for nothing ill, like ashen wood,
Or think him, like Herb John, for nothing good;°
Whether his valour they so much admire,
Or that for cowardice they all retire,
As heaven in storms, they call in gusts of state
On Monck and Parliament, yet both do hate.
All causes sure concur, but most they think
Under Herculean labours he may sink. 520
Soon then the independent troops would close,
And Hyde's last project would his place dispose.
  Ruyter the while, that had our ocean curbed,
Sailed now among our rivers undisturbed,
Surveyed their crystal streams and banks so green
And beauties ere this never naked seen.
Through the vain sedge, the bashful nymphs he eyed:
Bosoms, and all which from themselves they hide.
The sun much brighter, and the skies more clear,
He finds the air and all things sweeter here. 530
The sudden change, and such a tempting sight
Swells his old veins with fresh blood, fresh delight.°
Like amorous victors he begins to shave,
And his new face looks in the English wave.
His sporting navy all about him swim
And witness their complacence in their trim.
Their streaming silks play through the weather fair
And with inveigling colours court the air,
While the red flags breathe on their topmasts high
Terror and war, but want an enemy. 540
Among the shrouds the seamen sit and sing,

And wanton boys on every rope do cling.
Old Neptune springs the tides and water lent
(The gods themselves do help the provident),
And where the deep keel on the shallow cleaves,
With trident's lever, and great shoulder heaves.
Aeolus their sails inspires with eastern wind,
Puffs them along, and breathes upon them kind.
With pearly shell and Tritons all the while
Sound the sea-march and guide to Sheppey Isle. 550
 So have I seen in April's bud arise
A fleet of clouds, sailing along the skies;
The liquid region with their squadrons filled,
Their airy sterns the sun behind does gild;
And gentle gales them steer, and heaven drives,
When, all on sudden, their calm bosom rives
With thunder and lightning from each armed cloud;
Shepherds themselves in vain in bushes shroud.
Such up the stream the Belgic navy glides
And at Sheerness unloads its stormy sides. 560
 Spragge there, though practised in the sea command,°
With panting heart lay like a fish on land
And quickly judged the fort was not tenable—
Which, if a house, yet were not tenantable—
No man can sit there safe: the cannon pours
Thorough the walls untight and bullet showers,
The neighbourhood ill, and an unwholesome seat,
So at the first salute resolves retreat,
And swore that he would never more dwell there
Until the city put it in repair. 570
So he in front, his garrison in rear,
March straight to Chatham to increase the fear.
 There our sick ships unrigged in summer lay
Like moulting fowl, a weak and easy prey,
For whose strong bulk earth scarce could timber find,
The ocean water, or the heavens wind—
Those oaken giants of the ancient race,
That ruled all seas and did our Channel grace.
The conscious stag so, once the forest's dread,
Flies to the wood and hides his armless head. 580
Ruyter forthwith a squadron does untack;
They sail securely through the river's track.

An English pilot too (O shame, O sin!)
Cheated of pay, was he that showed them in.
Our wretched ships within their fate attend,
And all our hopes now on frail chain depend:
(Engine so slight to guard us from the sea,
It fitter seemed to captivate a flea).°
A skipper rude shocks it without respect,
Filling his sails more force to recollect.° 590
The English from shore the iron deaf invoke
For its last aid: 'Hold chain, or we are broke.'
But with her sailing weight, the Holland keel,
Snapping the brittle links, does thorough reel,
And to the rest the opened passage show;
Monck from the bank the dismal sight does view.
Our feathered gallants, which came down that day
To be spectators safe of the new play,
Leave him alone when first they hear the gun
(Cornbury the fleetest) and to London run. 600
Our seamen, whom no danger's shape could fright,
Unpaid, refuse to mount our ships for spite,
Or to their fellows swim on board the Dutch,
Which show the tempting metal in their clutch.
Oft had he sent of Duncombe and of Legge°
Cannon and powder, but in vain, to beg;
And Upnor Castle's ill-deserted wall,°
Now needful, does for ammunition call.
He finds, wheresoe'er he succor might expect,
Confusion, folly, treach'ry, fear, neglect. 610
But when the *Royal Charles* (what rage, what grief)°
He saw seized, and could give her no relief!
That sacred keel which had, as he, restored
His exiled sovereign on its happy board,
And thence the British Admiral became,°
Crowned, for that merit, with their master's name;
That pleasure-boat of war, in whose dear side
Secure so oft he had this foe defied,
Now a cheap spoil, and the mean victor's slave,
Taught the Dutch colours from its top to wave; 620
Of former glories the reproachful thought,
With present shame compared, his mind distraught.
Such from Euphrates' bank, a tigress fell

After the robbers for her whelps doth yell;
But sees enranged the river flow between,
Frustrate revenge and love, by loss more keen,
At her own breast her useless claws does arm:
She tears herself, since him she cannot harm.
   The guards, placed for the chain's and fleet's defence,°
Long since were fled on many a feigned pretence. 630
Daniel had there adventured, man of might,°
Sweet painter, draw his picture while I write.
Paint him of person tall, and big of bone,
Large limbs like ox, not to be killed but shown.°
Scarce can burnt ivory feign a hair so black,
Or face so red, thine ochre and thy lac.°
Mix a vain terror in his martial look,
And all those lines by which men are mistook;
But when, by shame constrained to go on board,
He heard how the wild cannon nearer roared, 640
And saw himself confined like sheep in pen,
Daniel then thought he was in lion's den.°
And when the frightful fireships he saw,
Pregnant with sulphur, to him nearer draw,
Captain, lieutenant, ensign, all make haste
Ere in the fiery furnace they be cast—
Three children tall, unsinged, away they row,
Like Shadrack, Meschack, and Abednego.°
   Not so brave Douglas, on whose lovely chin°
The early down but newly did begin, 650
And modest beauty yet his sex did veil,
While envious virgins hope he is a male.
His yellow locks curl back themselves to seek,
Nor other courtship knew but to his cheek.
Oft, as he in chill Esk or Seine by night
Hardened and cooled his limbs, so soft, so white,
Among the reeds, to be espied by him,
The nymphs would rustle; he would forward swim.
They sighed and said, 'Fond boy, why so untame
That fliest love's fires, reserved for other flame?' 660
Fixed on his ship, he faced that horrid day
And wondered much at those that run away.
Nor other fear himself could comprehend
Then, lest heaven fall ere thither he ascend,

But entertains the while his time too short
With birding at the Dutch, as if in sport,°
Or waves his sword, and could he them conjure°
Within its circle, knows himself secure.
The fatal bark him boards with grappling fire,
And safely through its port the Dutch retire. 670
That precious life he yet disdains to save
Or with known art to try the gentle wave.
Much him the honours of his ancient race
Inspire, nor would he his own deeds deface,
And secret joy in his calm soul does rise
That Monck looks on to see how Douglas dies.
Like a glad lover, the fierce flames he meets,
And tries his first embraces in their sheets.
His shape exact, which the bright flames enfold,
Like the sun's statue stands of burnished gold. 680
Round the transparent fire about him glows,
As the clear amber on the bee does close,°
And, as on angels' heads their glories shine,
His burning locks adorn his face divine.
But when in his immortal mind he felt
His altering form and soldered limbs to melt,
Down on the deck he laid himself and died,
With his dear sword reposing by his side,
And on the flaming plank, so rests his head
As one that's warmed himself and gone to bed. 690
His ship burns down, and with his relics sinks,
And the sad stream beneath his ashes drinks.
Fortunate boy, if either pencil's fame,
Or if my verse can propagate thy name,
When Oeta and Alcides are forgot,°
Our English youth shall sing the valiant Scot.°
Each doleful day still with fresh loss returns:
The *Loyal London* now a third time burns,°
And the true *Royal Oak* and *Royal James*,
Allied in fate, increase, with theirs, her flames. 700
Of all our navy none should now survive,
But that the ships themselves were taught to dive,
And the kind river in its creek them hides,
Fraughting their pierced keels with oozy tides.
Up to the bridge contagious terror struck:

The Tower itself with the near danger shook,
And were not Ruyter's maw with ravage cloyed,
E'en London's ashes had been then destroyed.
Officious fear, however, to prevent
Our loss does so much more our loss augment: 710
The Dutch had robbed those jewels of the crown;
Our merchantmen, lest they be burned, we drown.°
So when the fire did not enough devour,
The houses were demolished near the Tower.
Those ships that yearly from their teeming hole°
Unloaded here the birth of either Pole—
Furs from the north and silver from the west,
Wines from the south, and spices from the east;
From Gambo gold, and from the Ganges gems—°
Take a short voyage underneath the Thames, 720
Once a deep river, now with timber floored,
And shrunk, lest navigable, to a ford.°
  Now (nothing more at Chatham left to burn),
The Holland squadron leisurely return,
And spite of Ruperts and of Albemarles,
To Ruyter's triumph lead the captive *Charles*.
The pleasing sight he often does prolong:
Her masts erect, tough cordage, timbers strong,
Her moving shapes, all these he does survey,
And all admires, but most his easy prey. 730
The seamen search her all within, without:
Viewing her strength, they yet their conquest doubt;
Then with rude shouts, secure, the air they vex,
With gamesome joy insulting on her decks.
Such the feared Hebrew, captive, blinded, shorn,
Was led about in sport, the public scorn.°
  Black day accursed! On thee let no man hale
Out of the port, or dare to hoist a sail,
Nor row a boat in thy unlucky hour.
Thee, the year's monster, let thy dam devour, 740
And constant time, to keep his course yet right,
Fill up thy space with a redoubled night.
When aged Thames was bound with fetters base,
And Medway chaste ravished before his face,
And their dear offspring murdered in their sight,
Thou and thy fellows held'st the odious light.

Sad change since first that happy pair was wed,
When all the rivers graced their nuptial bed,
And Father Neptune promised to resign
His empire old to their immortal line! 750
Now with vain grief their vainer hopes they rue,
Themselves dishonoured, and the gods untrue,
And to each other, helpless couple, moan,
As the sad tortoise for the sea does groan.
But most they for their darling *Charles* complain,
And were it burnt, yet less would be their pain.
To see that fatal pledge of sea command,
Now in the ravisher De Ruyter's hand,
The Thames roared, swooning Medway turned her tide,
And were they mortal, both for grief had died. 760
  The court in farthing yet itself does please,
(And female Stuart there rules the four seas),
But fate does still accumulate our woes,
And Richmond here commands, as Ruyter those.°
  After this loss, to relish discontent,°
Someone must be accused by punishment.
All our miscarriages on Pett must fall:°
His name alone seems fit to answer all.
Whose counsel first did this mad war beget?
Who all commands sold through the navy? *Pett.* 770
Who would not follow when the Dutch were beat?
Who treated out the time at Bergen? *Pett.*°
Who the Dutch fleet with storms disabled met,
And rifling prizes, them neglected? *Pett.*
Who with false news prevented the Gazette,°
The fleet divided, writ for Rupert? *Pett.*
Who all our seamen cheated of their debt,
And all our prizes who did swallow? *Pett.*
Who did advise no navy out to set,
And who the forts left unrepaired? *Pett.* 780
Who to supply with powder did forget
Languard, Sheerness, Gravesend and Upnor? *Pett.*
Who all our ships exposed in Chatham's net?
Who should it be but the fanatic *Pett*?°
*Pett*, the sea-architect, in making ships
Was the first cause of all these naval slips:
Had he not built, none of these faults had been;
If no creation, there had been no sin.

But his great crime, one boat away he sent,°
That lost our fleet and did our flight prevent. 790
 Then (that reward might in its turn take place,
And march with punishment in equal pace),
Southampton dead, much of the treasure's care
And place in council fell to Duncombe's share.°
All men admired he to that pitch could fly:
Powder ne'er blew man up so soon so high,
But sure his late good husbandry in petre°
Showed him to manage the Exchequer meeter;
And who the forts would not vouchsafe a corn,°
To lavish the King's money more would scorn. 800
Who hath no chimneys, to give all is best,
And ablest Speaker, who of law has least;°
Who less estate, for Treasurer most fit,
And for a counsellor, he that has least wit.
But the true cause was that, in's brother May,°
The Exchequer might the Privy Purse obey.
 But now draws near the Parliament's return;
Hyde and the court again begin to mourn:
Frequent in council, earnest in debate,
All arts they try how to prolong its date. 810
Grave primate Sheldon (much in preaching there)°
Blames the last session and this more does fear:
With Boynton or with Middleton 'twere sweet,°
But with a Parliament abhors to meet,
And thinks 'twill ne'er be well within this nation,
Till it be governed by a convocation.
But in the Thames' mouth still De Ruyter laid;°
The peace not sure, new army must be paid.
Hyde saith he hourly waits for a dispatch;
Harry came post just as he showed his watch,° 820
All to agree the articles were clear—
The Holland fleet and Parliament so near—
Yet Harry must job back, and all mature,°
Binding, ere the Houses meet, the treaty sure,
And 'twixt necessity and spite, till then,
Let them come up so to go down again.°
 Up ambles country justice on his pad,
And vest bespeaks to be more seemly clad.°
Plain gentlemen are in stagecoach o'erthrown

And deputy-lieutenants in their own. 830
The portly burgess through the weather hot
Does for his corporation sweat and trot;
And all with sun and choler come adust
And threaten Hyde to raise a greater dust.
But fresh as from the mint, the courtiers fine
Salute them, smiling at their vain design,
And Turner gay up to his perch does march°
With face new bleached, smoothened and stiff with starch;
Tells them he at Whitehall had took a turn
And for three days thence moves them to adjourn. 840
'Not so!' quoth Tomkins, and straight drew his tongue,
Trusty as steel that always ready hung,
And so, proceeding in his motion warm,
The army soon raised, he doth as soon disarm.
True Trojan! While this town can girls afford,
And long as cider lasts in Hereford,
The girls shall always kiss thee, though grown old,
And in eternal healths thy name be trolled.
Meanwhile the certain news of peace arrives
At court, and so reprieves their guilty lives. 850
Hyde orders Turner that he should come late,
Lest some new Tomkins spring a fresh debate.
The King that day raised early from his rest,
Expects (as at a play) till Turner's dressed.°
At last together Ayton come and he:°
No dial more could with the sun agree.
The Speaker, summoned, to the Lords repairs,
Nor gave the Commons leave to say their prayers,
But like his prisoners to the bar them led,
Where mute they stand to hear their sentence read. 860
Trembling with joy and fear, Hyde them prorogues,
And had almost mistook and called them rogues.
Dear painter, draw this Speaker to the foot;
Where pencil cannot, there my pen shall do't:
That may his body, this his mind explain.
Paint him in golden gown, with mace's brain,
Bright hair, fair face, obscure and dull of head,
Like knife with ivory haft and edge of lead.
At prayers his eyes turn up the pious white,
But all the while his private bill's in sight. 870

In chair, he smoking sits like master cook,
And a poll bill does like his apron look.°
Well was he skilled to season any question
And made a sauce, fit for Whitehall's digestion,
Whence every day, the palate more to tickle,
Court-mushrooms ready are, sent in in pickle.°
When grievance urged, he swells like squatted toad,
Frisks like a frog, to croak a tax's load;
His patient piss he could hold longer than
A urinal, and sit like any hen; 880
At table jolly as a country host
And soaks his sack with Norfolk, like a toast;°
At night, than Chanticleer more brisk and hot,
And Sergeant's wife serves him for Pertelotte.°
Paint last the King, and a dead shade of night
Only dispersed by a weak taper's light,
And those bright gleams that dart along and glare
From his clear eyes, yet these too dark with care.
There, as in the calm horror all alone
He wakes, and muses of the uneasy throne; 890
Raise up a sudden shape with virgin's face,
(Though ill agree her posture, hour, or place),
Naked as born, and her round arms behind
With her own tresses, interwove and twined;
Her mouth locked up, a blind before her eyes,
Yet from beneath the veil her blushes rise,
And silent tears her secret anguish speak;
Her heart throbs and with very shame would break.
The object strange in him no terror moved:
He wondered first, then pitied, then he loved, 900
And with kind hand does the coy vision press
(Whose beauty greater seemed by her distress),
But soon shrunk back, chilled with her touch so cold,
And the airy picture vanished from his hold.
In his deep thoughts the wonder did increase,
And he divined 'twas England or the Peace.
Express him startling next with listening ear,°
As one that some unusual noise does hear.
With cannon, trumpets, drums, his door surround—
But let some other painter draw the sound. 910
Thrice did he rise, thrice the vain tumult fled,

But again thunders, when he lies in bed.
His mind secure does the known stroke repeat°
And finds the drums Louis's march did beat.
  Shake then the room, and all his curtains tear
And with the blue streaks infect the taper clear,
While the pale ghosts his eye does fixed admire
Of grandsire Harry and of Charles his sire.°
Harry sits down, and in his open side
The grisly wound reveals of which he died, 920
And ghastly Charles, turning his collar low,
The purple thread about his neck does show,
Then whispering to his son in words unheard,
Through the locked door both of them disappeared.
The wondrous night the pensive King revolves,
And rising straight on Hyde's disgrace resolves.
  At his first step, he Castlemaine does find,
Bennet, and Coventry, as't were designed;
And they, not knowing, the same thing propose
Which his hid mind did in its depths enclose. 930
Through their feigned speech their secret hearts he knew:
To her own husband, Castlemaine untrue;
False to his master Bristol, Arlington;°
And Coventry, falser than anyone,
Who to the brother, brother would betray,°
Nor therefore trusts himself to such as they.
His father's ghost, too, whispered him one note,
That who does cut his purse will cut his throat,
But in wise anger he their crimes forbears,
As thieves reprieved for executioners; 940
While Hyde provoked, his foaming tusk does whet,
To prove them traitors and himself the *Pett*.
  Painter, adieu! How will our arts agree,
Poetic picture, painted poetry;
But this great work is for our monarch fit,
And henceforth Charles only to Charles shall sit.
His master-hand the ancients shall outdo,
Himself the poet and the painter too.

*To the King*

So his bold tube, man to the sun applied°
And spots unknown to the bright star descried, 950

Showed they obscure him, while too near they please
And seem his courtiers, are but his disease.
Through optic trunk the planet seemed to hear,°
And hurls them off e'er since in his career.
  And you, great sir, that with him empire share,
Sun of our world, as he the Charles is there,
Blame not the Muse that brought those spots to sight,
Which in your splendour hid, corrode your light:
(Kings in the country oft have gone astray
Nor of a peasant scorned to learn the way).° 960
Would she the unattended throne reduce,
Banishing love, trust, ornament, and use,
Better it were to live in cloister's lock,
Or in fair fields to rule the easy flock.
She blames them only who the court restrain
And where all England serves, themselves would reign.
  Bold and accursed are they that all this while
Have strove to isle our monarch from his isle,
And to improve themselves, on false pretence,
About the Common-Prince have raised a fence; 970
The kingdom from the crown distinct would see
And peel the bark to burn at last the tree.
(But Ceres corn, and Flora is the spring,
Bacchus is wine, the country is the King.)
  Not so does rust insinuating wear,
Nor powder so the vaulted bastion tear,
Nor earthquake so a hollow isle o'erwhelm,
As scratching courtiers undermine a realm,
And through the palace's foundations bore,
Burrowing themselves to hoard their guilty store.° 980
The smallest vermin make the greatest waste,
And a poor warren once a city razed.
  But they, whom born to virtue and to wealth,
Nor guilt to flattery binds, nor want to stealth;
Whose generous conscience and whose courage high
Does with clear counsels their large souls supply;
That serve the King with their estates and care,
And, as in love, on Parliaments can stare,
(Where few the number, choice is there less hard):
Give us this court, and rule without a guard. 990

## *The loyal Scot*
## *Upon the Occasion of the Death of Captain Douglas burnt in one of his Majesty's Ships at Chatham*

*By Cleveland's Ghost*°

Of the old heroes when the warlike shades
Saw Douglas marching on the Elysian glades,
They straight consulting, gathered in a ring,
Which of their poets should his welcome sing,
And, as a favourable penance, chose
Cleveland, on whom they would the task impose.
He understood, and willingly addressed
His ready muse to court the warlike guest.
Much had he cured the tumour of his vein,°
He judged more clearly now, and saw more plain; 10
For those soft airs had tempered every thought,
And of wise Lethe he had took a draught,
Abruptly he began, disguising art,
As of his satire this had been a part.
  Not so brave Douglas, on whose lovely chin
The early down but newly did begin;
And modest beauty yet his sex did veil,
While envious virgins hope he is a male.
His shady locks curl back themselves to seek:
Nor other courtship knew but to his cheek. 20
Oft as he in chill Eske or Seine by night
Hardened and cooled those limbs so soft, so white,
Among the reeds, to be espied by him,
The nymphs would rustle; he would forward swim.
They sighed and said, 'Fond boy, why so untame
That fliest Love's fires, reserved for other flame?'
Fixed on his ship he faced the horrid day,
And wondered much at those that run away:
Nor other fear himself could comprehend
Than, lest heaven fall ere thither he ascend. 30
With birding at the Dutch, as though in sport,
He entertains the while his life too short
Or waves his sword, and could he them conjure

Within its circle, knows himself secure.
The fatal bark him boards with grappling fire,
And safely through its ports the Dutch retire.
That precious life he yet disdains to save,
Or with known art to try the gentle wave.
Much him the glories of his ancient race°
Inspire, nor could he his own deeds deface: 40
And secret joy in his calm soul does rise,
That Monck looks on to see how Douglas dies.°
Like a glad lover the fierce flames he meets,
And tries his first embraces in their sheets.
His shape exact, which the bright flames enfold,
Like the sun's statue stands of burnished gold.
Round the transparent fire about him glows,
As the clear amber on the bee doth close.
And as on angels' head their glories shine,
His burning locks adorn his face divine. 50
But when in his immortal mind he felt
His altering form and soldered limbs to melt,
Down on the deck he laid him down and died,
With his dear sword reposing by his side:
And on the flaming planks so rests his head
As one that hugs himself in a warm bed.
The ship burns down and with his relics sinks,
And the sad stream beneath his ashes drinks.
Fortunate boy, if e'er my verse may claim
That matchless grace to propagate thy fame, 60
When Oeta and Alcides are forgot
Our English youth shall sing the valiant Scot.
Skip-saddles Pegasus, thou needst not brag,°
Sometimes the Galloway proves the better nag.°
Shall not a death so generous now when told
Unite the distance, fill the breaches old?
Such in the Roman Forum, Curtius brave°
Galloping down closed up the gaping cave.
No more discourse of Scotch or English race,
Nor chant the fabulous hunt of Chevy Chase. 70
Mixed in Corinthian metal, at thy flame°
Our nations melting, thy colossus frame,
Shall fix a foot on either neighbouring shore,
And join those lands that seemed to part before.°

Prick down the point (whoever has the art),
Where Nature Scotland does from England part.
Anatomists may sooner fix the cells
Where life resides, or understanding dwells:
But this we know, though that exceed their skill,
That whosoever separates them doth kill.° 80
What ethic river is this wondrous Tweed,
Whose one bank virtue, th'other vice does breed?
Or what new perpendicular does rise
Up from her stream, continued to the skies,
That between us the common air should bar
And split the influence of every star?°
But who considers right will find indeed
'Tis Holy Island parts us, not the Tweed.°
Nothing but clergy could us two seclude,
No Scotch was ever like a bishop's feud. 90
All litanies in this have wanted faith.
There's no 'Deliver us from the bishop's wrath.'
Never shall Calvin pardoned be for Sales,°
Never for Burnet's sake the Lauderdales,°
For Becket's sake Kent always shall have tails.°
Who sermons e'er can pacify and prayers?°
Or to the joint stools reconcile the chairs?°
Nothing, not bogs, not sands, not seas, not alps,
Separate the world, so as the bishops' scalps.
Stretch for your line, their surcingle alone,° 100
'Twill make a more inhabitable zone.°
The friendly loadstone hath not more combined,
Than bishops cramped the commerce of mankind.
A bishop will like Mahomet tear the moon,
And slip one half into his sleeve as soon.
The juggling prelate on his hocus calls,°
Shows you first one, then makes that one two balls.
Instead of all the plagues, had bishops come,
Pharaoh at first would have sent Israel home.°
From church they need not censure men away, 110
A bishop's self is an anathema:
Where foxes dung, their earths the badgers yield,
At bishops' musk, even foxes quit the field.°
Their rank ambition all this heat has stirred,
A bishop's rennet makes the strongest curd.°

What reverend things (Lord) are lawn sleeves and ease!°
How a clean laundress and no sermons please!
They wanted zeal and learning, so forsook°
Bible and grammar for the service book.
Religion has too long the world depraved, 120
A shorter way's to be by clergy saved.
Believe, but only as the church believes,
And learn to pin your faith upon their sleeves.
(Ah, like Lot's wife they still look back and halt,
And, surpliced, show like pillars too of salt.)
Who that is wise would pulpit-toil endure?
A bishopric is a great *Sine-cure*.
Enough for them, God knows, to count their wealth,
To excommunicate, and study health.
A higher work is to their call annexed; 130
The nation they divide, their curates, text.
No bishop? Rather than it should be so,
No church, no trade, no king, no people, no.
All mischief's moulded by these state divines;
Aaron cast calves, but Moses them calcines.°
The legion-devil did but one man possess;
One bishop's fiend spirits a whole diocese.
That power alone can loose this spell that ties:
For only kings can bishops exorcise.
Will you be treated princes, here fall too: 140
Fish and flesh bishops are your ambigu.°
Howe'er insipid, yet the sauce will mend 'em,
Bishops are very good when *in commendam*.°
If wealth or vice can tempt your appetites,
These Templar Lords exceed the Templar Knights.°
And in the baron bishop you have both
Leviathan served up and behemoth.°
How can you bear such miscreants should live,
And holy ordure holy orders give?
None knows what god our flamen now adores:° 150
One mitre fits the heads of four Moors.°
No wonder if the orthodox do bleed,
While Arius stands at the Athanasian Creed.
What so obdurate pagan-heretic
But will transform for an archbishopric.
In faith erroneous and in life profane

These hypocrites their faith and linen stain.
Seth's pillars are no antique brick or stone;°
But of the choicest modern flesh and bone.
Who views but Gilbert's toils will reason find 160
Neither before to trust him nor behind.
How oft hath age his hallowing hands misled,
Confirming breasts and armpits for the head!
Abbot one buck, but he shot many a doe:°
Nor is our Sheldon whiter than his Snow.°
Their company's the worst that ever played,
And their religion all but masquerade.
The conscious prelate therefore did not err,
When for a church he built a theatre.
A congruous dress they to themselves adapt,° 170
Like smutty stories in clean linen wrapped.
Do but their piebald lordships once uncase
Of rochets, tippets, copes, and where's their Grace?
A hungry chaplain and a starved rat,
Eating their brethren, bishop turn and cat.
But an apocryphal Archbishop Bell
Like snake, by swallowing toads, doth dragon swell.
When daring Blood to have his rents regained°
Upon the English diadem distrained,
He chose the cassock, surcingle and gown,° 180
The fittest mask for one that robs a crown.
But his lay pity underneath prevailed,
And while he spared the keeper's life he failed.
With the priest's vestments had he but put on
A bishop's cruelty, the crown had gone.
Strange was the sight, that Scotch twin-headed man
With single body, like the two-necked swan;
And wild disputes betwixt those heads must grow
Where but two hands to act, two feet to go.
Nature in living emblem there expressed 190
What Britain was between two kings distressed.
But now when one head does both realms control,
The bishop's noddle perks up cheek by jowl.
They, though no poets, on Parnassus dream,
And in their causes think themselves supreme.
King's-head saith this, but bishop's-head that do;
Does Charles the Second reign, or Charles the Two?

Well that Scotch monster and our bishops sort,
(It was musician too, and dwelt at court).
  Hark, though at such a distance what a noise 200
Shattering the silent air disturbs our joys:
The mitred hubbub against Pluto moot,°
The cloven head must govern cloven foot.
Strange boldness! Bishops even there rebel,°
And plead their *jus divinum* though in Hell.°
Those whom you hear more clamorous yet and loud,
Of ceremonies wrangle in the crowd,
And would, like chemists fixing mercury,
Transfuse Indifference with Necessity.
To sit is necessary in Parliament, 210
To preach in diocese, indifferent;
To conform's necessary or be shent,
But to reform is all indifferent.
'Tis necessary bishops have their rent,
To cheat the plague-money, indifferent°
'Tis necessary to rebabel Paul's,
Indifferent to rob churches of their coals.
'Tis necessary Lambeth never wed,
Indifferent to have a wench in bed;
Such bishops are with all their complement, 220
Not necessary, nor indifferent.
  Incorrigible among all their pains,
Some sue for tithe of the Elysian plains.
Others attempt (to cool their fervent chine)
A second time to ravish Proserpine.
Even Father Dis, though so with age defaced,
With much ado preserves his postern chaste.
The innocentest mind there thirst alone,
And, uninforced, quaff healths in Phlegeton.
Luxury, malice, superstition, pride, 230
Oppression, avarice, ambition, Id-°
leness, and all the vice that did abound
While they lived here, still haunts them underground.
Had it not been for such a bias strong,
Two nations ne'er had missed the mark so long.
The world in all does but two nations bear,
The good, the bad, and those mixed everywhere:
Under each Pole place either of the two,

The good will bravely, bad will basely do;
And few indeed can parallel our climes 240
For worth heroic, or heroic crimes.
The trial would, however, be too nice,
Which stronger were, a Scotch or English vice,
Or whether the same virtue would reflect
From Scotch or English heart the same effect.
  Nation is all but name as shibboleth,°
Where a mistaken accent causes death.
In paradise names only Nature showed,
At Babel names from pride and discord flowed;
And ever since men with a female spite, 250
First call each other names, and then they fight.
Scotland and England! Cause of just uproar,
Does man and wife signify rogue and whore?
Say but a Scot, and straight we fall to sides,
That syllable like a Pict's wall divides.
Rational men's words pledges all of peace,
Perverted, serve dissensions to increase.
For shame, extirpate from each loyal breast,
That senseless rancour against interest.
One king, one faith, one language, and one isle, 260
English and Scotch, 'tis all but cross and pile.°
  Charles, our great soul, this only understands,
He our affection both and will commands.
And where twin sympathies cannot atone,°
Knows the last secret how to make them one.
Just so the prudent husbandman who sees
The idle tumult of his factious bees,
The morning dews, and flowers neglected grown,
The hive a comb-case, every bee a drone,°
Powders them o'er, till none discern their foes, 270
And all themselves in meal and friendship close;
The insect kingdom straight begins to thrive,
And each works honey for the common hive.
  Pardon, young hero, this so long transport,
(Thy death more noble did the same exhort).
My former satire for this verse forget,°
The hare's head 'gainst the goose giblets set.°
I single did against a nation write,
Against a nation thou didst singly fight.

My differing crime does more thy vertue raise, 280
And such my rashness best thy valour praise.
Here Douglas, smiling, said he did intend
After such frankness shown to be his friend,
Forewarned him therefore lest in time he were
Metempsychosed to some Scotch presbyter.

# THE REHEARSAL TRANSPROSED

## ANIMADVERSIONS UPON THE PREFACE TO BISHOP BRAMHALL'S VINDICATION, ETC.

---

THE author° of this Preface had first written *A Discourse of Ecclesiastical Policy*; after that, *A Defence and Continuation of the Ecclesiastical Policy*; and there he concludes his Epistle to the Reader in these words: 'But if this be the penance I must undergo for the wantonness of my pen, to answer the impertinent and slender exceptions of every peevish° and disingenuous caviller; reader, I am reformed from my incontinency of scribbling, and do here heartily bid thee an eternal farewell.' Now this expression lies open to his own dilemma against the Nonconformists confessing in their prayers to God such heinous enormities. For if he will not accept his own charge, his modesty is all impudent and counterfeit; or, if he will acknowledge it, why then he had been before, and did still remain upon record, the same lewd, wanton and incontinent scribbler.

But, however, I hoped he had been a clergyman of honour, and that when herein the world and he himself were now so fully agreed in the censure of his writings, he would have kept his word; or at least that his pen would not, so soon, have created us a disturbance of the same nature, and so far manifested how indifferent he is as to the business either of truth or eternity. But the author, alas, instead of his own, was fallen now into Amaryllis's dilemma. (I perceive the gentleman hath travelled by his remembering *Chi lava la testa al asino perde il sapone*;° and therefore hope I may without pedantry quote the words in her own whining Italian:)

> S'il peccar e si dolce e 'l non peccar si necessario,
> O troppo imperfetta natura che ripugni a la legge.
> O troppo dura legge che natura offendi.°
>
> If to scribble be so sweet, and not to scribble be so necessary;
> O too frail inclination, that contradictest obligation!
> O too severe obligation, that offendest inclination!

For all his promise to write no more, I durst always have laid ten pound to a crown on nature's side. And accordingly he hath now blessed us with, as he calls it, *A Preface, showing what Grounds there are of Fears and Jealousies of Popery.*

It will not be unpleasant to hear him begin his story: 'The ensuing treatise of Bishop Bramhall's° being somewhat superannuated, the bookseller was very solicitous to have it set off with some preface that might recommend it to the genius of the age, and reconcile it to the present juncture of affairs.' A pretty task indeed! That is as much as to say, to trick up° the good old Bishop in a yellow coif and a bull's head,° that he may be fit for the public, and appear in fashion. In the meantime, 'tis what I always presaged: from a writer of books, our author is already dwindled to a preface-monger, and from prefaces I am confident he may in a short time be improved to indite° tickets for the Beargarden.° But the bookseller I see was a cunning fellow, and knew his man. For who so proper as a young priest to sacrifice to the genius of the age; yea, though his conscience were the offering? And none more ready to nick a juncture of affairs° than a malapert° chaplain; though not one indeed of a hundred but dislocates them in the handling. And yet, our author is very maidenly, and condescends to his bookseller not without some reluctance, as being, forsooth, first of all 'none of the most zealous patrons of the press'.

Though he hath so lately forfeited his credit, yet herein I dare believe him: for the press hath owed him a shame a long time, and is but now beginning to pay off the debt. The press (that 'villainous' engine) invented much about the same time with the Reformation, that hath done more mischief to the discipline of our Church than all the doctrine can make amends for. 'Twas an happy time when all learning was in manuscript, and some little officer, like our author, did keep the keys of the library. When the clergy needed no more knowledge than to read the Liturgy, and the laity no more clerkship than to save them from hanging.° But now, since printing came into the world, such is the mischief, that a man cannot write a book but presently he is answered. Could the press but once be conjured to obey only an *Imprimatur*,° our author might not disdain perhaps to be one of its most zealous patrons. There have been ways found out to banish ministers, to fine not only the people, but even the grounds and fields where they assembled in conventicles;° but no art yet could prevent these seditious meetings of letters. Two or three brawny fellows in a corner, with mere° ink and elbow-grease, do more harm than 'a hundred systematical divines' with their 'sweaty preaching'. And, which is a strange thing, the very sponges,° which one would think should rather deface and blot out the whole book, and were anciently used to that purpose, are become now the instruments to make things legible. Their ugly printing letters, that look but like so many rotten teeth, how oft have they been pulled out by B. and L.° the public tooth drawers! and yet these rascally operators of the Press have got a trick to fasten them again in a few minutes, that they grow as firm a set and as biting and talkative as ever. O printing, how hast thou disturbed the peace of mankind! that lead, when moulded into bullets, is not so mortal as when founded into letters! There was a mistake sure in the story of Cadmus; and the serpent's teeth which he sowed° were nothing else but the letters which he invented. The first essay that was made towards this art was in single characters upon iron, wherewith of old they stigmatized slaves and remarkable offenders; and it was of good use sometimes to brand a schismatic. But a bulky Dutchman° diverted it quite from its first institution, and contriving those innumerable syntagms° of alphabets, hath pestered the world ever since with the 'gross bodies of their German divinity'. One would have thought in reason that a Dutchman at least might have contented himself only with the wine-press.

But, next of all, our author, beside his aversion from the press, alleges, that 'he is as much concerned as De Witt, or any of the high and mighty burgomasters, in matters of a closer and more comfortable importance to himself and his own affairs'. And yet whoever shall take the pains to read over his preface, will find that it intermeddles with the King, the Succession, the Privy-council, Popery, atheism, Bishops, ecclesiastical government, and above all with Nonconformity, and J.O.° A man would wonder what this thing should be of a 'closer importance'. But being 'more comfortable' too, I conclude it must be one of these three things: either his salvation, or a benefice, or a female. Now as to Salvation, he could not be so much concerned, for that care was over; there hath been a course taken to ensure all that are on his bottom. And he is yet surer of a benefice; or else his patrons must be very ungrateful. He cannot have deserved less than a Prebend for his first book, a sinecure for his second, and for his third a rectorship, although it were that of Malmesbury.° Why, then, of necessity it must be a female. For that, I confess, might have been a sufficient excuse from writing of Prefaces, and against the importunity of the bookseller. 'Twas fit that all business should have given place to the work of propagation. Nor was there anything that could more closely import him, than that the race and family of the railers° should be perpetuated among mankind. Who could in reason expect that a man should in the same moments undertake the labour of an author and a father? 'Nevertheless,' he saith, 'he could not but yield so far as to improve every fragment of time that he could get into his own disposal, to gratify the importunity of the bookseller.' Was ever civility graduated up and enhanced to such a value! His mistress herself could not have endeared a favour so nicely, nor granted it with more sweetness.

Was the bookseller more importunate, or the author more courteous?

The author was the pink of courtesy, the bookseller the burr of importunity.°

And so, not being able to shake him off, 'this,' he saith, 'hath brought forth this Preface, such a one as it is; for how it will prove, he himself neither is, nor (till 'tis too late) ever shall be a competent judge, in that it must be ravished out of his hands before his thoughts can possibly be cool enough to review or correct the indecencies either of its style or contrivance.' He is now growing a very enthusiast himself. No Nonconformist minister, as it seems, could have spoken more *extempore.* I see he is not so civil to his readers as he was to his bookseller: and so A.C. and James Collins° be gratified, he cares not how much the rest of the world be disobliged. Some man, that had less right to be fastidious and confident, would, before he exposed himself in public, both have cooled his thoughts, and corrected his indecencies; or would have considered whether it were necessary or wholesome that he should write at all. For as much as one of the ancient Sophists (they were a kind of orators of his form) killed himself with declaiming while he had a bone in his throat, and J.O. was still in being. 'Put up your trumpery, good noble marquess.'° But there was no holding him. Thus it must be and no better, when a man's fancy is up, and his breeches are down; when the mind and the body make contrary assignations, and he hath both a bookseller at once and a mistress to satisfy; like Archimedes, into the street he runs out naked with his invention. And truly, if at any time, we might now pardon this extravagance and rapture of our author, when he was perched upon the highest pinnacle of ecclesiastical felicity, being ready at once to assuage his concupiscence, and wreak his malice.

'But yet he knows not which way his mind will work itself and its thoughts.' This is Bays° the second.—''Tis no matter for the plot—the intrigo was out of his head.—But you'll apprehend it better when you see 't.'° Or rather, he is like Bays's actors, 'that could not guess what humour they were to be in whether angry, melancholy, merry, or in love'.° Nay, insomuch that he saith, 'he is neither prophet nor astrologer enough to foretell.' Never man certainly was so unacquainted with himself. And, indeed, 'tis part of his discretion to avoid his acquaintance and tell him as little of his mind as may be: for he is a dangerous fellow. But I must ask pardon if I treat him too homely.° It is his own fault that misled me at first, by concealing his quality under such vulgar comparisons as 'De Wit' and the 'Burgomasters'. I now see it all along: this can be no less a man than Prince Volscius° himself, in dispute betwixt his boots which way his mind will 'work itself'; whether love shall detain him with his 'closer importance', Parthenope, 'whose mother, sir, sells ale by the town wall':° or honour shall carry him 'to head the army that lies concealed for him at Knightsbridge', and to encounter J.O.

> Go on, cryes Honour: tender Love saith Nay.
> Honour aloud commands, Pluck both boots on.
> But safer Love doth whisper, Put on none.°

And so now, when it comes that he is 'not prophet nor astrologer enough to foretell' what will do 'tis just,

> For as bright day, with black approach of night,
> Contending, makes a doubtful puzzling light;
> So does my honour and my love together
> Puzzle me so, I am resolved on neither.°

Yet no astrologer could possibly have more advantage and opportunity to make a judgement. For he knew the very minute of the conception of his Preface, which was immediately upon his Majesty's issuing his Declaration of Indulgence° to tender consciences. Nor could he be ignorant of the moment when it was brought forth. And I can so far refresh his memory, that it came out in the dog-days:°

> . . . the season hot, and she too near:
> O mighty Love! J.O. will be undone;°

according to the rule in Davenant's Ephemerides:° 'But the heads which at this moment, and under present schemes and aspects of the heavens he intends to treat of (pure Sidrophel°) are these two: first, something of the treatise itself; secondly, of the seasonableness of its publication: and this, unless his humour jade° him ['tis come to a dog-trot already], will lead him further into the argument as it relates to the present state of things, and from thence 'tis odds but he shall take occasion to bestow some animadversions upon one J.O.' There's no trusting him: he doubtless knew from the beginning what he intended. And so too all his story of the bookseller, and all the '*volo nolo's*',° and 'shall I, shall I's' betwixt them, was nothing but fooling: and he now all along owns himself to be the publisher, and alleges the slighter and the main reasons that induced him. Would he had told us so at first; for then he had saved me thus much of my labour. Though, as it chances, it lights not amiss on our author, whose delicate stomach could not brook that J.O. should say, 'he had prevailed with himself, much against his inclination, to bestow a few (and those idle) hours upon examining his book':° and yet he himself stumbles so notoriously upon the very same fault at his own threshold.

But now from this Preamble he falls into his Preface to Bishop Bramhall: though indeed like Bays's prologue, that would have served as well for an epilogue, I do not see but the Preface might have passed as well for a postscript, or the headstall for a crupper.° And our author's divinity might have gone to push-pin° with the Bishop, which of their two treatises was the 'procatarctical cause'° of both their editions. For, as they are coupled together, to say the truth, 'tis not discernable, as in some animals, whether their motion begin at the head or the tail; whether the author made his preface for 'Bishop Bramhall's dear sake', or whether he published the Bishop's treatise for sake of his 'own dear Preface'. For my own part, I think it reasonable that the Bishop and our author should (like fair gamesters at leap-frog) stand and skip in their turns; and however our author got it for once, yet if the bookseller should ever be solicitous for a second edition, that then the Bishop's book should have the precedence.

But before I commit myself to the dangerous depths of his discourse, which I am now upon the brink of, I would, with his leave, make a motion; that, instead of author, I may henceforth indifferently call him Mr Bays, as oft as I shall see occasion. And that, first, because he hath no name, or at least will not own it, though he himself writes under the greatest security, and gives us the first letters of other men's names before he be asked them. Secondly, because he is, I perceive, a lover of elegancy of style, and can endure no man's tautologies but his own; and therefore I would not distaste him with too frequent repetition of one word. But chiefly, because Mr Bays and he do very much symbolize, in their understandings, in their expressions, in their humour, in their contempt and quarrelling of all others, though of their own profession. Because our divine, the author, manages his contest with the same prudence and civility which the players and poets have practised of late in their several divisions. And lastly, because both their talents do peculiarly lie in exposing and personating the Nonconformists. I would therefore give our author a name, the memory of which may perpetually excite him to the exercise and highest improvement of that virtue. For, our Cicero doth not yet equal our Roscius,° and one turn of Lacy's° face hath more 'Ecclesiastical Policy' in it than all the books of our author put together. Besides to say Mr Bays is more civil than to say 'villain' and 'caitiff', though these indeed are more tuant.° And to conclude, the Irrefragable Doctor° of School-divinity, page 460 of his *Defence*, determining concerning symbolical ceremonies, hath warranted me that not only governors, but anything else may have power to appropriate new names to things, without having absolute authority over the things themselves. And therefore henceforward, seeing I am on such sure ground, author, or Mr Bays, whether I please. Now, having had our dance, let us advance to our more serious counsels.

And first: our author begins with a panegyric upon Bishop Bramhall; a person whom my age had not given me leave to be acquainted with, nor my good fortune led me to converse with his writings: but for whom I had collected a deep reverence from the general reputation he carried, beside the veneration due to the place he filled in the Church of England. So that our author having a mind to show us some proof of his good nature, and that his eloquence lay not all in satire and invectives, could not, in my opinion, have fixed upon a fitter subject of commendation. And therefore I could have wished for my own sake that I had missed this occasion of being more fully informed of some of the Bishop's principles, whereby I have lost part of that pleasure which I had so long enjoyed in thinking well of so considerable a person. But, however, I recreate° myself with believing that my simple judgement cannot, beyond my intention, abate anything of his just value with others. And seeing he is long since dead, which I knew but lately, and now learn it with regret, I am the more obliged to repair in myself whatsoever breaches of his credit, by that additional civility which consecrates the ashes of the deceased. But by this means I am come to discern how it was possible for our author to speak a good word of any man. The Bishop was expired, and his writings jump° much with our author. So that if you have a mind to die, or to be of his party (there are but these two conditions), you may perhaps be rendered capable of his charity. And then write what you will, he will make you a Preface that shall recommend you and it to the genius of the age, and reconcile it to the juncture of affairs. But truly he hath acquitted himself herein so ill-favouredly to the Bishop, that I do not think it so much worth to gain his approbation; and I had rather live and enjoy mine own opinion than be so treated. For, beside his reflection on the Bishop, and the whole age he lived in, that 'he was, as far as the prejudice of the age would permit him, an acute philosopher' (which is a sufficient taste of Mr Bays's arrogance, that no man, no age can be so perfect but must abide his censure, and of the officious virulence of his humour, which infuses itself, by a malignant remark, that (but for this acuter philosopher) no man else would have thought of, into the praises of him whom he most intended to celebrate). If, I say, beside this, you consider the most elaborate and studious periods of his commendation, you find it at best very ridiculous. By the language he seems to transcribe out of the *Grand Cyrus*° and *Cassandra*,° but the exploits to have borrowed out of the *Knight of the Sun*° and *King Arthur*.° For in a luscious and effeminate style he gives him such a termagant character, as must either fright or turn the stomach of any reader: 'Being of a brave and enterprising temper, of an active and sprightly mind, he was always busied either in contriving or performing great designs.' Well, Mr Bays, I suppose by this, that he might have been an overmatch to the Bishop of Cologne° and the Bishop of Strasburg.° In another place, 'He finished all the glorious designs that he undertook.' This might have become the Bishop of Münster° before he had raised the siege from Groningen. 'As he was able to accomplish the most gallant attempts, so he was always ready not only to justify their innocence, but to make good their bravery.' I was too prodigal of my bishops at first, and now have never another left in the Gazette, which is too our author's Magazine. 'His reputation and innocence were both armour of proof against Tories° and Presbyterians.' But methinks, Mr Bays, having to do with such dangerous enemies, you should have furnished him too with some weapon of offence, a good old fox, like that of another hero, his contemporary in action upon the scene of Ireland, of whom it was sung:

> Down by his side he wore a sword of price,
> Keen as a frost, glazed like a new-made ice:
> That cracks men shelled in steel in a less trice
> Than squirrel's nuts, or the Highlander's lice.°

Then he saith, ''Tis true the Church of Ireland was the largest scene of his actions; but yet there in a little time he wrought out such wonderous alterations, and so exceeding all belief, as may convince us that he had a mind large and active enough to have managed the Roman Empire at its greatest extent.' This indeed of our author's is great; and yet it reacheth not a strain of his fellow pendets° in the history of the Mogol,° where he tells Dancehment Kan, 'when you put your foot in the stirrup, and when you march upon horseback in the front of the cavalry, the earth trembles under your feet, the eight elephants that hold it on their heads not being able to support it.' But enough of this trash.

Beside that it is the highest indecorum for a divine to write in such a style as this (part play-book and part romance) concerning a reverend

Bishop: these improbable eulogies too are of the greatest disservice to their own design, and do in effect diminish always the person whom they pretend to magnify. Any worthy man may pass through the world unquestioned and safe with a moderate recommendation; but when he is thus set off, and bedaubed with rhetoric, and embroidered so thick that you cannot discern the ground, it awakens naturally (and not altogether unjustly) interest, curiosity, and envy. For all men pretend a share in reputation, and love not to see it engrossed and monopolised and are subject to enquire (as of great estates suddenly got) whether he came by all this honestly, or of what credit the person is that tells the story. And the same hath happened as to this Bishop, while our Author attributes to him such achievements, which to one that could believe the *Legend of Captain Jones,*° might not be incredible. I have heard that there was indeed such a captain, an honest brave fellow; but a wag, that had a mind to be merry with him, hath quite spoiled his history. Had our author epitomized the legend of sixty-six books, *De Virtutibus Sancti Patricii*° (I mean not the ingenious writer of the *Friendly Debates,*° but St Patrick, the Irish Bishop), he could not have promised us greater miracles. And 'tis well for him that he hath escaped the fate of Secundinus,° who (as Jocelyn° relates it) acquainting Patrick that he was inspired to compose something in his commendation, the Bishop foretold the author should die as soon as it was perfected; which so done, so happened. I am sure our author had died no other death but of this his own Preface, and a surfeit upon Bishop Bramhall, if the swelling of truth could have choked him. He tells us, I remember somewhere, that this same Bishop of Derry said, the Scots had a civil expression for these 'improvers of verity', that 'they are good company'; and I shall say nothing severer, than that our author speaks the language of a lover, and so may claim some pardon, if the habit and excess of his courtship do as yet give a tincture to his discourse upon more ordinary subjects. For I would not by any means be mistaken, as if I thought our author so sharp set, or so necessitated that he should make a dead bishop his mistress; so far from that, that he hath taken such a course, that if the Bishop were alive, he would be out of love with himself. He hath, like those frightful looking-glasses made for sport, represented him in such bloated lineaments, as, I am confident, if he could see his face in it, he would break the glass. For, hence it falls out too, that men seeing the Bishop furbished up in so martial accoutrements, like another Odo, Bishop of Bayeux,° and having never before heard of his prowess, began to reflect what giants he defeated, and what damsels he rescued. Serious men consider whether he were engaged in the conduct of the

Irish army, and to have brought it over upon England, for the imputation of which the Earl of Strafford's° patron so undeservedly suffered. But none knows anything of it. Others think it not to be taken literally, but the wonderful and unheard-of alterations that he wrought out in Ireland are meant of some reformation that he made there in things of his own function. But then men ask again, how he comes to have all the honour of it, and whether all the while that great Bishop Ussher,° his metropolitan, were unconcerned? For even in ecclesiastical combats, how instrumental soever the Captain hath been, the General usually carries away the honour of the action. But the good Primate was engaged in designs of lesser moment, and was writing his *De Primordiis Ecclesiae Britannicae*,° and the story of Pelagius our countryman. He, honest man, was deep gone in Grub-street° and 'polemical divinity', and troubled with fits of 'modern orthodoxy'. He satisfied himself with being 'admired by the blue and white aprons', and 'pointed at by the more judicious tankard-bearers'. Nay, which is worst of all, he undertook to abate of our episcopal grandeur, and condescended indeed to reduce the ceremonious discipline in these nations to the primitive simplicity. What, then, was this that Bishop Bramhall did? Did he, like a Protestant apostle, in one day convert thousands of the Irish Papists? The contrary is evident by the Irish Rebellion° and massacre, which, notwithstanding his 'public employment and great abilities', happened in his time. So that, after all our author's bombast, when we have searched all over, we find ourselves bilked in our expectation; and he hath erected him, like a St Christopher, in the Popish churches, as big as ten porters, and yet only employed to sweat under the burden of an infant.

All that appears of him is, first, that he busied himself about a 'catholic agreement among the churches of Christendom'. But as to this, our author himself saith, that he was not 'so vain, or so presuming as to hope to see it effected in his days'. And yet but two pages before, he told us that 'the Bishop finished all the glorious designs which he undertook'. But this design of his he draws out in such a circuit of words, that 'tis better taking it from the Bishop himself, who speaks more plainly always and much more to the purpose. And he saith, page 87 of his *Vindication*, 'My design is rather to reconcile the Popish party to the Church of England, than the Church of England to the Pope.' And how he manages it, I had rather any man would learn by reading over his own book, than that I should be thought to misrepresent him, which I might, unless I transcribed the whole. But in sum, it seems to me that he is upon his own single judgement too liberal of the public, and that he retrenches both on our part more than he hath authority for, and grants more to the Popish than they can of right pretend to. It is, however, indeed a most glorious design, to reconcile all the churches to one doctrine and communion (though some that meddle in it do it chiefly in order to fetter men straiter under the formal bondage of fictitious discipline); but it is a thing rather to be wished and prayed for, than to be expected from these kind of endeavours. It is so large a field, that no man can see to the end of it: and all that have adventured to travel it have been bewildered. That man must have a vast opinion of his own sufficiency, that can think he may by his oratory or reason, either in his own time, or at any of our author's 'more happy junctures of affairs', so far persuade and fascinate the Roman Church, having by a regular contexture of continued Policy for so many ages interwoven itself with the secular interest, and made itself necessary to most princes, and having at last erected a throne of infallibility over their consciences, as to prevail with her to submit a power and empire so acquired and established in compromise to the arbitration of a humble proposer. God only in his own time, and by the inscrutable methods of his providence, is able to effect that alteration: though I think too He hath signified in part by what means he intends to accomplish it, and to range so considerable a Church, and once so exemplary, into primitive unity and Christian order. In the meantime such projects are fit for pregnant scholars that have nothing else to do, to go big with for forty years, and may qualify them to discourse with princes and statesmen at their hours of leisure; but I never saw that they came to use or possibility: no more than that of Alexander's architect,° who proposed to make him a statue of the mountain Athos (and that was no molehill); and among other things, that statue to carry in its hand a great habitable city. But the surveyor was gravelled,° being asked whence that city should be supplied with water. I would only have asked the Bishop, when he had carved and hammered the Romanists and Protestants into one colossal church, how we should have done as to matter of Bibles. For the Bishop, p. 117, complains that unqualified people should have a promiscuous licence to read the Scriptures: and you may guess thence, if he had moreover the Pope to friend, how the laity should have been used. There have been attempts in former ages to dig through the separating Isthmus of Peloponnesus,° and another to make communication between the Red Sea and the Mediterranean: both more easy than to cut this ecclesiastic canal, and yet both laid by partly upon the difficulty of doing it, and partly upon the inconveniences if it had been effected. I must confess freely, yet I ask pardon for the presumption, that I cannot look upon these undertaking churchmen, however otherwise of excellent prudence and learning, but as men struck with a notion, and crazed on that side of their head. And so I think even the Bishop had much better have busied himself in preaching in his own diocese, and disarming the Papists of their arguments, instead of rebating° our weapons, than in taking an ecumenical care upon him, which none called him to, and, as appears by the sequel, none conned him thanks for.° But if he were so great a politician as I have heard, and indeed believe him to have been, methinks he should in the first place have contrived how we might live well with our Protestant neighbours, and to have united us in one body under the King of England, as head of the Protestant interest, which might have rendered us more considerable, and put us into a more likely posture to have reduced the church of Rome to reason. For the most leading party of the English clergy in his time retained such a pontifical stiffness towards the foreign divines, that it puts me in mind of Austin° the monk, when he came into Kent, not deigning to rise up to the British or give them the hand, and could scarce afford their churches either communion or charity, or common civility. So that it is not to be wondered if they also on their parts looked upon our models of accommodation with the same jealousy that the British Christians had of Austin's design, to unite them first to (that is, under) the Saxons, and then deliver them both over, bound, to the Papal government and ceremonies. But seeing hereby our hands were weakened, and there was no probability of arriving so near the end of the work as to a consent among Protestants abroad; had the Bishop but gone that step, to have reconciled the ecclesiastical differences in our own nations, and that we might have stood firm at home before we had taken such a jump beyond sea, it would have been a performance worthy of his wisdom. For at that time the ecclesiastical rigours here were in the highest ferment, and the Church in being, arrayed itself against the peaceable Dissenters only in some points of worship. And what great undertaking could we be ripe for abroad, while so divided at home? or what fruit expected from the labour of those mediating divines in weighty matters, who were not yet past the sucking-bottle; but seemed to place all the business of Christianity in persecuting men for their consciences, differing from them in smaller matters? How ridiculous must we be to the Church of Rome to interpose in her affairs, and force our mediation upon her, when besides our ill correspondence with the foreign Protestants, she must observe our weakness within ourselves, that we could not, or would not step over a straw, though for the perpetual settlement and security of our Church and Nation! She might well look upon us as those that probably might be forced at some time by our folly to call her in to our assistance (for with no weapons or arguments but what are fetched out of her arsenals can the ceremonial-controversy be rightly defended), but never could she consider us of such authority or wisdom as to give balance to her counsels. But this was far from Bishop Bramhall's thoughts; who, so he might (like Caesar) 'manage the Roman empire at its utmost extent',° had quite forgot what would conduce to the peace of his own province and country. For, page 57, he settles this maxim as a truth, 'That second reformations are commonly like metal upon metal, which is false heraldry.' Where, by the way, it is a wonder that our author, in enumerating the Bishop's perfections in divinity, law, history, and philosophy, neglected this peculiar gift he had in heraldry; and omitted to tell us that his mind was large enough to have animated the kingdoms of Garter and Clarencieux° at their greatest dimensions. But, beside what I have said already in relation to this project upon Rome, there is this more, which I confess was below Bishop Bramhall's reflection, and was indeed fit only for some vulgar politician, or the commissioners of Scotland about the late Union: Whether it would not have succeeded, as in the consolidation of kingdoms, where the greatest swallows down the less; so also in church-coalition, that though the Pope had condescended (which the Bishop owns to be his right) to be only a patriarch, yet he would have swooped up the patriarchate of Lambeth to his morning's-draught, like an egg in muscadine. And then there is another danger always when things come once to a treaty, that beside the debates of reason, there is a better way of tampering to bring men over that have a power to conclude. And so who knows in such a treaty with Rome, if the Alps (as it is probable) would not have come over to England, as the Bishop designed it, England might not have been obliged, lying so commodious for navigation, to undertake a voyage to *Civita Vecchia*? But what though we should have made all the advances imaginable, it would have been to no purpose; and nothing less than an entire and total resignation of the Protestant cause would have contented her. For the Church of Rome is so well satisfied of her own sufficiency, and hath so much more wit than we had in Bishop Bramhall's days, or seem to have yet learned, that it would have succeeded just as at the Council of Trent. For there, though many divines of the greatest sincerity and learning endeavoured a reformation, yet no more could be obtained of her than the Nonconformists got of those of the Church of England at the conference of Worcester House.° But, on the contrary, all her excesses and errors were further riveted and confirmed, and that great machine of her Ecclesiastical Policy there perfected.

So that this enterprise of Bishop Bramhall's, being so ill laid and so unseasonable, deserves rather an excuse than a commendation. And all that can be gathered besides out of our author concerning him is of little better value; for he saith indeed, that 'he was a zealous and resolute assertor of the public rites and solemnities of the Church'. But those things, being only matters of external neatness, could never merit the trophies that our author erects him. For neither can a justice of peace for his severity about dirt-baskets deserve a statute. And as for 'his expunging some dear and darling articles from the Protestant cause', it is, as far as I can perceive, only his substituting some Arminian° tenets (which I name so, not for reproach, but for difference) instead of the Calvinian doctrines. But this too could not challenge all these triumphal ornaments in which he installs him: for, I suppose, these were but mere 'mistakes on either side, for want of being [as the Bishop saith, page 134] scholastically stated; and that he, with a distinction of school theology, could have smoothed over and plained away these knots, though they had been much harder.'

For the rest, which he leaves us to seek for, and I meet casually with in the Bishop's own book; I find him to have been doubtless a very good-natured gentleman. Page 160, 'He hath much respect for poor readers'; and page 161, he judges 'that if they come short of preachers in point of efficacy, yet they have the advantage of preachers as to point of security'. And page 163, he commends the care taken by 'the canons that the meanest cure of souls should have formal sermons at least four times every year'. Page 155, he 'maintains the public sports on the Lord's day by the proclamation to that purpose, and the example of the Reformed Churches beyond sea'; and 'for the public dances of our youth upon country greens on Sundays, after the duties of the day, he sees nothing in them but innocent, and agreeable to that under-sort of people.' And page 117 (which I quoted before), he 'takes the promiscuous licence to unqualified persons to read the Scriptures, far more prejudicial, nay, more pernicious, than the over-rigorous restraint of the Romanists.' And indeed, all along he complies much for peace-sake, and judiciously shows us wherein our separation from the Church of Rome is not warrantable. But although I cannot warrant any man who hence took occasion to traduce him of Popery, the contrary of which is evident, yet neither is it to be wondered, if he did hereby lie under some imputation, which he might otherwise have avoided. Neither can I be so hard-hearted as our author in the Nonconformists' case of discipline, to think it were better that 'he, or a hundred more divines of his temper, should suffer, though innocent, in their reputation, than that we should come under a possibility of losing our religion.' For as they (the Bishop, and, I hope, most of his Party) did not intend it so, neither could they have effected it. But he could not expect to enjoy his imagination without the annoyances incident to such as dwell in the middle storey—the pots from above, and the smoke from below. And those churches which are seated nearer upon the frontier of Popery did naturally and well if they took alarm at the march. For, in fact, that incomparable person, Grotius,° did yet make a bridge for the enemy to come over; or at least laid some of our most considerable passes open to them and unguarded: a crime something like what his son De Groot (here's *Gazette*° again for you) and his son-in-law Mombas have been charged with. And, as to the Bishop himself, his friend, an accusatory spirit, would desire no better play than he gives in his own vindication. But that's neither my business nor humour; and whatsoever may have glanced upon him was directed only to our author, for publishing that book, which the Bishop himself had thought fit to conceal, and for his impertinent efflorescence of rhetoric upon so mean topics, in so choice and copious a subject as Bishop Bramhall.

Yet though the Bishop prudently undertook a design, which he hoped not to accomplish in his own days, our author, however, was something wiser, and hath made sure to obtain his end. For the Bishop's honour was the furthest thing from his thoughts, and he hath managed that part so, that I have accounted it a work of some piety to vindicate his memory from so scurvy a commendation. But the author's end was only railing. He could never have induced himself to praise one man but in order to rail on another. He never oils his hone but that he may whet his razor, and that not to shave, but to cut men's throats. And whoever will take the pains to compare, will find that, as it is his only end, so his best, nay, his only talent is railing. So that he hath, while he pretends so much for the good Bishop, used him but for a stalking-horse till he might come within shot of the foreign divines and the Nonconformists. The other was only a copy of his countenance. But look to yourselves, my masters; for in so venomous a malice, courtesy is always fatal. Under colour of some men's having taxed the Bishop, he flies out into a furious debauch and breaks the windows; if he could, would raze the foundations of all the Protestant churches beyond sea: but for all men at home of their persuasion, if he meet them in the dark, he runs them through. He usurps to himself the authority of the Church of England, who is so well-bred, that if he would have allowed her to speak, she would doubtless have treated more civilly those over whom she pretends no jurisdiction: and under the names of Germany and Geneva, he rallies and rails at the whole Protestancy of Europe. For you are mistaken in our author (but I have worn him threadbare) if you think he designs to enter the lists where he hath but one man to combat. Mr Bays, ye know, 'prefers that one quality of fighting single with whole armies, before all the moral virtues put together'.° And yet I assure you, he hath several times obliged moral virtue so highly, that she owes him a good turn whensoever she can meet him. But it is a brave thing to be the ecclesiastical Drawcan sir;° he kills whole nations, he kills friend and foe; Hungary, Transylvania, Bohemia, Poland, Savoy, France, the Netherlands, Denmark, Sweden, and a great part of the Church of England, and all Scotland (for these, beside many more, he mocks under the title of Germany and Geneva) may perhaps rouse our mastiff, and make up a danger worthy of his courage. A man would guess that this giant had promised his 'comfortable importance' a cymar° of the beards of all the 'orthodox theologues' in Christendom. But I wonder how he comes to be prolocutor° of the Church of England! For he talks at that rate as if he were a *synodical individuum*;° nay, if he had a fifth council° in his belly, he could not dictate more dogmatically. There had been indeed, as I have heard, about the days of Bishop Bramhall, a sort of divines here of that leaven, who being dead, I cover their names, if not for health's sake, yet for decency, who never could speak of the first Reformers with any patience; who preened themselves in the peculiar virulency of their pens, and so they might say a tart thing concerning the foreign churches, cared not what obloquy they cast upon the history or the profession of religion. And those men undertook likewise to vent their wit and their choler under the style of the Church of England; and were indeed so far owned by her, that what preferments were in her own disposal, she rather conferred upon them. And now, when they were gone off the stage, there is risen up this spiritual Mr Bays, who, having assumed to himself an incongruous plurality of ecclesiastical offices, one of the most severe, of Penitentiary-Universal to the Reformed Churches; the other most ridiculous, of Buffoon-General to the Church of England, may be henceforth capable of any other promotion. And not being content to enjoy his own folly, he has taken two others into partnership, as fit for his design as those two° that clubbed with Mahomet in making the Koran; who by a perverse wit and representation might travesty the Scripture, and render all the careful and serious part of religion odious and contemptible. But, lest I might be mistaken as to the persons I mention, I will assure the reader that I intend not *Hudibras*;° for he is a man of the other robe, and his excellent wit hath taken a flight far above these whifflers:° that whoever dislikes the choice of his subject, cannot but commend his performance, and calculate if on so barren a theme he were so copious, what admirable sport he would have made with an ecclesiastical politician. But for a Daw-Divine not only to foul his own nest in England, but to pull in pieces the nests of those beyond sea, 'tis that which I think indecent and of very ill example. There is not indeed much danger (his book, his letter, and his Preface being writ in English) that they should pass abroad; but, if they be printed upon incombustible paper, or by reason of the many avocations of our Church, they may escape a censure, yet 'tis likely they may die at home, the common fate of such treatises, amongst the more judicious oilmen° and grocers. Unless Mr Bays be so far in love with his own whelp, that, as a modern lady, he will be at the charge of translating his works into Latin, transmitting them to the universities, and dedicating them in the Vatican. But, should they unhappily get vent abroad (as I hear some are already sent over for curiosity), what scandal, what heart-burning and animosity must it raise against our Church; unless they chance to take it right at first, and limit the provocation within the author. And then, what can he expect in return of his civility, but that the compliment which passed betwixt Arminius and Baudius° should concentre° upon him, that he is both *opprobrium academiae*° and *pestis Ecclesiae*?°

For they will see at the first that his books come not out under public authority or recommendation; but only as things of buffoonery do commonly, they carry with them their own *imprimatur*; (but I hope he hath considered Mr L.° in private, and payed his fees). Neither will the gravity therefore of their judgements take the measures I hope, either of the education at our universities, or of the spirit of our divines, or of the prudence, piety, and doctrine of the Church of England, from such an interloper. Those gardens of ours use to bear much better fruit. There may happen sometimes an ill year, or there may be such a crab-stock as cannot by all engrafting be corrected. But generally it proves otherwise. Once perhaps in a hundred years there may arise such a prodigy in the university (where all men else learn better arts and better manners), and from thence may creep into the Church (where the teachers at least ought to be well instructed in the knowledge and practice of Christianity); so prodigious a person, I say, may even there be hatched, as shall neither know or care how to behave himself to God or man; and who having never seen the receptacle of grace or conscience at an anatomical dissection, may conclude therefore that there is no such matter, or no such obligation among Christians; who shall persecute the Scripture itself, unless it will conform to his interpretation; who shall strive to put the world into blood, and animate princes to be the executioners of their own subjects for well-doing. All this is possible; but comes to pass as rarely and at as long periods in our climate as the birth of a false prophet. But unluckily, in this fatal year of seventy-two, among all the calamities that astrologers foretell, this also hath befallen us. I would not hereby confirm his vanity, as if I also believed that any scheme of heaven did influence his actions, or that he were so considerable as that the comet, under which they say we yet labour, had foreboded the appearance of his Preface. No, no: though he be a creature most noxious, yet his is more despicable. A comet is of far higher quality, and hath other kind of employment. Although we call it a hairy-star, it affords no prognostic of what breeds there: but the astrologer that would discern our author and his business must lay by his telescope, and use a microscope. You may find him still in Mr Calvin's° head. Poor Mr Calvin and Bishop Bramhall, what crime did you die guilty of, that you cannot lie quiet in your graves, but must be conjured up on the stage as oft as Mr Bays will ferret you? And which of you two are most unfortunate I cannot determine: whether the Bishop in being always courted, or the Presbyter in being always railed at. But in good earnest I think Mr Calvin hath the better of it. For, though an ill man cannot by praising confer honour, nor by reproaching fix an ignominy, and so they may seem on equal terms, yet there is more in it; for at the same time that we may imagine what is said by such an author to be false, we conceive the contrary to be true. What he saith of him indeed in this place did not come very well in; for Calvin writ nothing against Bishop Bramhall, and therefore here it amounts to no more than that his spirit forsooth had propagated an original waspishness and false orthodoxy amongst all his followers. But if you look in other pages of his book, and particularly page 663 of his *Defence*, you never saw such a scarecrow as he makes him: 'There sprang up a mighty bramble on the south side the Lake Leman, that (such is the rankness of the soil) spread and flourished with such a sudden growth, that partly by the industry of its agents abroad, and partly by its own indefatigable pains and pragmaticalness, it quite over-ran the whole Reformation.' You must conceive that Mr Bays was all this while in an ecstasy in Dodona's Grove;° or else here is strange work, worse than 'explicating a post', or 'examining a pillar'. A bramble that had 'agents abroad', and itself an 'indefatigable bramble'. But straight our bramble is transformed to a man, and he 'makes a chair of infallibility for himself' out of his own bramble-timber. Yet all this while we know not his name. One would suspect it might be a Bp. Bramble. But then 'he made himself both pope and emperor too of the greatest part of the reformed world'. How near does this come to his commendation of Bishop Bramhall before! For our author seems copious, but is indeed very poor of expression; and, as smiling and frowning are performed in the face with the same muscles very little altered, so the changing of a line or two in Mr Bays at any time, will make the same thing serve for a panegyric or a philippic. But what do you think of this man? Could Mistress Mopsa° herself have furnished you with a more pleasant and worshipful tale? It wants nothing of perfection, but that it doth not begin with 'Once upon a time'; which Master Bays, according to his accuracy, if he had though on't, would never have omitted. Yet some critical people, who will exact truth in falsehood, and tax up an old-wife's fable to the punctuality of History, were blaming him t'other day for placing this 'bramble' on the south side of Lake Leman. I said, it was well and wisely done, that he chose a south sun for the better and more sudden growth of such a fruit-tree. Ay, said they, but he means Calvin by the bramble; and the 'rank soil on the south side the Lake Leman' is the city of Geneva, situate (as he would have it) on the south side of that lake. Now it is strange that he, having travelled so well, should not have observed that the lake lies east and west, and that Geneva is built at the west end of it. Pish, said I, that's no such great matter; and as Mr Bays hath it on another occasion, 'Whether it be so or no, the fortunes of Caesar and the Roman empire are not concerned in 't.' One of the company would not let that pass, but told us if we looked in Caesar's Commentaries, we should find their fortunes were concerned; for it was the Helvetian Passage,° and many mistakes might have risen in the marching of the army. Why, then, replied I again, whether it be east, west, north, or south, there is neither vice nor idolatry in it, and the ecclesiastical politician may command you to believe it, and you are bound to acquiesce in his judgement, whatsoever may be your 'private opinion'. Another, to continue the mirth, answered, 'that yet there might be some religious consideration in building a town east and west, or north and south, and 'twas not a thing so indifferent as men thought it; but because in the Church of England, where the table is set altar-wise,° the minister is nevertheless obliged to stand at the north side (though it be the north end of the table), it was fit to place the Geneva presbyter in diametrical opposition to him upon the 'south side of the lake'. But this we all took for a cold conceit, and not enough matured. I, that was still upon the doubtful and excusing part, said, that to give the right situation of a town, it was necessary first to know in what position the gentleman's head then was when he made his observation, and that might cause a great diversity as much as this came to. Yes, replied my next neighbour; or, perhaps some roguing° boy that managed the puppets turned the city wrong, and so disoccidented our geographer. It was grown almost as good as a play among us; and at last they all concluded that Geneva had sold Mr Bays a bargain,° as the moon served the earth in *The Rehearsal*,° and in good sooth had 'turned her breech on him'. But this, I doubt not, Mr Bays will bring himself off with honour; but that which sticks with me is, that our author having undertaken to make Calvin and Geneva ridiculous, hath not pursued it to so high a point as the subject would have afforded. First, he might have taken the name of the beast Calvinus, and of that have given the anagram, 'Lucianus'. Next, I would have turned him inside outward, and have made him 'Usinulca'. That was a good hobgoblin name to have frightened children with. Then he should have been a 'bramble' still, ay, an 'indefatigable bramble' too: but after that he should have continued (for in such a book a passage in a play is clear gain, and a great loss if omitted), and upon that bramble 'reasons grew as plentiful as blackberries';° but both unwholesome, and they stained all the 'white aprons so' that there was no getting of it out. And then, to make a fuller description of the place, he should have added that near to the city of 'roaring lions'° there was a lake, and that lake was all of brimstone, but stored with over-grown trouts, which trouts spawned Presbyterians, and those spawned the Millecantons° of all other fanatics. That this shoal of Presbyterians landed at Geneva, and devoured all the bishops of Geneva's capons, which are of the greatest size of any in the Reformed world. And ever since their mouths have been so in relish, that the Presbyterians are in all parts the very cannibals of capons: insomuch that, if princes do not take care, the race of capons is in danger to be totally extinguished. But that the river Rhone was so sober and intelligent that its waters would not mix with this Lake Perilous, but ran sheer through, without ever touching it: nay, such is its apprehension lest the lake should overtake it, that the river dives itself underground, till the lake hath lost the scent: and yet when it rises again, imagining that the lake is still at its heels, it runs on so impetuously that it chooseth rather to pass through the roaring lions, and never thinks itself safe till it hath taken sanctuary at the popes' town of Avignon. He might too have proved that Calvin made himself 'Pope and Emperor', because the city of Geneva stamps upon its coin the two-headed imperial eagle. And, to have given us the utmost terror, he might have considered the alliance and vicinity of Geneva to the Canton of Berne, the arms of which city is the bear (and an argument in heraldry, even Bishop Bramhall himself being judge, might have also held in divinity), and therefore they keep under the town-house constantly a whole den of bears. So that there was never a more dangerous situation, nor anything so carefully to be avoided by all travellers in their wits, as Geneva—the lions on one side, and the bears on the other. This story would have been nuts to Mother Midnight,° and was fit to have been embellished with Mr Bays's allegorical eloquence. And all that he saith either by fits and girds° of Calvin, or in his justest narratives, hath less foundation in nature; and is indeed twice incredible, first in the matter related, and then because Mr Bays it comes from; or, to express it shorter, because of the tale and the tale's-man. He is not yet come to that authority, but that his dogmatical *ipse dixits*° may rather be a reason why we should not believe him. If Mr Bays will speak of controversy, let him enter into a regular disputation concerning these Calvinian tenets, and not write a history; or, if he will give us the history of Calvin, let him at the same time produce his authors. And whether history or controversy, let him be pleased so long to abate of the exuberancy of his fancy and wit, to dispense with his ornaments and superfluencies of invention and satire, and then a man may consider whether he may believe his story, and submit to his argument. But in the meantime (for all he pleads in page 97 of his *Defence*) it looks all so like subterfuge and inveigling; it is so nauseating and tedious a task, that no man thinks he owes the author so much service as to find out the reason of his own categoricalness for him. One may beat the bush a whole day, but after so much labour shall, for all game, only spring a butterfly, or start an hedgehog. Insomuch that I am ever and anon disputing with myself whether Mr Bays be indeed so ill-natured a person as some would have him, and do not rather innocently write things (as he professes page 4 of his *Preface*) so 'exceeding all belief' that he may make himself and the company merry. I sometimes could think that he intends no harm either to public or private, but only rails contentedly to himself and his Muses; that he seeks only his own diversion, and chargeth his gun with wind but to shoot at the air. Or that, like boys, so he may make a great paper-kite of his own letter of 850 pages, and his *Preface* of a hundred, he hath no further design upon the poultry of the village. But he takes care that I shall never be long deceived with that pleasing imagination: and though his hyperboles and impossibilities can have only a ridiculous effect, he will be sure to manifest that he had a felonious intention. He would take it ill if we should not value him as an enemy of mankind: and like a raging Indian (for in Europe it was never before practised), he runs amok (as they call it there), stabbing every man he meets, till himself be knocked on the head. This here is the least pernicious of all his mischiefs: though it be no less in this and all his other books than to make the 'German Protestancy' a reproachful proverb, and to turn Geneva and Calvin into a commonplace of railing. I had always heard that Calvin was a good scholar and an honest divine. I have indeed read that he spoke something contemptuously of our liturgy: *Sunt in illo libro quaedam tolerabiles ineptiae.*° But that was a sin which we may charitably suppose he repented of on his deathbed! And if Mr Bays had some just quarrel to him on that or other account, yet for divinity's sake he needed not thus have made a constant pissing-place of his grave. And as for Geneva, I never perceived before but that it was a very laudable city; that there grew an excellent grape on the south side of Lake Leman; that a man might make good cheer° there; and there was a pall-mall;° and one might shoot with the arbalet,° or play at court boule° on Sundays. What was here to enrage our author so, that he must raze the fort of St Katherine,° and attempt with the same success a second *escalade*?° but the difficulty of the enterprise doubtless provoked his courage, and the honour he might win made the justice of his quarrel. He knew that not only the commonwealth of Switzerland, but the King of France, the King of Spain, and the Duke of Savoy would enter the lists for the common preservation of the place; and therefore though it be otherwise but a petty town, he disdained not, where the race was to be run by monarchs, to exercise his footmanship. But is it not a great pity to see a man in the flower of his age and the vigour of his studies, to fall into such a distraction that his head runs upon nothing but Roman Empire and Ecclesiastical Policy? This happens by his growing too early acquainted with *Don Quixote*, and reading the Bible too late; so that the first impressions being most strong, and mixing with the last, as more novel, have made such a medley in his brain-pan that he is become a mad priest, which of all the sorts is the most incurable. Hence it is that you shall hear him anon instructing princes, like Sancho,° how to govern his island: as he is busied at present in vanquishing the Calvinists of Germany and Geneva. Had he no friends to have given him good counsel before his understanding were quite unsettled? or if there was none near, why did not men call in the neighbours, and send for the parson of the parish, to persuade with him in time, but let it run on thus till he is fit for nothing but Bedlam or Hoxton?° However, though it be a particular damage, it may tend to a general advantage; and young students will, I hope, by his example learn to beware henceforward of overweening presumption and preposterous ambition. For this gentleman, as I have heard, after he had read *Don Quixote* and the Bible, besides such school-books as were necessary for his age, was sent early to the university; and there studied hard, and in a short time became a competent rhetorician, and no ill disputant. He had learned how to erect a thesis, and to defend it *pro* or *con* with a serviceable distinction: while the truth (as his comrade Mr Bays hath it on another occasion),

> Before a full pot of ale you can swallow,
> Was here with a whoop and gone with a holla.°

And so thinking himself now ripe and qualified for the greatest undertakings and highest fortune, he therefore exchanged the narrowness of the university for the town; but coming out of the confinement of the square-cap and the quadrangle into the open air, the world began to turn round with him: which he imagined, though it were his own giddiness, to be nothing less than the quadrature° of the circle. This accident, concurring so happily to increase the good opinion which he naturally had of himself, he thenceforward applied to gain a like reputation with others. He followed the town life, haunted the best companies, and, to polish himself from any pedantic roughness, he read and saw the plays, with much care and more proficiency than most of the auditory. But all this while he forgot not the main chance, but hearing of a vacancy with a nobleman,° he clapped in,° and easily obtained to be his chaplain. From that day you may take the date of his preferments and his ruin. For having soon wrought himself dexterously into his patron's favour, by short graces and sermons, and a mimical way of drolling upon° the puritans, which he knew would take° both at chapel and at table; he gained a great authority likewise among all the domestics. They all listened to him as an oracle; and they allowed him by common consent to have not only all the divinity, but more wit too than all the rest of the family put together. This thing alone elevated him exceedingly in his own conceit, and raised his hypochondria into the region of the brain, that his head swelled like any bladder with wind and vapour. But after he was stretched to such a height in his own fancy, that he could not look down from top to toe but his eyes dazzled at the precipice of his stature, there fell out, or in, another natural chance which pushed him headlong. For being of an amorous complexion, and finding himself (as I told you) the cock-divine and the cock-wit of the family, he took the privilege to walk among the hens: and thought it was not impolitic to establish his new acquired reputation upon the gentlewomen's side. And they that perceived he was a rising man, and of pleasant conversation, dividing his day among them into canonical hours, of reading now the Common Prayer, and now the Romances, were very much taken with him. The sympathy of silk began to stir and attract the tippet° to the petticoat and the petticoat toward the tippet. The innocent ladies found a strange unquietness in their minds, and could not distinguish whether it were love or devotion. Neither was he wanting on his part to carry on the work; but shifted himself° every day with a clean surplice, and, as oft as he had occasion to bow, he directed his reverence towards the gentlewomen's pew. Till, having before had enough of the libertine, and undertaken his calling only for preferment, he was transported now with the sanctity of his office, even to ecstasy: and like the Bishop over Magdalen College altar, or like Maudlin de la Croix,° he was seen in his prayers to be lifted up sometimes in the air, and once particularly so high that he cracked his skull against the chapel ceiling. I do not hear, for all this, that he had ever practised upon the honour of the ladies, but that he preserved always the civility of a Platonic knight-errant. For all this courtship had no other operation than to make him still more in love with himself; and if he frequented their company, it was only to speculate° his own baby° in their eyes. But being thus, without competitor or rival, the darling of both sexes in the family, and his own minion, he grew beyond all measure elated, and that crack of his skull, as in broken looking-glasses, multiplied him in self-conceit and imagination.

Having fixed his centre in this nobleman's house, he thought he could now move and govern the whole earth with the same facility. Nothing now would serve him but he must be a madman in print, and write a book of *Ecclesiastical Policy*. There he distributes all the 'territories of conscience' into the Prince's province, and makes the 'hierarchy' to be but bishops of the air: and talks at such an extravagant rate in things of higher concernment, that the reader will avow that in the whole discourse he had not one 'lucid interval'. This book he was so bent upon, that he sat up late at nights, and wanting sleep, and drinking sometimes wine to animate his fancy, it increased his distemper. Beside that too he had the misfortune to have two friends, who being both also out of their wits, and of the same, though something a calmer, frenzy, spurred him on perpetually with commendation. But when his book was once come out, and he saw himself an author; that some of the gallants of the town laid by the new tune, and the 'tay, tay, tarry', to quote some of his impertinencies; that his title-page was posted and pasted up at every avenue next under the play for that afternoon at the King's or Duke's house:° the vain-glory of this totally confounded him. He lost all the little remains of his understanding, and his *cerebellum* was so dried up that there was more brains in a walnut, and both their shells were alike thin and brittle. The King of France° that lost his wits had not near so many unlucky circumstances to occasion it: and in the last of all there is some similitude. For, as a negligent page that rode behind and carried the King's lance, let it fall on his head, the King being in armour, and the day hot, which so disordered him that he never recovered it; so this gentleman in the dog days, straggling by Temple-bar, in a massy cassock and surcingle, and taking the opportunity at once to piss and admire the title-page of his book; a tall servant of his, one J.O., that was not so careful as he should be, or whether he did it of purpose, lets another book of four hundred leaves fall upon his head; which meeting with the former fracture in his cranium, and all the concurrent accidents already mentioned, has utterly undone him. And so, in conclusion, his madness hath formed itself into a perfect lycanthropy. He doth so verily believe himself to be a wolf, that his speech is all turned into howling, yelling, and barking; and if there were any sheep here, you should see him pull out their throats and suck the blood. Alas, that a sweet gentleman, and so hopeful, should miscarry! For want of cattle here, you find him raving now against all the Calvinists of England, and worrying the whole flock of them. For how can they hope to escape his chops and his paws better than those of Germany and Geneva; of which he is so hungry, that he hath scratched up even their dead bodies out of their graves to prey upon? And yet this is nothing if you saw him in the height of his fits: but he hath so beaten and spent himself before, that he is out of breath at present; and though you may discover the same fury, yet it wants of the same vigour. But however you see enough of him, my masters, to make you beware, I hope, of valuing too high, and trusting too far to your own abilities.

It were a wild thing for me to squire it after this knight, and accompany him here through all his extravagancies against our Calvinists. You find nothing but 'orthodoxy, systems, and syntagms, polemical theology, subtleties, and distinctions: Demosthenes; tankard-bearers; pragmatical; controversial': general terms without foundation or reason assigned. That they seem like words of Cabal,° and have no significance till they be deciphered; or, you would think he were playing at substantives and adjectives. All that rationally can be gathered from what he saith, is, that the man is mad. But if you would supply his meaning with your imagination, as if he spoke sense and to some determinate purpose; it is very strange that, conceiving himself to be the champion of the Church of England, he should bid such a general defiance to the Calvinists. For he knows, or perhaps I may better say he did know before this frenzy had subverted both his understanding and memory, that most of our ancient, and many of the later Bishops nearer our times, did both hold and maintain those doctrines which he traduces under that by-word.° And the contrary opinions were even in Bishop Prideaux's° time accounted so novel, that, being then public professor of divinity,° he thought fit to tax Doctor Heylin at the Commencement, for his new-fangled divinity: '*cujus*,' saith he, in the very words of promotion, '*te doctorem creo*.'° He knew likewise that of our present bishops, though one had leisure formerly to write a *Rationale*° of the Ceremonies and Liturgy, and another a treatise of the holiness of Lent; yet that most of them and 'tis to be supposed all, have studied other controversies, and at another rate than Mr Bays's lead can fathom. And as I know none of them that hath published any treatise against the Calvinian tenets, so I have the honour to be acquainted with some of them who are entirely of that judgement, and differ nothing but, as of good reason, in the point of Episcopacy. And as for that, Bishop Bramhall, page 61, hath proved that Calvin himself was of the Episcopal persuasion. So that I see no reason why Mr Bays should here and every where be such an enemy to 'controversial skill' or the Calvinists. But I perceive 'tis for Bishop Bramhall's sake here that all the tribe must suffer. This Bays is not a good dog: for he runs at a whole flock of sheep, when Mr B.° was the deer whom he had in view from the beginning. However, having soiled himself so long with everything he meets, after him now he goes, and will never leave till he hath run him down. Poor Mr B! I find that when he was a boy he plucked Bishop Bramhall's sloes and ate his bullace;° and now, when he is as superannuated as the Bishop's book, he must be whipped for't; there is no remedy. And yet I have heard, and Mr Bays himself seems to intimate as much, that however he might in his younger years have mistaken, yet that even as early as Bishop Bramhall's *Discourse*, he began to retract: and that as for all his sins against the Church of England, he hath in some late treatises cried *peccavi*° with a witness. But, Mr Bays, doth not this now look like sorcery and extortion, which of all crimes you purge yourself from so often without an accuser? For first, whereas the old Bishop was at rest, and had under his last pillow laid by all cares and contests of this lower world, you by your necromancy have disturbed him and raised his ghost to persecute and haunt Mr B., whom doubtless at his death he had pardoned. But if you called him up to ask some questions too concerning your Ecclesiastical Policy, as I am apt to suppose, I doubt you had no better answer than in the song:

Art thou forlorn of God, and com'st to me?
What can I tell thee, then, but misery?°

And then as for extortion; who but such a Hebrew Jew as you would, after an honest man had made so full and voluntary restitution, not yet have been satisfied without so many pounds of his flesh over into the bargain? Though J.O. be in a desperate condition, yet methinks Mr B., not 'being past grace' should not neither 'have been past mercy'. Are there no terms of pardon, Mr Bays? is there no time for expiation? but, after so ample a confession as he hath made, must he now be hanged too to make good the proverb?° It puts me in mind of a story in the time of the Guelphs and Ghibellines, whom I perceive Mr Bays hath heard of. There were two factions in Italy, of which the Guelphs were for the Pope, and the Ghibellines for the Emperor; and these were for many years carried on and fomented with much animosity, to the great disturbance of Christendom. Which of these two were the nonconformists in those days I can no more determine than which of our parties here at home is now schismatical. But so nonconformable they were to one another, that the historian said they took care to differ in the least circumstances of any human action: and as those that have the mason's word, secretly discern one another, so in the peeling or cutting but of an onion, a Guelph, and *vice versa*, would at first sight have distinguished a Ghibelline. Now, one of this latter sort, coming at Rome to confession upon Ash Wednesday, the Pope or the penitentiary sprinkling ashes on the man's head with the usual ceremony, instead of pronouncing *memento, homo, quod cinis es, et in cinerem reverteris*, changed it to *memento, homo, quod Ghibilinus es.*° *etc.* And even thus it fares with Mr B., who, though he should creep on his knees up the whole stairs of scholastic penitence, I am confident neither he, nor any of his party, shall by Mr Bays's good will ever be absolved. And therefore truly, if I were in Mr B.'s case, if I could not have my confession back again, yet it should be a warning unto me not without better grounds to be so coming and so good-natured for the future. But whatever he do, I hope others will consider what usage they are like to find at Mr Bays's hand, and not suffer themselves by the touch of his 'penitential rod' to be transformed into beasts, even into rats, as here he hath done with Mr B. I have indeed wondered often at this Bays's insolence, who summons in all the world and preacheth up only this repentance; and so frequently in his books he calls for 'testimonies, signal marks, public acknowledgment, satisfaction, recantation', and I know not what. He that hath made the passage to Heaven so easy that one may fly thither without grace (as Gonzales to the moon only by the help of his ganzas);° he that hath 'disintricated'° its narrow paths from those 'labyrinths' which J.O. and Mr B. have planted; this overseer of God's highways (if I may with reverence speak it), who hath paved a broad causeway with moral virtue through His kingdom; he methinks should not have made the 'process of loyalty' more difficult than that of salvation. What 'signal marks', what 'testimonies', would he have of this conversion? Every man cannot, as he hath done, write an *Ecclesiastical Policy*, a *Defence*, a *Preface*; and some, if they could, would not do it after his manner; lest instead of obliging thereby the King and the Church, it should be a testimony to the contrary. Neither, unless men have better principles of allegiance at home, are they likely to be reduced by Mr Bays's way of persuasion. He is the first minister of the gospel that ever had it in his commission 'to rail at all nations'. And though it hath been long practised, I never observed any great success by reviling men into conformity. I have heard that charms may even invite the moon out of heaven, but I never could see her moved by the rhetoric of barking. I think it ought to be highly penal for any man to impose other conditions upon his Majesty's good subjects than the King expects, or the law requires. When you have done all, you must yet appear before Mr Bays's tribunal, and he hath a new test yet to put you to. I must confess, at this rate the Nonconformists deserve some compassion; that after they have done or suffered legally and to the utmost, they must still be subjected to the wand of a verger or to the wanton lash of every pedant; that they must run the gantlope,° or down with their breeches as oft as he wants to prospect of a more pleasing nudity. But I think they may choose whether they will submit or no to his jurisdiction. Let them but (as I hope they do) fear God, honour the King, preserve their consciences, follow their trades, and look to their chimneys, and they need not fear Mr Bays and all his malice. But after he hath sufficiently insulted over Mr B.'s ignorance and vanity, with other compliments of the like nature, in recompense of that 'candour and civility' which he acknowledges 'him to have now learnt towards the Church of England', Mr Bays (forgetting what had past long since betwixt him and the bookseller) saith in excuse of his severity, that 'this treatise was not published to impair Mr B.'s esteem in the least, but for a correction of his scribbling humour, and to warn their rat-divines that are perpetually nibbling and gnawing other men's writings.' Now I must confess, Mr Bays, this is a very handsome welcome to Mr B., that was come so far to see you, and doubtless upon this encouragement he will visit you often. This is an admirable dexterity our author hath (I wish I could learn it) 'to correct a man's scribbling humour without impairing in the least his reputation'. He is as courteous as lightning, and can melt the sword without ever hurting the scabbard. But as for their rat-divines, I wonder they are not all poisoned with nibbling at his writings, he hath strewn so much arsenic in every leaf. But, however, methinks he should not have grudged them so slender a sustenance. For though there was a sow in Arcadia° so fat and insensible that she suffered a rat's nest in her buttock, and they had both diet and lodging in the same gammon, yet it is not every rat's good fortune to be so well provided. And for 'push-pin divinity', I confess it is a new term of art, and I shall henceforward take notice of it; but I am afraid in general it doth not tend much to the reputation of the faculty.

And now, though he told us at the beginning that the bookseller was the main reason of publishing this book of the Bishop and his own Preface, he tells us that the main reason of its publication was to give some check to their present disingenuity, that is to say, to that of J.O.; and J.O. be it at present. He is come so much nearer, however, to the truth, though we shall find ere we have done that there is still a mainer reason.

When I first took notice of this misunderstanding betwixt Mr Bays and J.O., I considered whether it were not execution-day with the whole Latin Alphabet; whether all the letters were not to suffer in the same manner, except C only, which (having been the mark of condemnation) might have a pardon to serve for the executioner. I began to repent of my undertaking, being afraid that the quarrel was with the whole criss-cross row,° and that we must fight it out through all the squadrons of the vowels, the mutes, the semi-vowels, and the liquids. I foresaw a sore and endless labour, and a battle the longest that ever was read of; being probable to continue as long as one letter was left alive, or there were any use of reading. Therefore, to spare mine own pains, and prevent ink-shed, I was advising the letters to go before Mr Bales,° or any other his Majesty's justices of peace, to swear that they were in danger of their lives, and desire that Mr Bays might be bound to the good behaviour. But after this I had another fancy, and that not altogether unreasonable; that Mr Bays had, only for health and exercise sake, drawn J.O. by chance out of the number of the rest, to try how he could rail at a letter, and that he might be well in breath upon any occasion of greater consequence. For, how perfect soever a man may have been in any science, yet without continual practice he will find a sensible decay of his faculty. Hence also, and upon the same natural ground, it is the wisdom of cats to whet their claws against the chairs and hangings, in meditation of the next rat they are to encounter. And I am confident that Mr Bays by this way hath brought himself into so good railing case, that pick what letter you will out of the alphabet, he is able to write an epistle upon it of 723 pages (I have now told them right) to the author of the *Friendly Debates*.

Now though this had very much of probability, I had yet a further conjecture; that this J.O. was a talisman, signed under some peculiar influence of the heavenly bodies, and that the fate of Mr Bays was bound up within it. Whether it be so or no I know not: but this I am assured of, without the help either of sidereal magic or judicial astrology, that when J and O are in conjunction, they do more certainly than any of the planets forbode that a great ecclesiastical politician shall that year run mad. I confess after all this, when I was come to the dregs of my fancy (for we all have our infirmities, and Mr Bays's *Defence* was but the blue-John° of his *Ecclesiastical Policy*, and this Preface the tap-droppings of his *Defence*), I reflected whether Mr Bays having no particular cause of indignation against the letters, there might not have been a mistake of the printer, and that they were to be read in one word *io*, that used to go before *paean*:° that is in English a triumph before the victory. Or whether it alluded to Io° that we read of at school, the daughter of Inachus; and that as Juno persecuted the heifer, so this was a he-cow, that is to say a bull, to be baited by Mr Bays the thunderer. But these being conceits too trivial, though a ragout° fit enough for Mr Bays's palate, I was forced moreover to quit them, remarking that it was a *J* consonant. And I plainly at last perceived that this J.O. was a very man as any of us are and had a head and a mouth with tongue and teeth in it, and hands with fingers and nails upon them: nay, that he could read and write, and speak as well as I or Mr Bays, either of us. When I once found this, the business appeared more serious, and I was willing to see what was the matter that so much exasperated Mr Bays, who is a 'person', as he saith himself, 'of such a tame and softly humour, and so cold a complexion, that he thinks himself scarce capable of hot and passionate impressions'. I concluded that necessarily there must be some extraordinary accident and occasion that could alter so good a nature. For I saw that he pursued J.O., if not from 'post to pillar', yet from 'pillar to post', and I discerned all along the footsteps of a most inveterate and implacable malice. As oft as he does but name those two first letters, he is, like the island of Fayal,° on fire in threescore and ten places.

You see, Mr Bays, that I too have improved my wit with reading the Gazettes. Were you of that fellow's° diet here about town, that epicurizes upon burning coals, drinks healths in scalding brimstone, scrunches the glasses for his dessert, and draws his breath through glowing tobacco pipes. Nay, to say a thing yet greater, had you never tasted other sustenance then the focus of burning-glasses, you could not show more flame than you do always upon that subject. And yet one would think that even from the 'little sports', with your 'comfortable importance' after supper, you should have learned when J.O. came into play, to 'love your love' with a J, because he is 'judicious' though you 'hate your love with a J' because he is 'jealous': and then to 'love your love with an O', because he is 'oraculous', though you 'hate your love with an O', because he is 'obscure'. Is it not strange that in those most benign minutes of a man's life, when the stars smile, the birds sing, the winds whisper, the fountains warble, the trees blossom, and universal nature seems to invite itself to the bridal; when the lion pulls in his claws and the aspic° lays by its poison, and all the most noxious creatures grow amorously innocent; that even then Mr Bays alone should not be able to refrain his malignity? As you love yourself, madam, let him not come near you. He hath been fed all his life with vipers instead of lampreys, and scorpions for crayfish; and if at any time he ate chickens, they had been crammed with spiders, till he hath so envenomed his whole substance that 'tis much safer to bed with a mountebank before he hath taken his antidote. But it cannot be any vulgar furnace that hath chafed so cool a salamander. 'Tis not the strewing of cowage° in his genial bed° that could thus disquiet him the first night. And therefore let's take the candle and see whether there be not somebody underneath that hath cut the bed cords. There was a worthy divine, not many years dead, who in his younger time, being of a facetious and unlucky humour, was commonly known by the name of Tom Triplet:° he was brought up at Paul's school, under a severe master, Dr Gill,° and from thence he went to the university. There he took liberty (as 'tis usual with those that are emancipated from school) to tell tales, and make the discipline ridiculous under which he was bred. But, not suspecting the doctor's intelligence, coming once to town, he went in full school to give him a visit, and expected no less than to get a playday for his former acquaintance. But, instead of that, he found himself horsed up in a trice; though he appealed in vain to the privileges of the university, pleaded *adultus*, and invoked the mercy of the spectators. Nor was he let down till the master had planted a grove of birch in his back side, for the terror and public example of all wags that divulge the secrets of Priscian,° and make merry with their teachers. This stuck so with Triplet, that all his lifetime he never forgave the doctor, but sent him every New-year's tide an anniversary ballad to a new tune, and so in his turn avenged himself of his jerking° pedagogue.

Now when I observed that of late years Mr Bays had regularly 'spawned' his books; in 1670 the *Ecclesiastical Policy*; in 1671 the *Defence of the Ecclesiastical Policy*; and now in 1672 this *Preface* to Bishop Bramhall; and that they were written in a style so vindictive and poignant that they wanted nothing but rhyme to be right Tom Triplet, and that their edge bore always upon J.O. either in broad meanings or in plain terms; I began to suspect that where there was so great resemblance in the effects, there might be some parallel in their causes. For though the piques of players among themselves, or of poet against poet, or of a Conformist divine against a Nonconformist, are dangerous, and of late times have caused great disturbance; yet I never remarked so irreconcileable and implacable a spirit as that of boys against their schoolmasters or tutors. The quarrels of their education have an influence upon their memories and understandings for ever after. They cannot speak of their teachers with any patience or civility; and their discourse is never so flippant, nor their wits so fluent, as when you put them upon that theme. Nay, I have heard old men, otherwise sober, peaceable, and good-natured, who never could forgive Osbolston, as the younger are still inveighing against Dr Busby.° It were well that both old and young would reform this vice, and consider how easy a thing it is upon particular grudges, and as they conceive out of a just censure, to slip either into juvenile petulancy or inveterate uncharitableness. And had there not been something of this in his own case, I am confident Mr Bays in his *Ecclesiastical Policy*, in order to the public peace and security of the government, could not have failed to admonish princes to beware of this growing evil, and to brandish the public rods, if not the axes, against the boys, to teach them better manners. And he would have assured them that they might have done it with all safety, notwithstanding that there were in proportion a hundred boys against one preceptor. But therefore is it not possible that J.O.° and Mr Bays have known one another formerly in the university; and that (as in seniority there is a kind of magistracy) Bays being yet young, J.O. conceived himself in those days to be his superior, and exercised an academical jurisdiction or dominion over him? Now whether J.O. might not be too severe upon him there (for all men are prone to be cogent° and supercilious when they are in office), or whether Mr Bays might not make some little escapes and excursions there (as young men are apt to do when they are got together), that I know not, and rather believe the contrary. But that is certain, that the young wits in the universities have always an animosity against the doctors, and take a peculiar felicity in having a lucky hit at any of them. I rather suppose that after Mr Bays had changed the place, and his condition, to be the nobleman's chaplain, that he might commit some exorbitance in J.O.'s opinion, or preach or write something to J.O.'s reproach, and published the secrets of the holy brotherhood:° and that J.O. having got him within his reach did therefore (figuratively speaking),

> Instead of maid Gillian,
> Take up his male pillion,
> And whipped him like a baggage . . .

as Tom Triplet expresses it. This might well raise Mr Bays's choler, who, considering himself to be now in holy orders, and conceiving that he had been as safe as in a sanctuary under his patron's protection, must needs take it ill to be handled so irreverently. If it were thus in fact, and that J.O. might presume too much upon his former authority to give him correction; yet it is the more excusable, if Mr Bays had on his part been guilty of so much disingenuity. For though a man may be allowed once in his life to change his Party, and the whole scene of his affairs, either for his safety or preferment; nay, though every man be obliged to change a hundred times backward and forward, if his judgement be so weak and variable; yet there are some drudgeries that no man of honour would put himself upon, and but few submit to if they were imposed. As, suppose one had thought fit to pass over from one persuasion of the Christian religion unto another; he would not choose to spit thrice at every article that he relinquished, to curse solemnly his father and mother for having educated him in those opinions, to animate his new acquaintances to the massacring of his former comrades. These are businesses that can only be expected from a renegade° of Algiers or Tunis; to overdo in expiation, and gain better credence of being a sincere Mussulman. And truly, though I can scarcely believe that Mr Bays hath so mean and desperate intentions, which yet his words seem too often to manifest; the offices, however, which he undertakes are almost as dishonourable. For he hath so studied and improved their jargon, as he calls it, heard their sermons and prayers so attentively, searched the scriptures so narrowly, that a man may justly suspect he had formerly set up J.O.'s profession, and having the language so perfectly, hath upon 'this juncture of affairs' betaken himself to turn spy and intelligencer;° and 'tis evident that he hath travelled the country for that purpose. So that I cannot resemble° him better than to that politic engine° who about two years ago was employed by some of Oxford as a missionary amongst the Nonconformists of the adjacent counties; and, upon design, either gathered a congregation of his own, or preached amongst others, till having got all their names, he threw off the vizard, and appeared in his own colours, an honest informer. But I would not have any man take Mr Bays's fanatical geography for authentic, lest he should be as far misled as in the situation of Geneva. It suffices that Mr Bays hath done therein as much as served to his purpose, and mixed probability enough for such as know not better, and whose ears are of a just bore for his fable.

But J.O. being of age and parts sufficient either to manage or to neglect this quarrel, I shall as far as possible decline the mentioning of him, seeing I have too, upon further intelligence and consideration, found that he was not the person whom Mr Bays principally intended. For, the truth of it is, the King was the person concerned from the beginning.

His Majesty, before his most happy and miraculous Restoration, had sent over a Declaration° of his indulgence to tender consciences in ecclesiastical matters. Which, as it was doubtless the real result of the last advice left him by his glorious father, and of his own consummate prudence and natural benignity; so at his return he religiously observed and promoted it as far as the passions and influence of the contrary Party would give leave. For whereas among all the decent circumstances of his welcome return, the providence of God had so co-operated with the duty of his subjects, that so glorious an action should neither be soiled with the blood of victory, nor lessened by any capitulations of treaty, so, not to be wanting on his part in courtesy, as I may say, to so happy a conjuncture, he imposed upon himself an oblivion of former offences, and this indulgence in ecclesiastical affairs. And to royal and generous minds no stipulations are so binding as their own voluntary promises: nor is it to be wondered if they hold those conditions that they put upon themselves the most inviolable. He therefore carried the Act of Oblivion and Indemnity° through; that party who had suffered so vastly in the late combustions° not refusing to imitate his generosity, but throwing all their particular losses and resentments into the public reckoning. But when it came to the ecclesiastical part, the accomplishment of which only remained behind to have perfected his Majesty's felicity, the business I warrant you would not go so (as I shall have occasion to say more particularly). For, though I am sorry to speak it, yet it is a sad truth, that the animosities and obstinacy of some of the clergy have in all ages been the greatest obstacle to the clemency, prudence, and good intentions of princes, and the establishment of their affairs. His Majesty therefore expected a better season, and having at last rid himself of a great minister of state° who had headed this interest, he now proceeded plainly to recommend to his parliament effectually, and with repeated instances, the consideration of tender consciences. After the King's last representing of this matter to the parliament, Mr Bays took so much time as was necessary for the maturing of so accurate a book, which was to be the standard of government for all future ages, and he was happily delivered in 1670 of his *Ecclesiastical Policy*. And, though he thought fit in this first book to treat his Majesty more tenderly than in those that followed, yet even in this he doth all along use great liberty and presumption. Nor can what he objects, page 282, to weak consciences, take place so justly upon them as upon himself: who, while his prince might expect his compliance, doth give him council, advises him how to govern the kingdom, blames and corrects the laws, and tells him how this and the other might be mended. But that I may not involve the thing in generals, but represent undeniably Mr Bays's performance in this undertaking, I shall without art write down his own words and his own *quod scripsi scripsi*,° as they lie naked to the view of every reader.

The grand thesis upon which he stakes not only all his own divinity and policy, his reputation, preferment and conscience (of most of which he hath no reason to be prodigal), but even the crowns and fate of princes, and the liberties, lives, and estates, and, which is more, the consciences of their subjects (which are too valuable to be trusted in his disposal), is this, page 10, 'That it is absolutely necessary to the peace and government of the world, that the supreme magistrate of every commonwealth should be vested with a power to govern and conduct the consciences of subjects in affairs of religion.' And page 12 he explains himself more fully, that 'unless princes have power to bind their subjects to that religion that they apprehend most advantageous to public peace and tranquillity, and restrain these religious mistakes that tend to its subversion, they are no better than statues and images of authority.' Page 13, 'A prince is endued with a power to conduct religion, and that must be subject to his dominion as well as all other affairs of state.' Page 27, 'If princes should forgo their sovereignty over men's consciences in matters of religion, they leave themselves less power than is absolutely necessary' etc. And in brief, 'The supreme government of every commonwealth, wherever it is lodged, must of necessity be universal, absolute, and uncontrollable in all affairs whatsoever that concern the interests of mankind and the ends of government.' Page 32, 'He in whom the supreme power resides, having authority to assign to every subject his proper function, and among others these of the priesthood; the exercise thereof as he has power to transfer upon others, so he may if he please reserve it to himself.' Page 33, 'Our saviour came not to unsettle the foundations of government, but left the government of the world in the same condition he found it.' Page 34, 'The government of religion was vested in princes by an antecedent right to Christ.' This being the magisterial and main point that he maintains, the rest of his assertions may be reckoned as corollaries to this thesis, and without which indeed such an unlimited maxim can never be justified. Therefore, to make a conscience fit for the nonce, he says, page 89, 'Men may think of things according to their own persuasions, and assert the freedom of their judgements against all the powers of the earth. This is the prerogative of the mind of man within its own dominions, its kingdom is intellectual, etc. Whilst conscience acts within its proper sphere, the civil power is so far from doing it violence, that it never can.' Page 92, 'Mankind have the same natural right to liberty of conscience in matters of religious worship as in affairs of justice and honesty; that is to say, a liberty of judgement, but not of practice.' And in the same page he determines Christian liberty to be 'founded upon the reasonableness of this principle'. Page 308, 'In cases and disputes of public concernment, private men are not properly *sui juris*;° they have no power over their own actions: they are not to be directed by their own judgements, or determined by their own wills, but by the commands and determinations of the public conscience; and if there be any sin in the command, he that imposed it shall answer for it, and not I whose whole duty it is to obey. The commands of authority will warrant my obedience, my obedience will hallow, or at least excuse my action, and so secure me from sin, if not from error; and in all doubtful and disputable cases 'tis better to err with Authority than to be in the right against it: not only because the danger of a little error (and so it is if it be disputable) is outweighed by the importance of the great duty of obedience, etc.'

Another of his corollaries is, 'That God hath appointed' (page 81) 'the magistrates to be his trustees upon earth, and his officials to act and determine in moral virtues and pious devotions according to all accidents and emergencies of affairs; to assign new particulars of the divine law; to declare new bounds of right and wrong, which the law of God neither doth nor can limit.' Page 69, 'Moral virtue being the most material and useful part of all religion, is also the utmost end of all its other duties.' Page 76, 'All religion must of necessity be resolved into enthusiasm or morality. The former is mere imposture, and therefore all that is true must be reduced to the latter.' Having thus enabled° the prince, dispensed with conscience, and fitted up a moral religion for that conscience; to show how much those moral virtues are to be valued, page 53 of the Preface to his *Ecclesiastical Policy*, he affirms that "'tis absolutely necessary to the peace and happiness of kingdoms that there be set up a more severe government over men's consciences and religious persuasions than over their vices and immoralities.' And page 55 of the same, that 'princes may with less hazard give liberty to men's vices and debaucheries than their consciences'. But for what belongs particularly to the use of their power in religion, he first (page 56 of his book) saith that 'the Protestant Reformation hath not been able to resettle princes in their full and natural rights in reference to its concerns'; and page 58, 'most Protestant princes have been frighted, not to say hectored out of the exercise of their ecclesiastical jurisdiction'. Page 271, 'If princes will be resolute (and if they will govern, so they must be), they may easily make the most stubborn conscience bend to their resolutions.' Page 221, 'Princes must be sure to bind on at first their ecclesiastical laws with the straitest knot, and afterwards keep them in force by the severity of their execution.' Page 223, speaking of honest and well-meaning men, 'so easy is it for men to deserve to be punished for their consciences, that there is no nation in the world in which, were government rightly understood and duly managed, mistakes and abuses of religion would not supply the galleys with vastly greater numbers than villainy.' Page 54 of the Preface to *Ecclesiastical Policy*: 'Of all villains the well-meaning zealot is the most dangerous.' Page 46, 'The fanatic party in country towns and villages ariseth not (to speak within compass) above the proportion of one to twenty. Whilst the public peace and settlement is so unluckily defeated by quarrels and mutinies of religion, to erect and create new trading combinations, is only to build so many nests of faction and sedition, etc. For it is notorious that there is not any sort of people so inclinable to seditious practices as the trading part of a nation.' And now, though many as material passages might be heaped-up out of his book on all these and other as tender subjects, I shall conclude this imperfect enumeration with one corollary more, to which indeed his grand thesis and all the superstructures are subordinate and accommodated. Page 166, 'Princes cannot pluck a pin out of the Church, but the state immediately shakes and totters.' This is the syntagm of Mr Bays's divinity, and system of his policy: the principles of which confine upon° the territories of Malmesbury, and the style, as far as his wit would give him leave, imitates that language: but the arrogance and dictature° with which he imposes it on the world surpasses by far the presumption either of *Gondibert*° or *Leviathan*.° For he had indeed a very politic fetch° or two that might have made a much wiser than he more confident. For he imagined, first of all, that he had perfectly secured himself from any man's answering him: not so much upon the true reason, that is, because indeed so paltry a book did not deserve an answer; as because he had so confounded the question with differing terms and contradictory expressions, that he might upon occasion affirm whatsoever he denied, or deny whatsoever he affirmed. And then besides, because he had so entangled the matter of conscience with the magistrates' power, that he supposed no man could handle it thoroughly without bringing himself within the statute of treasonable words, and at least a *praemunire*.° But last of all, because he thought that whoseover answered him must for certain be of a contrary judgement, and he that was of a contrary judgement should be a fanatic; and if one of them presumed to be meddling, then Mr Bays (as all divines have a *non-obstante*° to the *jejunium Cecilianum*,° and to the Act of Oblivion and Indemnity) would either burn that, or tear it in pieces. Being so well fortified on this side, upon the other he took himself to be impregnable. His Majesty must needs take it kindly that he gave him so great an accession of territory; and, lest he should not be thought rightly to understand government, nay, lest Mr Bays by virtue of page 271 should not think him fit to govern, he could not in prudence and safety but submit to his admonition and instructions. But if he would not, Mr Bays knew, ay that he did, how to be even with him, and would write another book that should do his business. For the same power that had given the prince that authority could also revoke it.

But let us see therefore what success the whole contrivance met with, or what it deserved. For, after things have been laid with all the depth of human policy, there happens lightly° some ugly little contrary accident from some quarter or other of heaven, that frustrates and renders all ridiculous.

And here, for brevity and distinction sake, I must make use of the same privilege by which I call him Mr Bays, to denominate also his several aphorisms or hypotheses: and let him take care whether or no they be significant.

First, The Unlimited Magistrate.
Secondly, The Public Conscience.
Thirdly, Moral Grace.
Fourthly, Debauchery Tolerated.
Fifthly, Persecution Recommended.
And lastly, Push-pin Divinity.

And now, though I intend not to be longer than the nature of animadversions requires (this also being but collateral to my work of examining the Preface, and having been so abundantly performed already), yet neither can I proceed well without some Preface. For, as I am obliged to ask pardon if I speak of serious things ridiculously, so I must now beg excuse if I should hap to discourse of ridiculous things seriously. But I shall, so far as possible, observe decorum, and, whatever I talk of, not commit such an absurdity as to be grave with a buffoon. But the principal cause of my apology is, because I see I am drawn in to mention kings and princes, and even our own; whom, as I think of with all duty and reverence, so I avoid speaking of either in jest or earnest, lest by reason of my private condition and breeding I should, though most unwillingly, trip in a word, or fail in the mannerliness of an expression. But Mr Bays, because princes sometimes hear men of his quality play their part, or preach a sermon, grows so insolent that he thinks himself fit to be their governor. So dangerous is it to let such creatures be too familiar. They know not their distance; and like the ass in the fable,° because they see the spaniel play with their master's legs, they think themselves privileged to paw and ramp° upon his shoulders. Yet though I must follow his track now I am in, I hope I shall not write after his copy.

As for his first hypothesis of the Unlimited Magistrate, I must for this once do him right, that after I had read in his 12th page, that 'princes have power to bind their subjects to that religion they apprehend most advantageous to public peace and tranquillity'; a long time after, not, as I remember, till page 82, when he bethought himself better, he saith, 'No rites nor ceremonies can be esteemed unlawful in the worship of God, unless they tend to debauch men either in their practices or their conceptions of the deity.' But no man is in ingenuity obliged to do him that service for the future; neither yet doth that limitation bind up or interpret what he before so loosely affirmed. However, take all along the power of the magistrate as he hath stated it; I am confident if Bishop Bramhall were alive (who could no more forbear° Grotius than Mr Bays could the Bishop, notwithstanding their friendship), he would bestow the same censure upon him that he doth upon Grotius, page 18, 'When I read his book° of the right of the sovereign magistrate in sacred things, he seemed to me to come too near an Erastian,° and to lessen the power of the keys too much, which Christ left as a legacy to his Church. It may be he did write that before he was come to full maturity of judgement: and some other things, I do not say after he was superannuated, but without that due deliberation which he useth at other times' (wherein a man may desire Mr Bays in Mr Bays); 'or, it may be, some things have been changed in his book, as I have been told by one of his nearest friends, and that we shall shortly see a more authentic edition of all his works. This is certain, that some of those things which I dislike were not his own judgement after he was come to maturity in theological matters.' And had Mr Bays (as he ought to have done) carried his book to any of the present bishops or their chaplains, for a licence to print it, I cannot conceive that he could have obtained it in better terms than what I have collected out of the 108 page of his answerer:° 'Notwithstanding the old pleas of the *jus divinum*° of episcopacy, of example and direction apostolical, of a parity of reason between the condition of the Church whilst under extraordinary officers, and whilst under ordinary, of the power of the Church to appoint ceremonies for decency and order, of the pattern of the churches of old' (all which, under protestation, are reserved till the first opportunity). I have upon reading of this book found that it may be of use for 'the present juncture of affairs', and therefore let it be printed. And as I think he hath disobliged the clergy of England in this matter, so I believe the favour that he doth his Majesty is not equivalent to that damage. For (that I may, with Mr Bays's leave, profane Ben Jonson) though the 'gravest divines should be his flatterers';° he hath a very quick sense and (shall I profane Horace too in the same period?):

*Hunc male si palpere, recalcitrat undique tutus.*°

If one stroke him ill-favouredly, he hath a terrible way of kicking, and will fling you to the stable door, but is himself safe on every side. He knows it's all but that you may get into the saddle again; and that the priest may ride him, though it be to a precipice. He therefore contents himself with the power that he hath inherited from his royal progenitors kings and queens of England, and as it is declared by parliament, and is not to be trepanned° into another kind of tenure of dominion to be held at Mr Bays's pleasure, and depend upon the strength only of his argument. But (that I may not offend in Latin too frequently) he considers that by not assuming a deity to himself, he becomes secure and worthy of his government. There are lightly about the courts of princes a sort of projectors for concealed lands,° to which they entitle the King to beg them for themselves; and yet generally they get not much by it, but are exceedingly vexatious to the subject. And even such a one is this Mr Bays, with his project of 'concealed power', that most princes, as he said, 'have not yet rightly understood'; but whereof the King is so little enamoured, that I am confident, were it not for prowling° and molesting the people, his Majesty would give Mr Bays the patent for it, and let him make his best on't, after he hath paid the fees to my Lord Keeper.

But one thing I must confess is very pleasant, and he hath passed a high compliment upon his Majesty in it: that he may, if he please, reserve the priesthood and the exercise of it to himself. Now this indeed is surprising; but this only troubles me, how his Majesty would look in all the sacerdotal habiliments, and the pontifical wardrobe. I am afraid the King would find himself incommoded with all that furniture upon his back, and would scarce reconcile him to wear even the lawn sleeves and the surplice. But what? even Charles the Fifth, as I have read, was, at his inauguration by the Pope, content to be vested, according to the Roman ceremonial, in the habit of a deacon; and a man would not scruple too much the formality of the dress in order to empire.

But one thing I doubt Mr Bays did not well consider; that, if the King may discharge the function of the priesthood, he may too (and 'tis all the reason in the world) assume the revenue. It would be the best subsidy that ever was voluntarily given by the clergy. But truly, otherwise, I do not see but that the King does lead a more unblamable conversation,° and takes more care of souls than many of them and understands their office much better, and deserves something already for the pains he hath taken.

The next is Public Conscience: for as to men's private consciences, he hath made them very inconsiderable, and reading what he saith of them with some attention, I only found this new and important discovery and great privilege of Christian liberty, that 'thought is free'. We are, however, obliged to him for that, seeing by consequence we may think of him what we please. And this he saith a man may assert against all the powers of the earth. And indeed with much reason and to great purpose, seeing, as he also alleges, the civil power is so far from doing violence to that liberty, that it never can. But yet if the freedom of thoughts be in not lying open to discovery, there have been ways of compelling men to discover them; or if the freedom consist in retaining their judgements when so manifested, that also hath been made penal. And I doubt not but, beside 'oaths' and 'renunciations', and 'assents' and 'consents', Mr Bays, if he were searched, hath twenty other tests and picklocks in his pocket. Would Mr Bays, then, persuade men to assert this against all the powers of the earth? I would ask, in what manner? To say the truth, I do not like him, and would wish the Nonconformists to be upon their guard, lest he trepan° them, first by this means into a plot, and then peach,° and so hang them. If Mr Bays meant otherwise in this matter, I confess my stupidity, and the fault is most his own, who should have writ to the capacity of vulgar readers. He cuts,° indeed, and falters in this discourse, which is no good sign, persuading men that they may and ought to practise against their consciences, where the commands of the magistrate intervene. None of them denies that it is their duty, where their judgements or consciences cannot comply with what is enjoined, that they ought in obedience patiently to suffer, but further they have not learned. I dare say that the casual divinity° of the Jesuits is all through as orthodox as this maxim of our author's: and as the opinion is brutish, so the consequences are devilish. To make it therefore go down more glibly, he saith, that ''tis better to err with authority, than to be in the right against it in all doubtful and disputable cases; because the great duty of obedience outweighs the danger of a little error (and little it is if it be disputable).' I cannot understand the truth of this reasoning, that whatsoever is disputable is little; for even the most important matters are subject to controversy, and besides, things are little or great according to the eyes or understandings of several men; and however, a man would suffer something rather than commit that little error against his conscience, which must render him a hypocrite to God and a knave amongst men. 'The commands, (he saith) and determinations of the public conscience ought to carry it; and if there be any sin in the command, he that imposed shall answer for it, and not I, whose duty it is to obey. (And mark) 'the commands of authority will warrant my obedience, my obedience will hallow,° or at least excuse, my action, and so secure me from sin, if not from error'; and so you are welcome gentlemen. Truly a very fair and conscionable reckoning! So far is this from hallowing the action, that I dare say it will, if followed home, lead only to all that 'sanctified villainy' for the invention of which we are beholden to the author. But let him have the honour of it, for he is the first divine that ever taught Christians how another man's sin could confer an 'imputative righteousness' upon all mankind that shall follow and comply with it. Though the subject made me serious, yet I could not read the expression without laughter: 'My obedience will hallow, or at least excuse, my action.' So inconsiderable a difference he seems to make betwixt those terms, that if ever our author come for his merits to be a bishop, a man might almost adventure, instead of 'consecrated', to say that he was 'excused'.

The third is Moral Grace. And whoever is not satisfied with those passages of his concerning it before quoted, may find enough where he discourseth it at large, even to surfeit. I cannot make either less or more of it than that he overturns the whole fabric of Christianity and power of religion. For my part, if Grace be resolved into morality, I think a man may almost as well make God too to be only a notional and moral existence.

And white-aproned Amaryllis was of that opinion:

*Ma tu sanctissima honesta, che sola sei*
*D'alma ben nata inviolabil nume.*°

'But thou most holy honesty, that only art the inviolable deity of the well-born soul.'

And so too was the moral poet; (for why may not I too bring out my Latin shreds as well as he his? *Quaesitum ad fontem solos deducere verpos*)°

*Nullum numen abest, si sit prudentia—*°

'There is no need of a deity where there is prudence; or, if you will, where there is Ecclesiastical Policy.'

But so far I must do Mr Bays right, that, to my best observation, if prudence had been God, Bays had been a most damnable atheist. Or, perhaps, only an idolator of their number, concerning whom he adds in the next line.

*. . . sed te*
*Nos facimus, fortuna, deam coeloque locamus.*

'But we make thee, Fortune, a goddess, and place thee in heaven.'

However, I cannot but be sorry that he hath undertaken this desperate vocation, when there are twenty other honest and painful ways wherein he might have got a 'living', and made Fortune propitious. But he cares not upon what argument or how dangerous he runs, to show his ambitious activity: whereas those that will dance upon ropes do lightly some time or other break their necks. And I have heard that even the Turk, every day he was to mount the high-rope, took leave of his 'comfortable importance', as if he should never see her more. But this is a matter foreign to my judicature, and therefore I leave him to be tried by any jury of divines: and that he may have all right done him, let half of them be school-divines and the other moiety systematical,° and let him except against as many as the law allows; and so God send him a good deliverance. But I am afraid he will never come off.°

The fourth is Debauchery Tolerated. For supposing, as he does, that 'tis better and 'safer to give a toleration to men's debaucheries than to their religious persuasions', it amounts to the same reckoning. This is a very ill way of discoursing, and that a 'greater severity ought to be exercised over men's consciences than over their vices and immoralities'. For it argues too much indiscretion, by avoiding one evil to run up into the contrary extreme. And debauched persons will be ready hence to conclude, although it be a perverse way of reasoning, that where the severity ought to be less, the crime is less also; nay even that the more they are debauched, it is just that the punishment should still abate in proportion; but, however, that it were very imprudent and unadvisable to reform and err on the religious hand, lest they should thereby incur the greater penalties. Mr Bays would have done much better had he singled out the theme of religion. He might have loaded it with all the truth which that subject would bear. I would allow him that 'rebellion is as the sin of witchcraft'° though that text of Scripture will scarce admit his interpretation. He could not have declaimed more sharply than I, or any honest man else, would upon occasion against all those who under pretence of conscience raise war, or create public disturbances. But comparisons of vice are dangerous, and though he should do this without design, yet, while he aggravates upon° religion, and puts it in balance, he doth so far alleviate and encourage debauchery. And moreover (which, to be sure, is against his design) he doth hereby more confirm the austerer sort of sinners, and furnishes them with a more specious colour and stronger argument. It had been better policy to instruct the magistrate that there is no readier way to shame these out of their religious niceties than by improving men's morals. But, as he handles it, never was there any point more unseasonably exposed; at such a time, when there is so general a depravation of manners, that even those who contribute towards it do yet complain of it; and though they cannot reform their practice, yet feel the effects, and tremble under the apprehension of the consequences. It were easy here to shew a man's reading, and to discourse out of history the causes of the decay and ruin of Mr Bay's Roman empire, when as the moralist has it,

> . . . saevior armis
> Luxuria incubuit, victumque ulciscitur orbem.°

And descending to those times since Christianity was in the throne, 'tis demonstrable that for one war upon a fanatical or religious account, there have been a hundred occasioned by the thirst of glory and empire that hath inflamed some great prince to invade his neighbours. And more have sprung from the contentiousness and ambition of some of the clergy; but the most of all from the corruption of manners and always fatal debauchery. It exhausts the estates of private persons, and makes them fit for nothing but the highway or an army. It debases the spirits and weakens the vigour of any nation; at once indisposing them for war, and rendering them incapable of peace. For if they escape intestine troubles, which would certainly follow when they had left themselves by their prodigality or intemperance no other means of subsistence but by preying upon one another; then must they either, to get a maintenance, pick a quarrel with some other nation, wherein they are sure to be worsted; or else (which more frequently happens) some neighbouring prince that understands government takes them at the advantage, and if they do not like ripe fruit fall into his lap, 'tis but shaking the tree once or twice, and he is sure of them. Where the horses are, like those of the Sybarites,° taught to dance, the enemy need only learn the tune and bring the fiddles. But therefore (as far as I understand) his Majesty, to obviate and prevent these inconveniences in his kingdoms, hath on the one hand never refused a just war; that so he might take down our grease and luxury, and keep the English courage in breath and exercise; and on the other (though himself most constantly addicted to the Church of England) hath thought fit to grant some liberty to all other sober people (and longer than they are so God forbid they should have it) thereby to give more temper and alloy to the common and notorious debauchery.

But Mr Bays nevertheless is for his fifth: Persecution Recommended; and he does it to the purpose. Julian° himself, who I think was first a reader, and held forth in the Christian churches before he turned apostate and then persecutor, could not have outdone him either in irony or cruelty. Only it is God's mercy that Mr Bays is not emperor. You have seen how he inveighs against trade: 'That whilst men's consciences are acted° by such peevish and ungovernable principles, to erect trading combinations is but to build so many nests of faction and sedition. 'Lay up your ships, my masters, set bills on your shop doors, shut up the Custom House; and why not adjourn the term, mure up° Westminster Hall, leave ploughing and sowing, and keep a dismal holiday through the nation? For Mr Bays is out of humour. But I assure you it is no jesting matter. For he hath in one place taken a list of the fanatic ministers, whom he reckons to be but about a hundred 'systematical divines'; though I believe the Bartholomew register° or the March licenses° would make them about a hundred and three or a hundred and four, or so: but this is but for rounder number, and breaks no square.° And then for their people, either 'they live in greater societies of men' (he means the City of London and the other cities and towns corporate, but expresses it so to prevent some inconvenience that might betide him) 'but there their noise is greater than their number. Or else in country towns and villages, where they arise not above the proportion of one to twenty.' It were not unwisely done indeed if he could persuade the magistrate that all the fanatics have but one neck,° so that he might cut off Nonconformity at one blow. I suppose the Nonconformists value themselves though upon their conscience, and not their numbers: but they would do well to be watchful, lest he have taken a list of their names as well as their number, and have set crosses upon all their doors against there should be occasion. But till that 'happy juncture', when Mr Bays 'shall be fully avenged of his new enemies, the wealthy fanatics' (which is soon done too, for he saith, 'there are but few of them men of estates or interest'), he is contented that they should only be exposed (they are his own expressions) to the 'pillories, whipping-posts, galleys, rods and axes'; and moreover and above, to all other punishments whatsoever, provided they be of a severer nature than those that are inflicted on men for their immoralities. O more than human clemency! I suppose the division betwixt immoralities and conscience is universal; and whatsoever is wicked or penal is comprehended within their territories. So that although a man should be guilty of all those heinous enormities which are not to be named among Christians, beside all lesser peccadillos expressly against the ten Commandments, or such other part of the divine law as shall be of the magistrates' making, he shall be in a better condition, and more gently handled, than a 'well-meaning zealot'; for this is the man that Mr Bays saith is 'of all villains the most dangerous' (even more dangerous, it seems, than a malicious and ill-meaning zealot): this is he whom in 'all kingdoms where government is rightly understood', he would have condemned 'to the galleys for his mistakes and abuses of religion'. Although the other punishments are more severe, yet this being more new and unacquainted, I cannot pass it by without some reflection. For I considered what princes make use of galleys. The first that occurred to me was the Turk, who, according to Bays's maxim, hath established Mahometism among his subjects, as the 'religion that he apprehends most advantageous to public peace and settlement'. Now in his empire the Christians only are guilty of those 'religious mistakes that tend to the subversion of Mahometism'; so that he understands government rightly in chaining the Christians to the oar. But then in Christendom, all that I could think of were the King of France, the King of Spain, the Knights of Malta, the Pope, and the rest of the Italian princes. And these all have bound their subjects to the Romish religion as most advantageous. But these people their galleys with immoral fellows and debauchees; whereas the Protestants, being their fanatics and mistakers in religion, should have been their *ciurma*.° But it is to be hoped these princes will take advice and understand it better for the future. And then at last I remembered that his Majesty too hath one galley lately built, but I dare say it is not with that intention: and our fanatics, though few, are so many, that one will not serve. But therefore if Mr Bays and his partners would be at the charge to build the King a whole squadron for this use, I know not but it might do very well (for we delight in novelties), and it would be a singular obligation to Sir John Baptist Dutel,° who might have some pretence to be general of his Majesty's galleys. But so much for that. Yet in the meantime I cannot but admire at Mr Bays's courage; who knowing how dangerous a villain a well-meaning zealot is, and having calculated to a man how many of them there are in the whole nation, yet dares thus openly stimulate the magistrate against them, and talk of nothing less, but much more, than 'pillories, whipping-posts, galleys, and axes', in this manner. It is sure some sign (and if he knew not so much he would scarce adventure) of the peaceableness of their principles, and of that restraint under which their tender consciences hold them when nevertheless he may walk night and day in safety; though it were so easy a thing to deify the divine after the ancient manner, and no man be the wiser. But that which I confess would vex me most, were I either an ill- or a well-meaning zealot, would be, after all, to hear him (as he frequently does) sneering at me in an ironical harangue to persuade me, forsooth, to take all patiently for conscience sake and the good example of mankind; nay, to wheedle one almost to make himself away, to save the hangman a labour. It was indeed near that pass in the primitive times, and the tired magistrates asked them, whether they had not halters and rivers and precipices, if they were so greedy of suffering?° But, by the good leave of your insolence, we are not come to that yet. *Non tibi, sed Petro*; or rather, *sed Regi*.° The Nonconformists have suffered as well as any men in the world, and could do so still if it were his Majesty's pleasure. Their 'duty to God hath hallowed', and their 'duty to the magistrate hath excused', both their pain and ignominy. To die by a noble hand is some satisfaction: but when his Majesty, for reasons best known to himself, hath been graciously pleased to abate of your rigors, I hope, Mr Bays, that we shall not see when you have a mind to junket with your 'comfortable importance', that the entremets° shall be of a fanatic's giblets, nor that a Nonconformist's head must be wiped off as oft as your nose drivels. 'Tis sufficient, Sir, we know your inclination, we know your abilities, and we know your lodging; and when there is any further occasion, you will doubtless be sent for. For, to say the truth, this Bays is an excellent tool, and more useful than ten other men. I will undertake that he shall rather than fail, be the trepanner, the informer, the witness, the attorney, the judge; and, if the Nonconformist need the benefit of his book, he shall be ordinary too, and say he is an ignorant fellow, *non legit*:° and then, to do him the last Christian office, he would be his hangman. In the meantime, let him enjoy it in speculation, secure of all the employments when they shall fall. For I know no gentleman that will take any of them out of his hands, although it be in an age wherein men cannot well support their quality without some accession from the public; and for the ordinary sort of people, they are, I know not by what disaster, besotted and abandoned to fanaticism. So that Mr Bays must either do it himself in person, or constitute the chief magistrate to be his deputy. But princes do indeed understand themselves better most of 'em, and do neither think it so safe to entrust a clergyman with their authority, nor decent for themselves to do the drudgery of the clergy. That would have passed in the days of Saint Dominic;° but when even the Inquisition hath lost its edge in the Popish countries, there is little appearance it should be set up in England. It were a worthy spectacle—were it not—to see his Majesty, like the governor in Synesius,° busied in his cabinet among those engines whose very names are so hard that it is some torture to name them; the Podostrabae, the Dactylethrae, the Otagrae, the Rhinolabides, the Cheilostrophia, devising, as they say there are particular diseases, so a peculiar rack for every limb and member of a Christian's body. Or, would he (with all reverence be it spoken) exchange his kingdom of England for that of Macassar? where the great *arcanum*° of Government is the cultivating of a garden of venomous plants, and preparing thence a poison, in which the prince dips a dart, that where it does but draw blood, rots the person immediately to pieces; and his office is with that to be the executioner of his subjects. God be praised, his Majesty is far of another temper; and he is wise, though some men be malicious.

But Mr Bays's sixth is that which I call his Push-Pin Divinity; for he would persuade princes that 'there cannot be a pin pulled out of the Church but the state immediately totters'. That is strange. And yet I have seen many a pin pulled out upon occasion, and yet not so much as the Church itself hath wagged. It is true indeed, and we have had sad experiments° of it, that some clergymen have been so opiniastre° that they have rather exposed the state to ruin than they would part with a pin, I will not say out of their church, but out of their sleeve. There is nothing more natural than for the ivy to be of opinion that the oak cannot stand without its support; or, seeing we are got into ivy, that the

Church cannot hold up longer than it underprops the walls; whereas it is a sneaking insinuating imp,° scarce better than bindweed, that sucks the tree dry and moulders the building where it catches. But what, pray, Mr Bays, is this pin in Pallas's buckler? Why, 'tis some ceremony or other that is 'indifferent in its own nature'; that 'hath no antecedent necessity' but 'only as commanded'; that 'signifies nothing in itself but what the commander pleases'; that even by the Church which commands it, is 'declared to have nothing of religion in it; and that is in itself of no great moment or consequence, only it is absolutely necessary that governors should enjoin it, to avoid the evils that would follow if it were not determined'. Very well, Mr Bays. This I see will keep cold: anon, perhaps I may have a stomach. But I must take care lest I swallow your pin.

Here we have had the titles, and some short rehearsal of Mr Bays's six plays. Not but that, should we disvalise° him, he hath to my knowledge a hundred more as good in his budget; but really I consult mine own repose. But now, among friends, was there ever anything so monstrous? You see what a man may come to with divinity and high-feeding. There is a scurvy disease° which, though some derive from America, others tell a story that the Genoese in their wars with Venice took some of their noblemen, whom they cut to pieces and barrelled up like tunny, and so maliciously vented it to the Venetians, who, eating it ignorantly, broke out in those nasty botches and ugly symptoms that are not curable but by mercury. What I relate it for is out of no further intention, nor is there any more similitude, than that the mind too hath its nodes° sometimes, and the style its buboes;° and that I doubt before Mr Bays can be rid of 'em, he must pass through the grand cure and a dry diet.

And now it is high time that I resume the thread of my former history concerning Mr Bays's books in relation to his Majesty. I do not find that the *Ecclesiastical Policy* found more acceptance than could be expected from so judicious a prince; nor do I perceive that he was ever considered of at a promotion of bishops, nor that he hath the reversion of the archbishopric of Canterbury. But if he have not by marriage barred his way, and if it should ever fall to his lot, I am resolved, instead of *his grace*, to call him always *his Morality*. But as he got no preferment that I know of at Court (though his patron doubtless having many things in his gift, did abundantly recompense him), so he missed no less of his aim as to the reformation of ecclesiastical government upon his principles. But still, what he complains of, page 20, 'the ecclesiastical laws were either weakened through want of execution, or in a manner cancelled by the opposition of civil constitutions'. For, beside what in

England, where all things went on at the same rate, in the neighbouring kingdom of Scotland there were I know not how many Mas Johns° restored in one day to the work of their ministry, and a door opened whereby all the rest might come in for the future, and all this by his Majesty's commission. Nay, I think there was (a thing of very ill example) an archbishop° turned out of his see for some misdemeanour or other. I have not been curious of his name nor his crime, because as much as possible I would not expose the nakedness of any person so eminent formerly in the Church. But henceforward the King fell into disgrace with Mr Bays, and anyone that had eyes might discern that our author did not afford his Majesty that countenance and favour which he had formerly enjoyed. So that a book too of J.O.'s° happening mischievously to come out at the same season, upon pretence of answering that, he resolved to make his Majesty feel the effects of his displeasure. He therefore set pen to paper again, and having kept his midwife of the 'Friendly Debate' by him all the time of his pregnancy for fear of miscarrying, he was at last happily delivered of his second child, the *Defence of the Ecclesiastical Policy* in the year 1671. It was a very lusty baby, and twice as big as the former, and (which some observed as an ill sign, and that if it lived it would prove a great tyrant) it had, when born, all the teeth,° as perfect as ever you saw in any man's head. But I do not reckon much upon those ominous criticisms. For there was partly a natural cause in it, Mr Bays having gone so many months more than the civil law allows for the utmost term of legitimation, that it was no wonder if the brat were at its birth more forward than others usually are. And indeed, Mr Bays was so provident against abortion, and careful for some reason that the child should cry, that the only question in town (though without much cause, for truly 'twas very like him) was whether it was not spurious or supposititious. But, allegories and raillery and hard words apart: in this his second book, and what I quoted before out of Bishop Bramhall, page 18, with allusion to our author, is here fallen out as exactly true as if it had been expressly calculated for Bays's meridian. He finds himself to have come too near, nay to have far outgone an Erastian; that he had writ his *Ecclesiastical Policy* before he was come to maturity of judgement; that one might desire Mr Bays in Mr Bays; that something had been changed in his book. That a more authentic edition was necessary; that some things which he had said before were not his judgement after he was come to maturity in theological matters.

I will not herein too much insist upon his Reply, where his answerer asks him pertinently enough to his grand thesis, what was then become of their old plea of *Jus Divinum*?° Why, saith he, must you prescribe me what I shall write? Perhaps my next book shall be of that subject. For, perhaps he said so only for evasion, being old excellent° at parrying and fencing. Though I have good reason to believe that we may shortly see some piece of his upon that theme, and in defence of an aphorism of a great prelate° in the last king's time, 'that the King had no more to do in ecclesiastical matters than Jack that rubbed his horses' heels.' For Mr Bays is so enterprising, you know, 'look to't, I'll do't.'° He has face enough to say or unsay anything, and 'tis his privilege, what the school divines deny to be even within the power of the almighty, to make contradictions true. An evidence of which (though I reserve the further instances to another occasion that draws near) does plainly appear in what I now principally urge, to show how dangerous a thing it is for his Majesty and all other princes to lose Mr Bays's favour. For whereas he had all along in his first book treated them like a company of ignorants, and that did not understand government (but that is pardonable in Mr Bays), in this his second, now that they will not do as he would have them, when he had given them power and instructions how to be wiser for the future, he casts them quite off, like men that were desperate. He had, you know, page 35 of his first book, and in other places, vested them with a universal and unlimited power, and uncontrollable in the government of religion (that is, over men's consciences); but now in his second, to make them an example to all incorrigible and ungrateful persons, he strips and disrobes them again of all these regal ornaments that he had superinduced upon them and leaves them good princes in *querpo*° as he found 'em, to shift for themselves in the wide world as well as they can. Do but read his own words, page 237 of his *Defence*, paragraph 5, and sure you will be of my mind. 'To vest the supreme magistrate in an unlimited and uncontrollable power, is clearly to defeat the efficacy and obligatory force of all his laws, that cannot possibly have any binding virtue upon the minds of men, when they have no other inducement to obedience but only to avoid the penalty. But if the supreme power be absolute and unlimited, it doth for that very reason remove and evacuate all other obligations, for otherwise it is restrained and conditional; and if men lie under no other impulsion than of the law itself, they lie under no other obligation than that of prudence and self-interest, and it remains entirely in the choice of their own discretion whether they shall or shall not obey, and then there is neither government nor obligation to obedience; and the principle of men's compliance with the mind of their superiors is not the declaration of their will and pleasure, but purely the determination of their own judgements; and therefore necessary for the security of the government, though for nothing else, to set bounds to its jurisdiction; otherwise, like the Roman empire, etc.' I know it would be difficult to quote twenty lines in Mr Bays but we should encounter with the Roman empire. But observe how laboriously here he hath asserted and proved that all he had said in his first book was a mere mistake before he were come to years of discretion. For, as in law a man is not accounted so till he hath completed twenty-one, and 'tis but the last minute of that time that makes him his own man (as to all things but conscience I mean, for as to that he saith men are never *sui juris*),° so though the distance of Bays's books was but betwixt 1670 and 1671, yet a year, nay an instant at any time of a man's life may make him wiser, and he hath, like all other fruits, his annual maturity. It was so long since as 1670, page 33, that this 'universal unlimited and uncontrollable power was the natural right of princes antecedent to Christ, firmly established by the unalterable dictates of natural reason, universal practice, and consent of nations, that the scripture rather supposes than asserts the ecclesiastical (and so the civil) jurisdiction of princes.' 'Twas in 1670, page 10, that it was 'absolutely necessary'; and page 12 'that princes have that power to bind their subjects to that religion that they apprehend most advantageous to public peace, etc.' So that they derive their title from eternal necessity, which the moralists say the gods themselves cannot impeach. His Majesty may lay by his *dieu*, and make use only of his *mon droit*. He hath a patent for his kingdom under the broad seal of nature, and next under that, and immediately *before* Christ, is over all persons and in all causes as well ecclesiastical as civil (and over all men's consciences) within his Majesty's realms and dominions supreme head and governor. 'Tis true, the author sometimes for fashion's sake speaks in that book of religion and of a deity; but his principles do necessarily, if not in terms, make the prince's power paramount to both those, and if he may by his uncontrollable and unlimited universal authority introduce what religion he may, of consequence what deity also he pleases. Or if there were no deity, yet there must be some religion, that being an engine most advantageous for public peace and tranquillity. This was in 1670; but by 1671 you see the case is altered. Even one night hath made some men grey. And now p. 238 of his second book, beside what before p. 237, he hath made princes accountable, ay and to so severe an auditor as God himself. 'The thrones of princes are established upon the dominion of God.' And page 241, ''Tis no part of the prince's concernment to institute rules of moral good and evil; that is the care and the prerogative of a superior lawgiver.' And page 260, he owns, that if the subjects can plead a clear and undoubted pre-engagement to that higher authority, they have liberty to remonstrate to° the equity of their laws. I do not like this remonstrating nor these remonstrants.° I wish again that Mr Bays would tell us what he means by the term, and where it will end, whether he would have the fanatics remonstrate: but they are wary, and ashamed of what they have done in former times of that nature: or whether he himself hath a mind to remonstrate, because the fanatics are tolerated. That is the thing, that is the business of this whole book; and knowing that there is a clear and undoubted pre-engagement to the higher authority of nature and necessity, if the King will persist in tolerating these people, who knows, after remonstrating, what Mr Bays will do next? But now in sum, what shall we say of this man, and how had the King been served if he had followed Bays's advice, and assumed the power of this first book? He had run himself into a fine *premunire*, when now, after all, he comes to be made accountable to God, nay even to his subjects. And by this means it happens, though it were beyond Mr Bays's forecast, and I dare say he would rather have given the prince again a power antecedent to Christ, and to bring in what religion he please: he hath obliged him to as tender a conscience as any of his Christian subjects, and then good night to 'Ecclesiastical Policy'. I have herein endeavoured the utmost ingenuity° toward Mr Bays, for he hath laid himself open but to too many disadvantages already, so that I need not, I would not press him beyond measure, but to my best understanding, and if I fail, I even ask him pardon, I do him right. 'Tis true, that being distracted betwixt his desire that the consciences of men should be persecuted, and his anger at princes that will not be advised, he confounds himself everywhere in his reasonings, that you can hardly distinguish which is the 'whoop' and which is the 'holla', and he makes indentures° on each side of the way wheresoever he goes. But no man that is sober will follow him, lest some Justice of Peace should make him pay his five shillings, beside the scandal; and it is apparent to everyone what he drives at. But were this otherwise, I can spare it, and 'tis sufficient to my purpose that I do thus historically deduce the reason of his setting forth his books, and show that it was plainly to 'remonstrate' against the power of his prince, and the measures that he hath taken of governing; to set his Majesty at variance not only with his subjects, but with himself, and to raise a Civil War in his intellectual Kingdom betwixt his controllable and his uncontrollable jurisdiction. And because, having to do with a wise man, as Mr Bays is, one may often gather more of his mind out of a word that drops casually, than out of his whole watchful and serious discourse, when he is talking of matters of policy and that require caution,—I cannot slight one passage of Mr Bays, page 656, where, raging bitterly against all the Presbyterians and other sects, and as much against the allowing them any tenderness, liberty, toleration, or indulgence, he concludes thus: 'Tenderness and indulgence to such men were to nourish vipers in our own bowels, and the most sottish neglect of our own quiet and security and we should deserve to perish with the dishonour of Sardanapalus.'° Now this of Sardanapalus I remember some little thing ever since I read, I think it was my Justine;° and I would not willingly be such a fool as to make a dangerous similitude that has no foundation. For if Mr Bays in the Preface of his *Defence*, to excuse his long teeming° before it were brought forth, places it partly upon his recreations, I know not why much more a prince should not 'be willing to enjoy the innocent comforts of this life, as well as to do the common drudgeries'. But I am thinking what Mr Bays meant by it; for every similitude must have, though not all, yet some likeness. Now I am sure there were no Nonconformists and Presbyterians in Sardanapalus's days. I am sure also that Sardanapalus was no clergyman, that he was no subject; but he was one of the uncontrollable creatures that, instead of exercising his ecclesiastical power, delighted in spinning; till somebody came in on the sudden, and catching him at it, cut his thread. Come, 'tis better we left this argument and the company too, for you see the crime, you see the sentence: and whoever it be, there is some prince or other whom Mr Bays will have to perish. That page 641 is indeed not so severe, but 'tis pretty well; where, on the same kind of subject, whetting the prince against those people, he saith, 'that prince that hath felt the pounces° of these ravening vultures, if after that he shall be persuaded to regard their fair speeches at such time as they want power, without other evident and unquestionable tokens of their conversation, deserves to be king of the night.' Now, for this matter, I believe Mr Bays knows that his Majesty hath received such evident and unquestionable tokens of loyalty° from the Nonconformists; otherwise his own loyalty would have hindered him from daring to use that expression.

And now I should continue my history to his third book in hand, the *Preface* to Bishop Bramhall. But having his second book still before me, I could not but look a little further into it, to see how he hath left matters standing betwixt himself and his answerer. And first I lighted on that place where he strives to disentangle himself from what he had said about trade in his former book. Here therefore he defies the whole fanatic world to discover one syllable that tends to its discouragement. Let us put it upon that issue, and by this one example take the pattern of his ingenuity in all his other contests. Whoop, Mr Bays, page 49: 'with what conscience does the answerer tell the people that I have represented all tradesmen as seditious, when 'tis so notorious I only suppose that some of them may be tainted with seditious principles? If I should affirm that when the nobility or clergy are possessed with principles that incline to rebellion and disloyal practices, they are of all rebels the most dangerous, should I be thought to impeach them of treason and rebellion?' Holla, Mr Bays, but in the 49th page of your first book° you say expressly, 'For 'tis notorious that there is not any sort of people so inclinable to seditious practices as the trading part of a nation.' Is this the same thing now? and how does this defence take off the objection? And yet he tears and insults and declaims as if he had the truth on his side. At last he strives to bring himself off and salve the matter in the same page 49 with, 'In brief, it is not the rich citizen, but the wealthy fanatic that I have branded for an ungovernable beast, and that not as wealthy but as fanatic.' Subtle distinguisher! I see, if we give him but rope enough what he will come to. Mr Bays, many as prosper° a man as yourself hath marched up Holborn° for distinguishing betwixt the wealth and the fanatic; and moreover let me tell you, fanatic money hath no ear-mark.°

So concerning the magistrate's power in religion, wherein his answerer had remarked some unsafe passages: Whoop, Mr Bays! p. 12 of his first book before quoted: 'Unless princes have power to bind their subjects to what religion they apprehend most advantageous, etc. they are no better than statues of authority.' Holla, Bays. Page 467 of the second book: 'this bold calumny I have already I hope competently enough discovered and detested. Yet he repeats this fundamental forgery in all places, so that his whole book is by one huge lie 400 pages long.' Judge now who is the forger; and yet he roars too here as if he would mix heaven and earth together. But you may spare your raving; you will never claw it off ° as long as your name is Bays.

So his answerer,° it seems, having, p. 85, said that Bays confines the whole duty of conscience to the inward thoughts and persuasions of the mind, over which the magistrate hath no power at all: Whoop, Bays! page 89 of his first book: 'Let all matters of mere conscience, whether purely moral or religious, be subject to conscience only; i.e. let men think of things according to their own persuasions, and assert the freedom of their judgements against all the powers of the earth. This is the prerogative of the mind of man within its own dominions; its kingdom is intellectual, etc.' Page 91, 'Liberty of conscience is internal and invisible, and confined to the minds and judgements of men; and while conscience acts within its proper sphere, the civil power is so far from doing it violence, that it never can.' Holla, Bays! Page 299 of his second book: 'This in downright English is a shameless lie. Sir, you must pardon my rudeness, for I will assure you after long meditation, I could not devise a more pertinent answer to so bold a one as this.' I believe you, Mr Bays: you meditated long, some twelve months at least; and you could not devise any other answer; and in good earnest he hath not attempted to give any other answer. 'I confess 'tis no extraordinary conceit; but 'tis the best repartee my barren fancy was able to suggest to me upon so rude an occasion.' Well, Mr Bays! I see it must come to a quarrel; for thus the hectors use to do, and to give the lie at adventure, when they have a mind to try a man's courage. But I have often known them die on the spot.

So his answerer, page 134, having taxed him for his speaking against an expression in the act of parliament of 5to Eliz. concerning the Wednesday fast: Whoop, Bays! Page 59 of his first book: 'The act for the Wednesday fast, the *Jejunium Caecilianum*' (our ecclesiastical politician is the better statesman of the two by far, and may make sport with Cecil when he pleases), 'was enjoined with this clause of exception, that if any person should affirm it to be imposed with an intention to bind the conscience, he should be punished as a spreader of false news.' So careful was the supreme magistrate in those days not to impose upon the conscience; and the wisdom of it is confirmed by the experience of our time: when so eminent a divine, as I mentioned before, thought fit to write a whole volume concerning the holiness of Lent; though, if I be not deceived, this doctrine too is prohibited by Act of Parliament under the same penalty. But, saith Bays there, 'the matter indeed of this law was not of any great moment, but this declaration annexed to it proved of a fatal and mischievous consequence.' 'Tis very well worth reading at large: but in short the consequence (or the occasion, 'tis no matter when I have to do with Bays) was, that 'princes, how peremptory soever they have been in asserting the rights of their supreme power, in civil affairs they have been forced to seem modest and diffident in the exercise of their ecclesiastical supremacy.' Now, Holla, Bays! p. 298 of his second book: 'To what purpose does he so briskly taunt me for thwarting mine own principles, because I have censured the impertinency of a needless provision in an Act of Parliament?' Observe, these are not the answerer's but Bays's own words; whereby you may see with what reverence and duty he uses to speak of his superiors and their actions, when they are not so happy as to please him. 'I may obey the law, though I may be of a different persuasion from the lawgivers in an opinion remote and impertinent to the matter of the law itself: nay, I may condemn the wisdom of enacting it, and yet at the same time think myself to lie under an indispensable obligation to obey it: for the formal reason of its obligatory power [as any casuist will inform him] is not the judgement and opinion of the lawgiver, but the declaration of his will and pleasure.' Very good and sound, Mr Bays: but here you have opened a passage; and this is as impertinent in you and more dangerous than what you blamed in that Act, that the Nonconformists may speak against your ecclesiastical laws; for their casuists then tell them that, they lying under an indispensable obligation not to conform to some of them do fulfil and satisfy their obedience in submitting to the penalty.

I looked further into what he saith in defence of the magistrate's assuming the priesthood; what for his scheme of moral grace; what to palliate his irreverent expressions concerning our blessed saviour and the Holy Spirit; what of all other matters objected by his answerer: and if you will believe me, (but I had much rather the reader would take the pains to examine all himself) there is scarce anything but slender trifling, unworthy of a logician, and beastly railing unbecoming any man, much more a divine. At last, having read it all through with some attention, I resolved, having failed so of anything material, to try my fortune whether it might be more lucky, and to open the book in several places as it chanced. But whereas they say that in the *Sortes Virgilianae*,° wheresoever you light you will find something that will hit and is proper to your intention; on the contrary here, there was not any leaf that I met with but had something impertinent, so that I resolved to give it over. This only I observed upon the whole, that he does treat his answerer the most basely and ungratefully that ever man did. For, whereas in his whole first book there was not one sound principle, and scarce anything in the second but what the answerer had given him occasion to amend and rectify if he had understanding; after so great an obligation he handles him with more rudeness than is imaginable. I know it may be said in Mr Bays's defence, that in this his second book he hath made his matters in many places much worse than they were before. But I say that was Bays's want of understanding: and that he knew not how to take hold of so charitable an opportunity as was offered him, and 'twas none of the answerer's fault. There are amongst men some that do not study always the true rules of wisdom and honesty, but delight in a perverse kind of cunning, which sometimes may take for a while and attain their design; but most usually it fails in the end and hath a foul farewell. And such are all Mr Bays's plots. In all his writings he doth so confound terms, he leaps cross, he hath more doubles (nay triples and quadruples) than any hare, so that he thinks himself secure of the hunters. And in the second book, even the length of it was some Policy. For you must know it is all but an epistle to the author of the *Friendly Debate*; and thought he with himself, who hath so much leisure from his own affairs, that he will read a letter of another man's business of eight hundred pages? But yet, thought he again (and I could be content they did read it), in all matters of argument I will so muddle myself in ink, that there shall be no catching, no finding me; and besides, I will speak always with so magisterial a confidence, that no modest man (and most ingenious persons are so) shall so much as quetch° at me, but be beat out of countenance; and plain men shall think that I durst not talk at such a rate but that I have a commission. I will first, said he in his heart, like a stout vagrant, beg; and if that will not do, I will command the question, and as soon as I have got it, I will so alter the property and put on another periwig, that I defy them all for discovering me or ever finding it again. This, beside all the lock° and advantage that I have the Nonconformists upon since the late times; and though they were born since, and have taken more sober principles, it shall be all one for that matter. And then for oratory and railing, let Bays alone. This contrivance is indeed all the strength of Mr Bays's argument, and as he said (how properly let the reader judge), page 69 before quoted, 'that moral virtue is not only the most material and useful part of all religion, but the ultimate end of all its other duties': so, railing is not only the most material and useful part of his religion, his reason, his oratory, and his practice, but the ultimate end of this and all his other books. Otherwise he is neither so strongly fortified nor so well guarded but that, without any ceremony of trenches or approaches, you may at the very first march up to his counterscarp° without danger. He puts me in mind of the incorrigible scold,° that though she was ducked over head and ears under water, yet stretched up her hands with her two thumb-nails in the nit-cracking posture, or with two fingers divaricated,° to call the man still in that language lousy rascal and cuckold. But indeed, when I consider how miserable a wretch his answerer has rendered him, and yet how he persists still and more to rail and revile him, I can liken it to nothing better betwixt them than to what I have seen with some pleasure, the hawking at the magpie. The poor bird understands very well the terrible pounces of that vulture; but therefore she chatters amain° most ruefully, and spreads and cocks her tail, so that one that first saw and heard the sport would think that she insulted over° the hawk in that chatter, and she huffed her train in token of courage and victory; when, alas, 'tis her fear all, and another way of crying the hawk mercy; and to the end that, the hawk finding nothing but tail and feather to strike at, she may so perhaps shelter her body.

Therefore I think there is nothing in my way that hinders me, but that I may now go on to the history of this Mr Bays's third book, the *Preface* to Bishop Bramhall, and to what 'juncture of affairs' it was reconciled. His Majesty (perhaps upon Mr Bays's frequent admonitions, both in his first and second book, that princes should be more attentive and confident in exercising their ecclesiastical jurisdiction, though I rather believe he never deigned to read a line in him, but what he did herein was only the result of his own good understanding) resolved to make some clear trial how the Nonconformists could bear themselves under some liberty of conscience. And accordingly he issued on March the 15th, 1671, his gracious Declaration of Indulgence, of which I wish his Majesty and the kingdom much joy, and as far as my slender judgement can divine, dare augurate° and presage mutual felicity and that whatever human accident may happen (I fear not what Bays foresees), they will, they can never have cause to repent this action or its consequences. But hereupon Bays finding that the King had so vigorously exerted his ecclesiastical power, but to a purpose quite contrary to what Mr Bays had always intended, he grew terribly angry at the King and his Privy Council; so that hereupon 'he started,' as himself says, 'into many warm and glowing meditations: his heart burnt and the fire kindled, and that heated him into all this wild and rambling talk (as some will be forward enough to call it), though he hopes it is not altogether idle, and whether it be or be not, he hath now neither leisure nor patience to examine.' This he confesses upon his best recollection, in the last page of this Preface: whereupon I cannot but animadvert, as in my first page, that this too lies open to his dilemma against the Nonconformists' prayers; for if he will not accept his own charge, his modesty is all impudent and counterfeit; if he does acknowledge it, he is a hot-headed incendiary, and a wild rambling talker, and in part, if not altogether, an idle fellow. Really I cannot but pity him, and look upon him as under some great disturbance and dispondency of mind—that this, with some other scattering passages here and there, argues him to be in as ill a case as Tiberius was in his distracted letter° to the Senate: there wants nothing of it but the *Dii deaeque me perdant*,° wishing, Let the gods and goddesses confound him worse than he finds himself to be every day confounded. But that I may not lose my thread. Upon occasion of this his Majesty's gracious Declaration, and against it, he writes this his third book, the Preface to Bishop Bramhall, and accordingly was unhappily delivered of it in June

(I have forgot) or July, in 1672. For he did not go his full time of it, but miscarried; partly by a new fright from J.O., and partly by a fall he had upon a 'closer importance'. But of all his three bolts this was the soonest shot, and therefore 'tis no wonder if he missed his mark, and took no care where his arrow glanced. But what he saith of his Majesty and his council, being toward the latter end of his Discourse, I am forced to defer that a little, because, there being no method at all in his wild rambling talk, I must either tread just on in his footsteps, or else I shall be in a perpetual maze, and never know when I am come to my journey's end.

And here I cannot altogether escape the mentioning of J.O. again, whom, (though I have shown that he was not the main cause of publishing Bays's books) yet he singles out, and on his pretence runs down all the Nonconformists; this being, as he imagined, the safest way by which he might proceed first to undermine, and then blow up his Majesty's gracious Declaration. And this indeed is the least unmethodical part in the whole Discourse. For first he undertakes to defend, that railing is not only lawful, but expedient. Secondly, that though he had railed, the person he spoke of ought not to have taken notice of it. And thirdly, that he did not rail. As to these things I do not much trouble myself, nor interest myself in the least in J.O.'s quarrel; no otherwise than if he were John a Nokes, and I heard him railed at by John a Styles.° Nor yet would I concern myself unnecessarily in any man's behalf; knowing that 'tis better being at the beginning of a feast, than to come in at the latter end of a fray. For if so I should, as often it happens in such rencounters,° not only draw Mr Bays, but J.O. too, upon my back, I should have made a sweet business on't for myself.

Now as to the lawfulness and expedience of railing: were it not that I do really make conscience of using Scripture with such a drolling° companion as Mr Bays, I could overload him thence both with authority and example. Nor is it worth one's while to teach him out of other authors, and the best precedents of the kind, how he, being a Christian and a divine, ought to have carried himself. But I cannot but remark his insolence, and how bold he makes upon this argument, p. 88 of his second book, with the memories of those great persons there enumerated, several of whom, and particularly my Lord Verulam,° I could quote to his confusion, upon a contrary and much better account. 'So far am I from repenting my severity towards them, that I am tempted rather to applaud it by the glorious examples of the greatest wits of our nation, King James, Archbishop Whitgift, Archbishop Bancroft, Bishop Andrews, Bishop Bilson, Bishop Montague, Bishop Bramhall, Sir

Walter Raleigh, Lord Bacon, etc.' and he might have added Mr Tarlton,° with as good pretence to his honour as himself. The niches are yet empty in the Old Exchange; pray let us speak to the statuary, that, next to King James's we may have Bays's *effigies*.° For such great wits are princes' fellows, at least when dead. At this rate there is not a scold at Billingsgate but may defend herself by the pattern of King James and Archbishop Whitgift, etc.; yet this is passable, if you consider our man. But that is most intolerable, p. 7 of the Preface to his first book, where he justifies his debauched way of writing by parallel to our blessed Saviour. And I cannot but with some awe reflect how near the punishment was to the offence; when, having undertaken so profane an argument, he was in the very instant so infatuated as to say that Christ was not only 'in a hot fit of zeal, but in a seeming fury too, and transport of passion'. But however, seeing he hath brought us so good vouchers,° let us suppose what is not to be supposed, that railing is lawful. Whether it be expedient or no, will yet be a new question. And I think Mr Bays, when he hath had time 'to cool his thoughts', may be trusted yet with that consideration, and to compute whether the good that he hath done by railing do countervail the damage which both he in particular and the cause he labours, have suffered by it. For in my observation, if we meet with an argument in the streets, both men, women, and boys, that are the auditory, do usually give it on the modester side, and conclude that she that rails most has the least reason.

For the second: where he would prove that though he had railed, yet his answerer, J.O., ought not have taken notice of it, nor those of the party who are under the same condemnation; but that he should have abstracted and kept close to the argument,—I must confess it is a very secure and wholesome way of railing. And allowing this, he hath good reason to find fault with his answerer, as he does, for turning over his book; though without turning it over, I know not how he could have answered him, but with his hat, or with mum.° But for aught I can see in that only answer which is to his first book, he hath been obedient, and abstracted the argument sufficiently: and if he hath been anywhere severe upon him, he hath done it more cleanly, and much more like a gentleman, and it hath been only in showing the necessary inferences that must follow upon the author's maxims and unsound principles. But as to any answer to Bays's second book or this third, for aught I can see, J.O. sleeps upon both ears.°

To this third undertaking, to show that he hath not railed, I shall not say anything more, but let it be judged by the company, and to them let it be referred. But in my poor opinion, I never saw a man through all his three books in so high a salivation.

And therefore, till I meet with something more serious, I will take a walk in the garden and gather some of Mr Bays's flowers; or I might more properly have said, I will go see Bedlam,° and pick straws with our madman. First he saith, that 'some that pretend a great interest in the holy brotherhood, upon every slight accident are beating up the drums against the Pope and Popish Plots; they descry Popery in every common and usual chance; and a chimney cannot take fire in the city or suburbs, but thcy are immediately crying, Jesuits and fireballs'. I understand you, sir. This, Mr Bays, is your prologue, that is to bc spoke by Thunder and Lightning: 'I am loud Thunder; brisk Lightning I. I strike men down. I fire the town. Look to't. We'll do't.'° Mr Bays, it is something dangerous meddling with those matters. As innocent persons as yourself have felt the fury of the wild multitude, when such a calamity hath disordered them; and after your late severity against tradesmen, it had been better you had not touched the fire. Take heed lest the reasons which sparkle, forsooth, in your discourse have not set their chimneys on fire. None accuses you, what you make sport with, of burning the ships at Chatham,° much less of blowing up the Thames. But you ought to be careful, lest having so newly distinguished betwixt the fanatic and his wealth, they should say that you are distinguishing now betwixt the fanatics and their houses. These things are too edged to be jested with, if you did but consider that not only the 'holy brotherhood', but the 'sober and intelligent citizens', are equally involved in these sad accidents. And in that lamentable conflagration° (which was so terrible, that, though so many years ago, it is yet fresh in men's memories; and besides, is yearly, by Act of Parliament, observed with due humiliation and solemnity)° it was not trade only and merchandise suffered, which you call their Diana, and was not so much to be considered; but St Paul's too was burnt, which the historians tell us was Diana's temple.°

The next thing is more directly levelled at J.O. for having in some later book used those words, 'We cannot conform to Arminianism or Socinianism on the one hand, or Popery on the other.' What the answerer meant by those words, I concern not myself; only I cannot but say that there is a very great neglect somewhere, wheresoever the inspection of books is lodged, that at least the Socinian° books are tolerated and sell as openly as the Bible. But Bays turns all into mirth: 'He might as well have added all the *-isms* in the Old Testament, Perizzitism, Hivitism, Jebuzitism, Hittitism, etc.'

No, Mr Bays, that need not; and though this indeed is a very pretty conceit, and 'twere pity it should have been lost, yet I can tell you a better way. For, if rhyming be the business, and you are so good at 'tagging° of points in a garret', there is another word that will do it better, and for which, I know not how truly, you tax your answerer too here, as if he said, 'The Church of England were desperately schismatical, because the Independents are resolved, one and all, to continue separate from her communion.' Therefore let schism, if you please, rhyme to *-ism*. And though no man is obliged to produce the authority of the greatest wits of the nation to justify a rhyme, yet for your 'dear sake', Mr Bays, I will this once supererogate.° The first shall be your good friend Bishop Bramhall; who among many other memorable passages, which I believe were the reason that he never thought fit to print his own book, page 101 teaches us, not absurdly, that 'it was not the erroneous opinions of the Church of Rome, but the obtruding them by laws upon other churches, which warranted a separation'. But if this will not do, *vous avez* Doctor Thorndike's° deposition in print; for he, I hear, is lately dead. 'The Church of England in separating from the Church of Rome is guilty of schism before God.' I have not the book by me, but I am sure 'tis candidly recited as I have read it. Then (to show too that there is a King on this side) his present Majesty's father, in his Declaration, 4t° *Caroli*, 1628, affirms that a Book, entitled *Appello Caesarem*, or *An Appeal to Ceasar*, and 'published in the year 1625, by Richard Montague,° then Bachelor of Divinity, and now Bishop of Chichester, had opened the way to these schisms and divisions which have since ensued in the church; and that therefore, for the redress and remedy thereof, and for the satisfaction of the consciences of his good people, he had not only by public proclamation called in that book, which ministered matter of offence, but, to prevent the like danger for the future, reprinted the Articles of Religion established in the time of Queen Elizabeth, of famous memory: and by a Declaration, before those Articles, did restrain all opinions to the sense of those Articles, that nothing might be left for private fancies and innovations, etc.' And if this will not amount fully, I shall conclude with a villainous pamphlet that I met with t'other day, but of which a great wit indeed was the Author. And whereas Mr Bays is always defying the Nonconformists with Mr Hooker's° *Ecclesiastical Polity*, and the *Friendly Debate*: I am of opinion, though I have a great reverence for Mr Hooker, who in some things did answer himself, that this little book, of not full eight leaves, hath shut that *Ecclesiastical Polity*, and Mr Bays's too, out of doors; but for the *Friendly Debate*, I must confess that is unanswerable. 'Tis one

Mr Hales,° of Eton; a most learned divine, and one of the Church of England, and most remarkable for his sufferings in the late times, and his Christian patience under them. And I reckon it not one of the least ignominies of that age, that so eminent a person should have been by the iniquity of the times reduced to those necessities under which he lived; as I account it no small honour to have grown up into some part of his acquaintance, and conversed a while with the living remains of one of the clearest heads and best-prepared breasts in Christendom. That which I speak of is his little *Treatise of Schism*,° which, though I had read many years ago, was quite out of my mind, till I occasionally lighted upon't at a bookseller's stall. I hope it will not be tedious, though I write of some few (and yet whatsoever I omit I shall have left behind more) material passages. 'Schism is one of those theological scarecrows with which they who use to uphold a party in religion, use to fright away such, as making enquiry into it are ready to relinquish and oppose it, if it appear either erroneous or suspicious. Schism is, if we would define it, an unnecessary separation of Christians from that part of the visible church of which they were once members. Some, reverencing antiquity more than needs, have suffered themselves to be scared with imputation of schism more than needs. Nothing absolves men from the guilt of schism but true and unpretended conscience. But the judgements of the ancients many times (to speak most gently) are justly to be suspected. Where the cause of schism is necessary, there not he that separates, but he that is the cause of separation, is the schismatic. Where the occasion of separation is unnecessary, neither side can be excused from guilt of schism. But who shall be the judge? That is a point of great difficulty, because it carries fire in the tail of it; for it brings with it a piece of doctrine which is seldom pleasing to superiors. You shall find that all schisms have crept into the Church by one of these three ways—either upon matter of fact, or upon matter of opinion, or point of ambition. For the first, I call that matter of fact, when something is required to be done by us which either we know or strongly suspect to be unlawful.' Where he instances in the old great controversy about Easter, 'For it being upon error taken for necessary that an Easter must be kept, and upon worse than error (for it was no less than a point of Judaism forced upon the Church) thought further necessary that the ground of the time for the feast must be the rule left by Moses to the Jews: there arose a stout question, whether it was to be celebrated with the Jews on the fourteenth Moon, or the Sunday following. This caused as great a combustion as ever was; the West separating and refusing communion with the East for many years together.

Here I cannot see but all the world were schismatics, excepting only that we charitably suppose to excuse them from it, that all parties did what they did out of conscience. A thing which befell them by the ignorance, for I will not say the malice, of their guides; and that through the just judgement of God, because, through sloth and blind obedience, men examined not the things they were taught, but like beasts of burden patiently couched down, and indifferently underwent all whatsoever their superiors laid upon them. If the discretion of the chiefest guides of the church did, in a point so trivial, so inconsiderable, so mainly fail them, can we, without the imputation of great grossness and folly, think so poor-spirited persons competent judges of the questions now on foot betwixt the churches? Where, or among whom, or how many the church shall be, it is a thing indifferent. What if those to whom the execution of the public service is committed, do something either unseemly or suspicious, or peradventure unlawful; what if the garments they wear be censured, nay, indeed be suspicious. What if the gesture or adoration be used to the altars, as now we have learned to speak? What if the homilist have preached or delivered any doctrine, of the truth of which we are not well persuaded (a thing which very often falls out); yet, for all this, we may not separate, except we be constrained personally to bear a part in it ourselves. Nothing can be a just cause of refusing communion in schism that concerns fact, but only to require the execution of some unlawful or suspected act. For, not only in reason, but in religion too, that maxim admits of no release, *Cautissimi cujusque praeceptum, quod dubitas, ne feceris*:° That whatsoever you doubt of, that you in no case do.' He instances then in the second Council of Nice,° where, saith he, 'the Synod itself was the schismatical party in the point of using the images, which,' saith he, 'all acknowledge unnecessary, most do suspect, and many hold utterly unlawful: can then the enjoining of such a thing be aught else but an abuse? Can the refusal of communion here be thought any other thing than duty? Here, or upon the like occasion, to separate may peradventure bring personal trouble or danger, against which it concerns any honest man to have *pectus praeparatum*."° Then of schism from opinion: 'Prayer, confession, thanksgiving, reading of scripture, administration of sacraments in the plainest and simplest manner, were matter enough to furnish out a sufficient liturgy, though nothing either of private opinion or of Church pomp, of garments, of prescribed gestures, of imagery, of music, of matter concerning the dead, of many superfluities which creep into the church under the name of order and decency, did interpose itself. To charge churches and liturgies with things unnecessary, was the first beginning of superstition. If the Fathers and special guides of the Church would be a little sparing in encumbering churches with superfluities, or not over-rigid either in reviving obsolete customs or imposing new, there would be far less cause of schism or superstition; and all the inconvenience likely to ensue would be but this—they should in so doing yield a little to the imbecility of their inferiors; a thing which Saint Paul would never have refused to do. It is alike unlawful to make profession of known or suspected falsehood, as to put in practice unlawful or suspected actions. The third thing I named for matter of schism was ambition, I mean episcopal ambition; one head of which is one bishop's claiming supremacy over another, which, as it hath been from time to time a great trespass against the Church's peace, so it is now the final ruin of it. For they do but abuse themselves and others, who would persuade us that Bishops by Christ's institution have any superiority over other men further than that of reverence, or that any Bishop is superior to another further than positive order agreed upon among Christians hath prescribed. Time hath taken leave, sometimes, to fix this name of CONVENTICLES upon good and honest meetings. Though open assemblies are required, yet at all times, while men are really pious, all meetings of men for mutual help of piety and devotion, wheresoever and by whomsoever celebrated, were permitted without exception. In times of manifest corruption and persecution, wherein religious assembling is dangerous, private meetings, howsoever besides public order, are not only lawful, but they are of necessity and duty. All pious assemblies, in time of persecution and corruption howsoever practised, are indeed, or rather alone, the lawful congregations; and public assemblies, though according to form of law, are indeed nothing else but RIOTS and CONVENTICLES, if they be stained with corruption and superstition.' Do you not see now, Mr Bays, that you needed not to have gone so far for a word, when you might have had it in the neighbourhood? If there be any coherence left in your skull, you cannot but perceive that I brought you authority enough to prove that schism (for the reason we may discourse another time) does at least rhyme to *-ism*. But you have a peculiar delight and felicity (which no man envies you) in scripture-drollery; nothing less will taste to your palate; whereas otherwise you have travelled so far in Italy, that you could not escape the titles of some books which would have served your turn as well, *cardinalism, nepotism, putanism,*° if you were in a paroxysm of the *-isms*.

When I had writ this, and undergone so grateful a penance for no less than that I had transcribed before out of our author, I could not, upon comparing them both together, but reflect most seriously upon the difference of their two ways of discoursing. I could not but admire that majesty and beauty which sits upon the forehead of masculine Truth and generous Honesty; but no less detest the deformity of Falsehood disguised in all its ornaments. How much another thing it is to hear him speak, that hath cleared himself from froth and groans, and who suffers neither sloth, nor fear, nor ambition, nor any other tempting spirit of that nature to abuse him, from one who, as Mr Hales expresseth it, makes Christianity lackey to ambition! How wretchedly, the one, to uphold his fiction, must incite princes to persecution and tyranny, degrade grace to morality, debauch conscience against its own principles, distort and misinterpret the Scripture, fill the world with blood, execution, and massacre; while the other needs and requires no more but a peaceable and unprejudicate° soul, and the native simplicity of a Christian spirit! And methinks, if our author had any spark of virtue unextinguished, he should upon considering these together, retire into his closet, and there lament and pine away for his desperate folly; for the disgrace he hath, as far as in him is, brought upon the Church of England by such an undertaking, and for the eternal shame to which he has hereby condemned his own memory.

I ask you heartily pardon, Mr Bays, for treating you against decorum here, with so much gravity. 'Tis possible I may not trouble you above once or twice more in the like nature; but so often at least, I hope, one may in the writing of a whole book , have leave to be serious. Your next flower, and that indeed is a sweet one, 'Dear heart, how could I hug and kiss thee for all this love and sweetness!' Fie, Fie, Mr Bays! Is this the language of a divine, and to be used, as you sometimes express it, in the face of the sun? Who can escape from thinking that you are adreamed° of your 'comfortable importance'? These are (as the moral satirist calls them in the cleanliest manner the thing would bear) 'words left betwixt the sheets'.° Somebody might take it ill that you should misapply your courtship to an enemy. But in the Roman Empire it was the privilege of the hangman to deflower a virgin before execution. But, sweet Mr Bays (for I know you do nothing without a precedent of some of the greatest wits of the nation), whose example had you for this 'seeming transport' of a gentler 'passion'?

Then comes 'Well fare poor Macedo° for a modest fool.' This I know is matter of *Gazette*, which is as canonical as *Ecclesiastical Policy*. Therefore I have the less to say to't. Only, I could wish that there were some severer laws against such villains who raise so false and scandalous reports of worthy gentlemen, and that those laws were put in execution; and that men might not be suffered to walk the streets in so confident a garb, who commit those assassinates° upon the reputation of deserving persons.

Here follows a sore charge: that the answerer had 'without any provocation, in a public and solemn way, undertaken the defence of the fanatic cause'. Here indeed, Mr Bays, you have reason,° and you might have had as just a quarrel against whosoever had undertaken it. For your design and hope was from the beginning, that no man would have answered you in a public and solemn way; and nothing would vex a wise man, as you are, more than to have his intention and counsel frustrated. When you have ranged all your forces in battle, when you have placed your cannon, when you have sounded a charge, and given the word to fall upon the whole party,—if you could then persuade every particular person of 'em, that you gave him no provocation, I confess, Mr Bays, this were an excellent and a new way of your inventing, to conquer single ('tis your moral virtue) whole armies.° And so the 'admiring drove' might stand gaping, till, one by one, you had cut all their throats. But, Mr Bays, I cannot discern but you gave him as much provocation in your first Book as he has you in his *Evangelical Love, Church Peace and Unity*,° which is the pretence of your issuing this preface.

For, having for your 'dear sake' (beside many other troubles that I have undertaken, without your giving me any provocation) sought out and perused that book too, I do not find you anywhere personally concerned, but as you have, it seems upon some conviction, assumed to yourself some vices or errors against which he speaks only in general, and with some modesty. But for the rest, you say upon full perusal, 'you find not one syllable to the purpose beside a perpetual repetition of the old out-worn story of unscriptural ceremonies, and some frequent whinings, and sometimes ravings', etc. Now to see the dulness of some men's capacities above others. I, upon this occasion, began, I know not how it came, at page 127, and thence read on to the end of his book. And from thence I turned to the beginning and continued to page 127, and could not all along observe anything but what was very pertinent to the matter in hand. But this is your way of excusing yourself from replying to things that yet you will be meddling with and nibbling at; and 'tis besides a pretty knack (the Nonconformists have it not alone) of frightening or discouraging sober people from reading those dangerous treatises, which might contribute to their better information. I cannot but observe, Mr Bays, this admirable way (like fat Sir John Falstaff's singular dexterity in sinking°) that you have of answering whole books or discourses, how pithy and knotty soever, in a line or two, nay sometimes with a word. So it fares with this book of the answerer's. So with a book or discourse of his, I know not, of the *Morality of the Lord's Day*; which is answered by a 'Septenary Portion in the Hebdomadal Revolution'. So, whether book or discourse, I also know not of the 'self-evidencing light of the scripture,' where Bays offers (and it seems strange) to produce as good proofs for it out of the Koran. So I showed you where he answers demonstration with the lie. And one thing more comes into my mind; where, after he has blundered a great while to bring himself off the magistrate's exercising the priesthood in his own person, he concludes with an irresistible defence against his answerer: 'this is suitable to the genius of his ingenuity, and betrays him as much as the word ENTANGLEMENT, which is the shibboleth° of all his writings.' So he defeats all the 'gross bodies of orthodoxy' with calling them 'systems and syntagms'. So you know he answers all the controversial books of the Calvinists that ever had been written, with the tale of Robin Hood, and the 'mighty bramble on the south side of the lake Leman'. Mr Bays, you cannot enough esteem and cherish this faculty. For, next to your single beating whole armies, I do not know any virtue that you have need of so often, or that will upon trial be found more useful.

And to this succeeds another flower, I am sure, though I can scarce smell out the sense of it. But it is printed in a distinct character, and that is always a certain sign of a flower. For our booksellers have many arts to make us 'yield to their importunity': and among the rest, they promise us, that it shall be printed on fine paper, and in a very large and fair letter; that it shall be very well examined, that there be no *errata*; that wheresoever there is a pretty conceit, it shall be marked out in another character; that the sentences shall be boxed up in several paragraphs and more drawers than in any cabinet; that the books shall all be bound up in calves' leather. But my greatest care was, that when I quoted any sentence or word of our author's, it might be so discernable, lest I should go for a plagiarist. And I am much offended to see that in several places he hath not kept touch with me. The word of Mr Bays's that he has here made notorious is 'categoricalness': and I observe that wheresoever there comes a word of that termination he shows it the same honour, as if he had a mind to make Bays a collar of *-nesses*.° What the mystery is, I cannot so easily imagine; no more than of 'shibboleth' and 'entanglement'. But I doubt Mr Bays is sick of many complicated diseases; or, to keep to our rhyme, *sicknesses*. He is troubled not only with the *-isms* but the *-nesses*. He might, if he had pleased, here too to have shown his wit, as he did in the others, and have told us of Sheerness, Dungeness, Inverness, and Caithness. But he omitted it perhaps in this place, knowing how well he had acquitted himself in another, and out of the scripture too, which gives his wit the highest relish. 'Tis page 72 of his first book, where, to prove that the fruits of the spirit are no more than morality, he quotes Saint Paul, Gal. 5: 22, where the apostle enumerates them: 'love, joy, peace, patience, gentleness, goodness, faith, meekness, and temperance': but our author translates 'joy' to 'cheerfulness', 'peace' to 'peaceableness', and 'faith' to 'faithfulness'. What ignorance, or rather what forgery is this of scripture and religion! Who is there of the 'systematical German, Geneva, orthodox Divines' but could have taught him better? Who is there of the 'sober, intelligent, Episcopal Divines' of the Church of England but would abhor this interpretation? Yet when his answerer, I see, objects this to him, page 220, Bays, like a dexterous scholastical disputant, it being told him that joy is not cheerfulness, but that 'spiritual joy which is unspeakable'; that peace is not peaceableness in his sense, but 'that peace of God which through Jesus Christ is wrought in the hearts of believers by the Holy Ghost'; and that faith in God is there intended, not faithfulness in our duties, trusts, or offices: what does he do? Page 337, he very ingenuously and wisely, when he is to answer, quite forgets that faith was once named; and having suppressed that, as to the rest he wipes his mouth, and rubs his forehead, and saith the 'cavil is but a little one, and the fortune of Caesar and the Roman Empire depend not upon it, and therefore he will not trouble the reader with a critical account of the reason of his translation.' No, don't, Mr Bays. 'Tis very well; let it alone. But, though not the fortunes of Caesar and the Roman Empire, I doubt there is something more depends upon it, if it be matter of salvation. And I am afraid besides, that there may a curse too belong to him who shall knowingly add or diminish in the scripture.° Do you think Bishop Bramhall himself, if he had seen this, could have abstained (p. 117 before quoted) from telling our author, 'That the promiscuous licence given to people qualified or unqualified, not only to read but to interpret the scriptures according to their private spirits or particular fancies, without regard either to the analogy of faith, which they understand not, or to the interpretation of the doctors of former ages, is more prejudicial, I might better say pernicious, both to particular Christians and to whole societies, than the over-rigorous restraint of the Romanists'?

The next is a piece of mirth, on occasion of some discourse of the answerer's about the Morality of the Lord's-day; where it seems he useth some hard words, which I am naturally an enemy to; but might be done of purpose to keep the controversy from the white-aprons! within the white-surplices, to be more learnedly debated. But this fares no better than all the rest. There is no kind of morality, I see, but Bays will try to debauch it. 'O what edifying doctrine,' saith he, 'is this to the white-aprons! and doubtless they would, with the Jews, sooner roast themselves than a small joint of mutton upon the sacred day of rest.' Now I do not, neither, I believe, does Bays himself, know any of them that are thus superstitious. So that Mr Bays might, if he had pleased, have spared his gibing at that day, which hath more sacredness in it by far than many, nay than any of those things he pleads for. But when men are once *adepti*° and have attained Bays's height, and 'divinity' at least is 'rightly understood', they have a privilege, it seems, not only to play and make merry *on* the Sabbath day, but *with* it.

After this I walked a great way through bushes and brambles before I could find another flower; but then I met with two upon one stalk on occasion of his answerer's having said something of the day of judgement, when men should be accountable. 'Oh,' saith he, 'we shall be sure to be accounted with at the day of judgement'; and again, 'Ah sweet day when these people of God shall once for all, to their unspeakable comfort and support, wreak their eternal revenge upon their reprobate enemies.' This puts me in mind of another expression of our author's alluding too this way: ''Tis an easy matter by this dancing and capering humour to perpetuate all the controversies in the world, how plainly soever determinable, to the coming of Elias;° and after this rate shall the barber's basin remain Mambrino's helmet,° and the ass's panel° a furniture° for the great horse, till the day of judgement.' Now, good Mr Bays, I am one that desire to be very well resolved in these things; and though not much indeed, yet I attribute something to your judgement. Pray tell us in good earnest, what you think of these things, that we may know how to take our measures of living accordingly. For if indeed there be no judgement, no account for what is done here below, I have lost a great deal of precious time, that I might have enjoyed in one of the fruits of your spirit, that is 'cheerfulness'. How many good jests have I balked, even in writing this book, lest I should be brought to answer for every profane and idle word. How frequent opportunities have I missed in my life of geniality and pleasure, and fulfilling nature in all its ends! How have you frighted the magistrate in vain from exercising his uncontrollable ecclesiastical power, with the fear of an after-reckoning to God Almighty! And how have you, p. 238, defeated the obligatory force of all his laws, and set his subjects at liberty from all obligations to the duty of obedience for they lie under no obligation, you say then, but of prudence and self-interest. But unless there hath been some error in our education, and we have been seasoned with ill books at first, so that we can never lose the impression, there is some such matter, and the governor had reason when he trembled to hear Saint Paul° discoursing of that subject. The fanatical 'Book of Martyrs'° (for we will not with some call the Bible so) tells us some old stories of persons that have been cited by some of them to appear at such a day, and that by dying at the time prefixed, they have saved their recognisances. And in the Scotch history° we read of a great cardinal that was so summoned by poor Mr Wishart,° and yet could not help it, but he must take that long and sad journey of death to answer at the Grand Assizes. If therefore there be such a thing, I would not for fear, and if there be not, yet I would not for good luck's sake, set that terrible day at defiance, or make too merry with it. 'Tis possible that the Nonconformists many of them may be too censorious of others, and too confident of their own integrity. Others of them are more temperate, and perhaps destitute of all human redress against their sufferings. Some of those make rash challenges, and the other just appeals to appear at that dreadful tribunal. In the meantime, 'tis not for you to be both the enemy and their judge. Much less does it befit you, because perhaps they speak too sillily or demurely of it, or too braving and confidently, therefore to make a mere mockery of the whole business of that Supreme Judge and judicature. And one thing I will say more, though slighter: that, though I am not so far gone as Campanella° was in the efficacy of words, and the magic of the face, and pronunciation, yet I marked how your answerer looked when he spoke of the day of judgement. Very gravely, I assure you, and yet without any dressing or adorning his *superciliums*;° and I have most often observed that serious words have produced serious effects.

I have by this time, methinks, gathered enough: nor are there many more left, unless I should go for a flower to the dunghill, which he saith 'is his only magazine'. And this being an expression which he has several times used (for no Nonconformist repeats so often), I cannot but remark, that besides his natural talent, Mr Bays hath been very industrious, and neglected no opportunity of acquiring a perfection of railing. For this is a phrase borrowed from a modern author lately dead,° and I suppose Bays had given him a bond for repayment at the day that he spoke of so lately.

There are, indeed, several others at which I am forced to stop my nose. For by the smell, any man may discern they grew upon a ranker soil than that on the south side of the Lake Leman, even upon the bank of the Thames in the meadow of Billingsgate: as that of the lie,° which, he saith, no gentleman, much less a divine, ought to put up. Now if this were to be tried by a court-martial of the brothers of the blade,° 'tis to be considered whether it were the downright lie, or whether it were only the lie by interpretation. For in the disputes of the Schools there is nothing more usual than *hoc est verum, hoc est falsum*.° But this passes without any blemish of honour on either side, and so far it is from any obligation to a challenge or a duel, that it never comes to be decided so much as by the study-door key. But *quod restat probandum*° does the business without demanding other satisfaction. Then, if it were the downright lie, it is to be examined who gave the lie first; for that alters the case. And last of all (but which is indeed upon a quarrel the least material point, yet, it too comes under some consideration), which of the two was in the right, and which of them spoke truth, and which lied. These are all things to be discussed in their proper places. For I do not observe that the answerer gave Bays the downright lie. But I find that Bays gave him the lie first in terms. And as to the truth of the things controverted and alleged, there needs no more than the depositions that I formerly transcribed concerning Bays's own words. But all this is only a scene out of Bays's *Rehearsal*.

> Villain, thou liest!
> Arm, arm Valerio, arm!
> The lie no flesh can bear, I trow.°

And then as to the success of the combat,

> . . . They fly, they fly,
> Who first did give the lie.°

For that of caitiff and other provocations that are proper for the same Court, I will not meddle further. And for the being epast grace, and so past mercy', I shall only observe that the Church of England is much obliged to Mr Bays for having proved that Nonconformity is the sin against the Holy Ghost.

There remains but one flower more that I have a mind to. But that indeed is a rapper.° 'Tis a 'flower of the sun', and might alone serve both for a staff and a nosegay for any nobleman's porter. 'Symbolicalness is the very essence of paganism, superstition, and idolatry. They will and ought sooner to broil in Smithfield than submit to such abominations of the strumpet and the beast.° 'Tis the very potion wherewith the scarlet whore made drunk the kings of the earth. Elagabalus° and Bishop Bonner° loved it like clary° and eggs, and always made it their morning's draught upon burning-days; and it is not to be doubted but the seven vials of wrath that were to be poured out upon the nations of the earth under the reign of Antichrist, were filled with symbolical extracts and spirits.' This I confess a pretty posy for the nose of such a divine. Doctor Bayly's° romance of the wall-flower had nothing comparable to't. And I question whether, as well as Mr Bays loves preferment, yet though he had lived in the primitive Church, he would not, as Heliodorus Bishop of Trissa,° I take it, that renounced his bishopric rather than his title to the History of Theagenes and Chariclea, have done in like manner; nay, and have delivered up his Bible too into the bargain, before he would quit the honour of so excellent a piece of drollery. This is surely the bill of fare, not at the ordination dinner° at the Nags Head, but of the excusation dinner° at the Cock; and never did divine make so good cheer of Owen's peas porridge° and scripture. I know no dainty wanting, or that could have pleased his tooth so well, except the leg of a pheasant at the Dog and Partridge; for he is of Thomas a Becket's diet, who ate, he said, *phaesianum sicut alii muluellum*,° and can mortify himself upon pheasant as well as others with saltfish. Good Mr Bays, or Mr Thunder, or Mr Cartwright° (not the Nonconformist Cartwright° that was, you say—as some others too of your acquaintance—converted; but the player in *The Rehearsal*), this divinity I doubt was the Bacchus of your thigh, and not the Pallas of your brain.

Here it is that, after so great an excess of wit, he thinks fit to take a julep and resettle his brain and the government. He grows as serious as 'tis possible for a madman, and pretends to sum up the whole state of the controversy with the Nonconformists. And to be sure he will make the story as plausible for himself as he may; but therefore it was that I have before so particularly quoted and bound him up with his own words as fast as such a Proteus could be pinioned. For he is as waxen as the first matter,° and no form comes amiss to him. Every change of posture does either alter his opinion or vary the expression by which we should judge of it; and sitting he is of one mind, and standing of another. Therefore I take myself the less concerned to fight with a windmill like Don Quixote; or to whip a gig° as boys do; or with the lackeys at Charing Cross or Lincoln's Inn Fields to play at the Wheel of Fortune;° lest I should fall into the hands of my Lord Chief Justice, or Sir Edmond Godfrey.° The truth is, in short, and let Bays make more or less of it if he can, Bays had at first built up such a stupendous magistrate as never was of God's making. He had put all princes upon the rack to stretch them to his dimension. And as a straight line continued grows a circle, he had given them so infinite a power, that it was extended unto impotency. For though he found it not till it was too late in the cause, yet he felt it all along (which is the understanding of brutes) in the effect. For hence it is that he so often complains that princes knew not aright that supremacy over consciences, to which they were so lately, since their deserting the Church of Rome, restored; that in most nations government was not rightly understood, and many expressions of that nature: whereas indeed the matter is, that princes have always found that uncontrollable government over CONSCIENCE to be both unsafe and impracticable. He had run himself here to a stand, and perceived that there was a God, there was scripture; the magistrate himself had a conscience, and must 'take care that he did not enjoin things apparently evil'. Being at a stop here, he would therefore try how he could play the broker on the subjects' side: and no pimp did ever enter into a more serious disputation to vitiate° an innocent virgin than he to debauch their consciences. And to harden their unpractised modesty, he emboldens them by his own example, showing them the experiment upon his own conscience first. But after all, he finds himself again at the same stand here, and is run up to the wall by an angel.° God, and scripture, and conscience will not let him go further; but he owns, that if the magistrate enjoins things apparently evil, the subject may have liberty to remonstrate. What shall he do, then? for it is too glorious an enterprise to be abandoned at the first rebuff. Why, he gives us a new translation of the Bible, and a new commentary! He saith that tenderness of conscience might be allowed in a Church to be constituted, not in a Church constituted already. That tenderness of conscience and scandal are ignorance, pride, and obstinacy. He saith the Nonconformists should communicate with him till they have clear evidence that it is evil. This is a civil way indeed of gaining the question, to persuade men that are unsatisfied, to be satisfied till they be dissatisfied. He threatens, he rails, he jeers them, if it were possible, out of all their consciences and honesty; and finding that will not do, he calls out the magistrate, tells him these men are not fit to live; there can be no security of government while they are in being: bring out the pillories, whipping-posts, galleys, rods, and axes (which are *ratio ultima cleri*, a clergyman's last argument, ay and his first too), and pull in pieces all the Trading Corporations, those nests of faction and sedition. This is a faithful account of the sum and intention of all his undertaking, for which, I confess, he was as picked° a man as could have been employed or found out in a whole kingdom; but it is so much too hard a task for any man to achieve, that no goose but would grow giddy with it.

For whereas he reduces the whole controversy to a matter of two or three symbolical ceremonies (and if there be nothing else, more the shame of those that keep such a pudder° for them), it is very well worth observing how he hath behaved himself, and how come off in this dispute. It seems that the Conformists define a sacrament to be an outward visible sign of an inward spiritual grace. It seems that the sacraments are usually called in the Greek *symbola*. It seems further that some of the Nonconformists, under the name therefore of symbolical ceremonies, dispute the lawfulness of those that are by our Church enjoined, whereby the Nonconformists can only intend that these ceremonies are so applied as if they were of a sacramental nature and institution, and that therefore they are unlawful. Our author's answerer, handling this argument, does among other things make use of a pertinent passage in Saint Austin, *Signa cum ad res divinas pertinent sacramenta appellantur.*° What does Mr Bays in this case? for it went hard with him. Why, as good luck would have it, not being willing that so great a politician, to the irreparable damage of the Church, should yet be destroyed, J.O. had forgot to quote the book and the page. Now though you send a man the length of your weapon, and name your second, yet Mr Bays, being, as you see often, admirably read in the laws of duelling, knew that unless the time and place be appointed, there is no danger. He saith therefore, page 452 of his second book, that he 'should have advantage on his side, if he should lay odds with him that there is no such passage in all the volumes of Saint Austin'. But however, that is neither civil nor ingenuous to trouble him with such objections, as he cannot answer without reading over eight or ten large volumes in folio. It was too much to expect from one of so much business, good Augustulus:°

> Quum tot sustineas et tanta negotia solus,
> Res sacras armis tuteris, moribus ornes,
> Legibus emendes . . . °

Which may be thus translated: 'When you alone have the ceremonies to defend with whipping-posts, rods, and axes; when you have grace to turn into morality; when you have the Act of Oblivion and Indemnity and the Ecclesiastical Declaration of March to tear in pieces; it were unreasonable and too much to the damage of the public to put you in such an employment'. I ask your pardon, Mr Bays, for this paraphrase and digression; for I perceive I am even hardened in my Latin, and am prone to use it without fear or reverence. But, Mr Bays, there might have been a remedy for this, had you pleased. Where, then, were all your leaf-turners? a sort of poor readers 'that you as well as Bishop Bramhall ought to have some reverence for', having made so much use of them to gather materials for your structures and superstructures. I cannot be persuaded, for all this, but that he knows it well enough, the passage being so remarkable in itself, and so dirtied with the Nonconformists' thumbs, that he could not possibly miss it; and I doubt he does but laugh at me now, when, to save him a labour, I tell him in the simplicity of my heart, that even I myself met with it in *Ep. 5ta ad Marcellinum,*° and the words these, *Nimis autem longum est convenienter disputare de varietate signorum, quae cum ad res divinas pertinent Sacramenta appellantur.*° But, whether there be such a place or no, he hath no mind that his answerer should make use of it; nor of the schoolmen, whom before he had owned for the authors of the Church of England's divinity; but would bind up the answerer to the law only and the Gospel. And now Mr Bays saith he will be of the schoolmen's opinion 'as long as they speak sense', and no longer (and so I believe of Saint Austin's), that is to say, so long as they will serve his turn; for all politicians shake men off when they have no more use of 'em, or find them to thwart the design. But, Mr Bays, why may not your answerer or any man else quote Saint Austin, as well as you may the Scriptures? I am sure there is less danger of perverting the place, or of misinterpretation. And though perhaps a Nonconformist may value the authority of the Bible above that of the Fathers, yet the Welsh have a proverb, that the Bible and a stone do well together; meaning, perhaps, that if one miss, the other will hit. You, that are a duellist, know how great a bravery 'tis to gain an enemy's sword, and that there is no more home thrust in disputation than the *argumentum ad hominem*. So that if your adversary fell upon you with one of your own Fathers, it was gallantly done on his part; and no less wisely on yours to fence in this manner, and use all your shifts to put it by. For you too, Mr Bays, do know—no man better—that it is not at all times safe nor honourable to be of a father's opinion.

Having escaped this danger, he grows, nor can I blame him, exceeding merry; and insults heavily over 'symbolical' wheresoever he meets with it; for in his answerer I find it not. But wheresoever 'twas, it serves to good purpose. For no man would imagine that he could have received so universal a defeat, and appear in so good humour. A terrible disputant he is, when he has set up a hard word to be his opponent. 'Tis a very wholesome thing he knows, and prolongs life; for all the while he can keep up this ball he may decline the question. But the poor word is sure to be mumbled and mousled° to purpose, and to be made an example. But let us, with Mr Bays's leave, examine the thing for once a little closer. The Nonconformists, as I took notice before, do object to some of the rites of the Church of England, under the name of symbolical or significant ceremonies. They observe the Church of England does, in the discourse of ceremonies printed before the Common Prayer Book, declare that the retaining of those ceremonies is not only 'as they serve for decent order and godly discipline, but as they are apt to stir up the dull mind of man to the remembrance of his duty to God, by some special and notable significancy, whereby he may be edified'. They further observe the Church of England's definition of a sacrament: that it is 'an outward visible sign of an inward spiritual grace'. They find these ceremonies, so constituted, imposed upon them by authority; and moreover, according to our author's principle, made a new part of the divine law. They therefore quarrel and except against these under the notion of sacraments, and insist that the Church is not empowered to institute such ceremonies under such obligations and penalties as are imposed. Or, if you will, instead of Church you may say rather the magistrate; forasmuch as our author hath *pro hac vice*° delivered the keys and the whole power of the house into his hands.

Now the author having got them at this lock, cries victory. Nothing less will serve him than a three days' triumph, as if he had conquered Europe, Asia, and Africa, and let him have a fourth day added, if he please, over the *terra incognita*° of Geneva. There is no end of his ostentation and pageantry; and the dejected Nonconformists follow the wheels of his chariot, to be led afterwards to the prison and there executed. He had said, page 446 of his second book, 'Here Cartwright began his objection, and here he was immediately checked in his career by Whitgift' (you might, Mr Author, for respect sake, have called him at least Mr if not Archbishop Whitgift), 'who told him plainly, he could not be ignorant that to the making of a sacrament, besides the external element, there is required a commandment of God in his word that it should be done, and a promise annexed to it, whereof the sacrament is a seal'. And in pursuance hereof, page 447, our author saith, 'Here then I fix my foot, and dare him to his teeth to prove that anything can be capable of the nature or office of sacraments that is not established by divine institution and upon promise of divine acceptance.' Upon the confidence of this argument 'tis that he *Hectors* and *Achillizes* all the Nonconformists out of the pit in this Preface. This is the sword that was consecrated first upon the altar, and thence presented to the champions of the Church in all ages. This is that with which Archbishop Whitgift gave 'Cartwright his death's wound; and laid the Puritan reformation a-gasping'. This is the weapon wherewith Master Hooker 'gained those lasting and eternal trophies over that baffled° cause'. This is that with which Bishop Bramhall 'wrought those wonderful things that exceeded all belief'. This hath been transmitted sucessively to the writer of the *Friendly Debate*, and to this our author. It is, in conclusion, the *curtana*° of our Church. 'Tis Sir Solomon's sword;° cock of as many men as it hath been drawn against. Woe worth the man that comes in the way of so dead-doing a tool and when wielded with the arm of such a Scanderbag° as our author! The Nonconformists had need desire a truce to bury their dead. Nay, there are none left alive to desire it; but they are slain, every mother's son of them. Yet perhaps they are but stounded,° and may revive again. For I do not see, all this while, that any of them have written, as a great prelate° of ours, a book of 'Seven Sacraments', or attempted to prove that these symbolical ceremonies are indeed sacraments. Nothing less. 'Tis that which they most labour against, and they complain that these things should be imposed on them with so high penalty, as want nothing of a sacramental nature but divine institution. And because a human institution is herein made of equal force to a divine institution, therefore it is that they are aggrieved. All that they mean, or could mean, as far as I or any man can perceive, is only that these ceremonies are a kind of anti-sacraments, and so obtruded upon the Church, that without condescending to these additional inventions, no man is to be admitted to partake of the true sacraments which were of Christ's appointing. For, without the sign of the cross our Church will not receive anyone to baptism; as also without kneeling no man is suffered to come to the communion. So that, methinks, our author and his partners have wounded themselves only with this argument; and have had as little occasion here to sing their *Te Deums* as the Roman Emperor° had to triumph over the ocean, because he had gathered periwinkles and scallop shells on the beach. For the author may transform their reasonings as oft as he pleases (even as oft as he doth his own, or the scriptures'): but this is indeed their fort, out of which I do not see they are likely to be beat with all our author's cannon: that no such new conditions ought to be imposed upon Christians by a less than divine authority, and unto which if they do not submit, though against their consciences, they shall therefore be deprived of communion with the Church. And I wonder that our author 'could not observe anything in the discourse of evangelical love, that was to the purpose, beside a perpetual repetition' of the outworn story of unscriptural ceremonies, and a peculiar uncouthness and obscurity of style; when as this plea is there for so many pages distinctly and vigorously insisted on. For it is a childish thing (how high soever our author magnifies himself in this way of reasoning) either to demand from the Nonconformists a pattern of their worship from the scripture, who affect therein a simplicity free from all exterior circumstances but such as are natural or customary; or else to require of them some particular command against the cross, or kneeling, and suchlike ceremonies, which in the time of the Apostles, and many ages after, were never thought of. But therefore general and applicable rules of scripture they urge as directions to the conscience; unto which our author gives no satisfactory solution, but by superseding and extinguishing the conscience, or exposing it to the severest penalties. But here I say, then, is their main exception: that things indifferent, and that have no proper signature or significancy to that purpose, should by command be made necessary conditions of Church-communion. I have many times wished, for peaceableness-sake, that they had a greater latitude; but if, unless they should stretch their consciences till they tear again, they cannot conform, what remedy? For I must confess that Christians have a better right and title to the Church, and to the ordinances of God there, than the author hath to his surplice. And that right is so undoubted and ancient, that it is not to be innovated upon° by human restrictions and capitulations.

Bishop Bramhall, page 141, saith, 'I do profess to all the world, that the transforming of indifferent opinions into necessary articles of faith hath been that *insana laurus*,° or cursed bay-tree, the cause of all our brawling and contention.' That which he saw in matter of doctrine he would not discern in discipline; whereas this among us—the transforming of things at best indifferent into necessary points of practice—hath been of as ill consequence. And (to reform a little my seriousness) I shall not let this pass without taking notice that you, Mr Bays, being the most extravagant person in this matter that ever I heard of, as I have shown you are mad, and so the *insana laurus*; so I wish you may not prove 'that cursed bay-tree, too', as the Bishop translates it. If you had thought of this, perhaps we might have missed both the Bishop's book and your Preface; for you see that sometimes no man hath a worse friend than he brings from home.

It is true, and very piously done, that our Church does declare that the kneeling at the Lord's Supper is not enjoined for adoration of those elements, and concerning the other ceremonies as before. But the Romanists (from whom we have them, and who said of old we would come to feed on their meat as well as eat of their porridge) do offer us here many a fair declaration, and distinction in very weighty matters, to which nevertheless the conscience of our Church hath not complied. But in this particular matter of kneeling, which came in first with the doctrine of transubstantiation, the Romish Church do reproach us with flat idolatry, in that we, not believing the real presence in the bread and wine, do yet pay to something or other the same adoration. Suppose the ancient pagans had declared to the primitive Christians that the offering of some grains of incense was only to perfume the room, or that the delivering up of their Bibles was but for preserving the book more carefully. Do you think the Christians would have palliated so far, and colluded with their consciences? Men are too prone to err on that hand. In the last king's time, some eminent persons of our clergy made an open defection to the Church of Rome. One, and he yet certainly a Protestant, and that hath deserved well of that cause, wrote the book of *Seven Sacraments*. One in the Church at present, though certainly no less a Protestant, could not abstain from arguing the *Holiness of Lent*: Doctor Thorndike, lately dead, left for his epitaph, *Hic jacet corpus Herberti Thorndike Praebendarii hujus Ecclesiae, qui vivus veram Reformatae Ecclesiae rationem et modum precibus studiisq; prosequebatur* and nevertheless he adds, *Tu, Lector, requiem ei et beatam in Christo resurrectionem precare.*° Which things I do thus sparingly set down, only to show the danger of inventive piety; and if men once come to add new devices to the Scripture, how easily they slide on into superstition. Therefore, although the Church do consider herself so much as not to alter her mode unto the fancy of others, yet I cannot see why she ought to exclude those from communion whose weaker consciences cannot for fear of scandal step further. For the Nonconformists, as to these Declarations of our Church against the reverence to the creatures of bread and wine, and concerning the other ceremonies as before, will be ready to think they have as good a plea as that so much commended by our author against the clause, 'that whosoever should affirm the Wednesday fast to be imposed with an intention to bind the conscience', should be punished 'like the spreaders of false news'; which is, saith a learned Prelate, 'plainly to them that understand it, to evacuate the whole law. For all human power being derived from God, and bound upon our consciences by his power, not by man, he that saith it shall not bind the conscience, saith it shall be no law, it shall have no authority from God, and then it hath none at all; and if it be not tied upon the conscience, then to break it is no sin, and then to keep it is no duty. So that a law without such an intention is a contradiction. It is a law only which binds as we please, and we may obey when we have a mind to it, and to so much we are tied before the constitution. But then if by such a Declaration it was meant, that to keep such fasting-days was no part of a direct commandment from God, that is, God had not required them by himself immediately, and so it was abstracting from that law no duty evangelical, it had been below the wisdom of the contrivers of it; for no man pretends it, no man saith it, no man thinks it; and they might as well have declared that that law was none of the Ten Commandments.' Page 59 of his first book. So much pains does that learned Prelate of his take (whoever he was) to prove a whole parliament of England coxcombs. Now I say that those ecclesiastical laws, with such declarations concerning the ceremonies by them enjoined, might, *mutatis mutandis*,° be taxed upon the same topic. But I love not that task, and shall rather leave it to Mr Bays to paraphrase his learned Prelate; for he is very good at correcting the impertinence of laws and lawgivers; and though this work, indeed, be not for his turn at present, yet it may be for the future. And I have heard a good engineer say, that he never fortified any place so, but that he reserved a feeble point, whereby he knew how to take it, if there were occasion.

I know a medicine for Mr Bays's hiccough (it is but naming J.O.), but I cannot tell certainly, though I have a shrewd guess, what is the cause of it. For indeed all his arguments here are so abrupt and short, that I cannot liken them better, considering too that frequent and perpetual repetition. Such as this, 'Why may not the sovereign power bestow this privilege upon ceremony as well as use and custom, by virtue of its prerogative? What greater immortality is there in them when determined by the command and institution of the prince than when by the consent and institution of the people?' This is the tap-lash° of what he said, page 110: 'When the civil magistrate takes upon him to determine any particular forms of outward worship, 'tis of no worse consequence than if he should go about to define the signification of all words used in the worship of God.' And page 108 of his first book: 'So that all the magistrate's power of instituting significant ceremonies etc. can be no more usurpation upon the CONSCIENCES of men than if the sovereign authority should take upon itself, as some princes have done, to define the signification of words.' And afterwards: 'The same gesture and actions are indifferently capable of signifying either honour or contumely, and so words; and therefore 'tis necessary their signification should be determined, etc.' 'Tis all very well worth reading. Page 441 of his second book: ''Tis no other usurpation upon their subjects' consciences than if he should take upon him to refine their language and determine the proper signification of all phrases employed in divine worship, as well as in trades, arts, and sciences.' Page 461 of the same: 'Once we will so far gratify the tenderness of their consciences and curiosity of their fancies, as to promise never to ascribe any other significancy to things than what himself is here content to bestow upon words.' And 462 of the same: 'So that you see my comparison between the signification of words and ceremonies stands firm as the pillars of the earth and the foundations of our faith.' Mr Bays might, I see, have spared Sir Solomon's sword of the divine institution of the sacraments. Here is the terriblest weapon in all his armory; and therefore, I perceive, reserved by our duellist for the last onset. And I, who am a great well-wisher to the pillars of the earth, or the eight elephants, lest we should have an earthquake; and much more a servant to the king's prerogative, lest we should all fall into confusion; and perfectly devoted to the foundations of our faith, lest we should run out into Popery or paganism,—have no heart to this encounter, lest, if I should prove that the magistrate's absolute, unlimited, and uncontrollable power doth not extend to define the signification of all words, I should thereby not only be the occasion of all those mischiefs mentioned, but, which is of far more dismal importance, the loss of two or three so significant ceremonies. But, though I therefore will not dispute against that flower of the prince's crown, yet I hope that, without doing much harm, I may observe that, for the most part, they left it to the people, and seldom themselves exercised it. And even Augustus Caesar,° though he was so great an emperor and so valiant a man in his own person, was used to fly from a new word, though it were single, as studiously as a mariner would avoid a rock for fear of splitting. The difference of one syllable in the same word hath made as considerable a controversy as most have been in the church, betwixt the *homousians* and the *homoiousians*.° One letter in the name of *beans* in Languedoc, one party calling them *faves* and the other *haves*; as the transposition only of a letter a, another time in the name of a goat, by some called *crabe*, and by others *cabre*, was the loss of more men's lives than the distinguishing but by an aspiration in *shibboleth* upon the like occasion. So that if a man would be learnedly impertinent, he might enlarge here, to show that 'tis as dangerous to take a man by the tongue as a bear by the tooth. And had I a mind to play the politician, like Mr Bays, upon so pleasant and copious a subject, I would demonstrate that though the imposition of ceremonies hath bred much mischief in the world, yet (shall I not venture too on a word once for trial?) such a penetration or transubstantiation of language would throw all into rebellion and anarchy, would shake the crowns of all princes, and reduce the world into a second Babel. Therefore, Mr Bays, I doubt you were not well advised to make so close an analogy betwixt imposing of significant words and significant ceremonies: for I fear the argument may be improved against you, and that princes, finding that of words so impracticable and of ill consequence, will conclude that of ceremonies to be no less pernicious. And the Nonconformists (who are great traders, you know, in scripture and therefore thrown out of the Temple) will be certainly on your back; for they will appropriate your pregnant text of 'Let all things be done decently and in order'° to preaching or praying in an unknown tongue, which such an imposition of words would be; and then, to keep you to your similitude, they will say too that yours are all Latin ceremonies, and the congregation does not understand them. But were not this dominion of words so dangerous (for how many millions of men did it cost your Roman Empire to attain it!) yet it was very unmannerly in you to assign to princes, who have enough beside, so mean a trouble. When you gave them leave to exercise the priesthood in person, that was something to the purpose; that was both honourable, and something belongs to it that would have helped to bear the charge. But this mint of words will never quit cost,° nor pay for the coinage. This is such a drudgery, that rather than undergo it, I dare say there is no prince but would resign to you so pedantical a sovereignty. I cannot but think how full that prince's head must be of proclamations; for if he published but once a proclamation to that purpose, he must forthwith set out another to stamp and declare the signification of all the words contained in it, and then another to appoint the meaning of all the words in this, and so on—that here is work cut out in one Paper of State for the whole Privy Council; both Secretaries of State, and all the clerks of the Council, for one king's reign, and *in infinitum*. But I cannot but wonder, knowing how ambitious Mr Bays is of the power over words, and jealous of his own prerogative of refining language, how he came to be so liberal of it to the prince; why, the same thing that induced him to give the prince a power antecedent and independent to Christ, and to establish what religion he pleased, etc. Nothing but his spite against the Nonconformists. I know not that thing in the world, except a jest, that he would not part with to be satisfied in that particular. He hoped, doubtless, by holding up this maxim, to obtain that the words of the Declaration of the 15th March should be understood by contraries. You may well think he expected no less an equivalent; he would never else have permitted the prince even to define the signification of all words used in the worship of God, and to determine the proper signification of all phrases employed in divine worship. Nay, Mr Bays, if it be come to that, and you will surrender your Liturgy to the prince, I know not what you mean, for 'tis bound up with your Bible. Was it ever heard that the Book so sacred, and in which there could not one error be found by all the Presbyterians at the Worcester-House conference, should, upon so uncertain a prospect, be now abandoned so far as that every word and phrase in it may receive a new and contrary signification? But the king, for aught I see, likes it well enough as it is, and therefore I do so too. Yet in case his Majesty should ever think fit to reform it, and because such kind of work is usually referred back to some of the clergy, I would gladly put in a caveat, that our author may in no case be one of them. For 'tis known that Mr Bays is subject to a distemper; and who knows but when he is in a fit, as he made such mad alterations of the fruit of the Spirit, in the Epistle for the day, he may as well insert in some other part of the service, 'Well fare poor Macedo for a modest fool'; and then, 'Oh how I hug thee, dear heart, for this!' and pretend that the supreme magistrate should stamp upon it a signification sacred and serious. I would not have spoken so severely of him, but that his 'more laboured periods', as he calls them, are so often filled with much bolder and more unwholesome translations. But however, that he may not at his better intervals be wholly unemployed in the work of uniformity, I should recommend to him rather to turn the Liturgy and the *Rationale* into the universal language, and so in time the whole world might come to be of his parish.

When he was drawn thus low, did not he, think you, stand in need of tilting?° He had done much more service to the cause had he laid by all those cheating argumentations, and dealt candidly, like the good Archdeacon not long since dead; who went about both court and country, preaching upon the 'cloak left at Troas, and the books, but especially the parchments'.° The honest man had found out there the whole liturgy, the canonical habits, and all the equipage of a Conformist. This was something to the matter in hand, to produce apostolical example and authority, and much more to the purpose than that beaten° text of 'doing all things decently and in order'.

One argument I confess remains still behind, and that will justify anything. 'Tis that which I called lately *rationem ultimam cleri*; force, law, execution, or what you will have it. I would not be mistaken, as though I hereby meant the body of the English clergy, who have been ever since the Reformation (I say it without disparagement to the foreign Churches) of the most eminent for divinity and piety in all Christendom. And as far am I from censuring, under this title, the Bishops of England, for whose function, their learning, their persons I have too deep a veneration to speak anything of them irreverently. But those that I intend only are a particular bran° of persons, who will, in spite of fate, be accounted the Church of England, and to show they are pluralists, never write in a modester style than 'We, We'; nay, even these, several of them, are men of parts sufficient to deserve a rank among the teachers and governors of the Church. Only what Bishop Bramhall saith of Grotius's defect in school-divinity,

> *Unum hoc maceror et doleo tibi deesse,*°

I may apply to their excess and rigour in matter of discipline. They want all consideration, all moderation in those things; and I never heard of any of them at any time, who, if they got into power or office, did ever make the least experiment or overture towards the peace of the Church and nation they lived in. They are the 'Politic Would-be's'° of the clergy. Not Bishops, but men that have a mind to be Bishops, and that will do anything in the world to compass it. And though princes have always a particular mark upon these men, and value them no more than they deserve, yet I know not very well, or perhaps I do not know, how it oftentimes happens that they come to be advanced. They are men of a fiery nature, that must always be uppermost; and, so they may increase their own splendour, care not though they set all on flame about them. You would think the same day that they took up divinity they divested themselves of humanity, and, so they may procure and execute a law against the Nonconformists, that they had forgot the Gospel. They cannot endure that humility, that meekness, that strictness of manners and conversation, which is the true way of gaining reputation and authority to the clergy; much less can they content themselves with the ordinary and comfortable provision that is made for the ministry; but having wholly calculated themselves for preferment and grandeur, know or practise no other means to make themselves venerable but by ceremony and severity. Whereas the highest advantage of promotion is the opportunity of condescension, and the greatest dignity in our Church can but raise them to the title of 'Your Grace', which is in the Latin *vestra clementia.*° But of all these, none are so eager and virulent, as some who having had relation to the late times, have got access to 'ecclesiastical fortune', and are resolved to make their best of her. For so, of all beasts, none are so fierce and cruel as those that have been taught once by hunger to prey upon their own kind; as, of all men, none are so inhuman as the cannibals. But whether this be the true way of ingratiating themselves with a generous and discerning prince, I meddle not; nor whether it be an ingenuous practice towards those whom they have been formerly acquainted with: but whatsoever they think themselves obliged to for the approving of their new loyalty, I rather commend. That which astonishes me, and only raises my indignation is, that of all sorts of men this kind of clergy should always be, and have been for the most precipitate, brutish, and sanguinary counsels. The former Civil War cannot make them wise, nor his Majesty's happy return, good-natured; but they are still for running things up unto the same extremes. The softness of the universities where they have been bred, the gentleness of Christianity in which they have been nurtured, hath but exasperated their nature; and they seem to have contracted no idea of wisdom but what they learnt at school, the pedantry of whipping. They take themselves qualified to preach the Gospel, and no less to intermeddle in affairs of state; though the reach of their divinity is but to persecution, and an inquisition is the height of their policy.

And you, Mr Bays, had you lived in the days of Augustus Caesar (be not scandalized; for why may you not bring sixteen hundred years, as well as five hours, into one of your Plays?), would not you have made, think you, an excellent privy counsellor? His father° too was murdered. Or (to come nearer both to our times and your resemblance of the late war, which you trumpet always in the ear of his Majesty) had you happened in the time of Henry the fourth of France, should you not have done well in the cabinet? His predecessor too was assassinated. No, Mr Bays, you would not have been for their purpose: they took other measures of government, and accordingly it succeeded with them. And his Majesty, whose genius hath much of both those princes, and who derives half of the blood in his veins from the latter, will in all probability not be so forward to hearken to your advice as to follow their example. For these kings, Mr Bays, how negligent soever or ignorant you take 'em to be, have, I doubt, a shrewd understanding with them. 'Tis a trade that, God be thanked, neither you nor I are of; and therefore we are not so competent judges of their actions. I myself have oftentimes seen them, some of them, do strange things, and unreasonable in my opinion; and yet a little while, or sometimes many years after, I have found that all the men in the world could not have contrived any thing better. 'Tis not with them as with you. You have but one cure of souls, or perhaps two, as being a nobleman's chaplain, to look after; and if you make conscience of discharging them as you ought, you would find you had work sufficient without writing your *Ecclesiastical Policies*. But they are the incumbents of whole kingdoms, and the rectorship of the common people, the nobility, and even of the clergy, whom you are prone to 'affirm when possessed with principles that incline to rebellion and disloyal practices, to be of all rebels the most dangerous', page 49. The care, I say, of all these, rests upon them. So that they are fain to condescend to many things for peace sake, and the quiet of mankind, that your proud heart would break before it would bend to. They do not think fit to require anything that is impossible, unnecessary, or wanton, of their people; but are fain to consider the very temper of the climate in which they live, the constitution and laws under which they have been formerly bred; and upon ill occasions to give them good words, and humour them like children. They reflect upon the histories of former times, and the present transactions, to regulate themselves by in every circumstance. They have heard that one of your Roman emperors,° when his captain of the Life Guard came for the word,° by giving it unhandsomely, received a dagger. They observe how the Parliament of Poland° will be their King's tailor, and among other reasons, because he would not wear their mode, have suffered the Turk to enter, as coming nearer their fashion. Nay, that even Alexander the Great had almost lost all he had conquered, by forcing his subjects to conform to the Persian habit. That the King of Spain, when upon a progress he enters Biscay, is pleased to ride with one leg naked; and above all, to take care that there be not any Bishop in his retinue. So their people will pay their taxes in good gold and silver, they demand no subsidy of so many bushel of fleas, lest they should receive the same answer with the tyrant,° that the subject could not furnish that quantity; and besides, they would be leaping out still before they could be measured, and should they fine the people for non-payment, they reckon there would be little got by distraining. They have been told that a certain queen being desired to give a town seal to one of her cities, lighting from horse, sat down naked on the snow, and left them that impression; and though it cause no disturbance, but all the town-leases are letters patents, kings do not approve the example. That the late Queen of Sweden did herself no good with saying, *io non voglio governar le bestie*,° but afterwards resigned. That the occasion of the revolt of Switzerland from the Emperor, and its turning commonwealth, was only the imposing of a civil ceremony by a capricious governor, who set up a pole in the highway, with a cap upon the top of it, to which he would have all passengers be uncovered, and do obeisance. One sturdy Swiss° that would not conform, thereupon overturned the government, as 'tis at large in history. That the King of Spain lost Flanders chiefly upon introducing the Inquisition. And you now, Mr Bays, will think these, and a hundred more that I could tell you, but idle stories; and yet kings can tell how to make use of 'em. And hence 'tis that instead of assuming your unhoopable° jurisdiction, they are so satisfied with the abundance of their power, that they rather think meet to abate of its exercise by their discretion. The greater their fortune is, they are content to use the less extravagancy. But because I see, Mr Bays, you are a little deaf on this ear, I will talk somewhat closer to you. In this very matter of ceremonies, which you are so bent upon, that your mind is always running on't, when you should be hearkening to the sermon; do not you think that the King knows every word you said, although he never gave your book the reading? That you say, that the clause 5° Eliz. of the Wednesday fast° has been the original of all the Puritan disorders. That the controversy is now reduced only to two or three 'symbolical ceremonies'. That these ceremonies are things indifferent in their own nature, and have no antecedent necessary, but only bind as they are commanded. That they signify nothing in themselves but what the commander pleases. That the Church itself declares that there is nothing of religion or adoration in them. That they are no parts of religious worship. That they are only circumstances. That the imposing of a significant ceremony is no more than to impose significancy upon a word. That there is not a word of any of these ceremonies in the scriptures. That they are in themselves of no great moment and consequence, but 'tis absolutely necessary that government should enjoin them, to avoid the evil that would follow if they were not determined; and that there cannot be a pin pulled out of the Church, but the State immediately totters. Do not you think that the King has considered all these things? I believe he has; and perhaps, as you have minced the matter, he may well think the Nonconformists have very nice stomachs, that they cannot digest such chopped hay; but, on the other side, he must needs take you to be very strange men to cram these in spite down the throats of any Christian. If a man have an antipathy against anything, the company is generally so civil as to refrain the use of it, however not to press it upon the person. If a man be sick or weak, the Pope grants a dispensation from the Lent, or fasting days; aye and from many a thing that strikes deeper in his religion. If one have got a cold, their betters will force them to be covered. There is no end of similitudes; but I am let into them by your calling these ceremonies 'pins of the Church'. It would almost tempt a prince that is curious, and that is settled (God be praised) pretty fast in his throne, to try for experiment whether the pulling out of one of these pins would make the state totter. But, Mr Bays, there is more in it. 'Tis a matter of conscience; and if kings do, out of discretion, connive at the other infirmities of their people; if great persons do out of civility condescend to their inferiors; and if all men out of common humanity do yield to the weaker; will your clergy only be the men who, in an affair of conscience, and where perhaps 'tis you are in the wrong, be the only hard-hearted and inflexible tyrants; and not only so, but instigate and provoke princes to be the ministers of your cruelty? But, I say, princes, so far as I can take the height of things so far above me, must needs have other thoughts, and are past such boy's play to stake their crowns against your pins. They do not think fit to command things unnecessary, and where the profit cannot countervail the hazard. But above all, they consider that God has instated them in the government of mankind, with that encumbrance (if it may so be called) of reason, and that encumbrance upon reason of conscience. That he might have given them as large an extent of ground and other kind of cattle for their subjects; but it had been a melancholy empire to have been only supreme grazers and sovereign shepherds. And therefore, though the laziness of that brutal magistracy might have been more secure, yet the difficulty of this does make it more honourable. That men therefore are to be dealt with reasonably, and conscientious men by conscience. That even law is force, and the execution of that law a greater violence; and therefore with rational creatures not to be used but upon the utmost extremity. That the body is in the power of the mind; so that corporal punishments do never reach the offender, but the innocent suffers for the guilty. That the mind is in the hand of God, and cannot correct those persuasions which upon the best of its natural capacity it hath collected; so that it too, though erroneous, is so far innocent. That the prince therefore, by how much God hath endued him with a clearer reason, and by consequence with a more enlightened judgement, ought the rather to take heed, lest by punishing conscience he violate not only his own, but the Divine Majesty. But as to that, Mr Bays, which you still inculcate of the late war, and its horrid catastrophe, which you will needs have to be upon a religious account: 'tis four and twenty years ago, and after an Act of Oblivion; and for aught I can see, it had been as seasonable to have shown Ceasar's bloody coat, or Thomas à Becket's bloody rochet.° The chief of the offenders have long since made satisfaction to justice; and the whole nation hath been swept sufficiently of late years by those terrible scourges of heaven;° so that methinks you might in all this while have satiated your mischievous appetite. Whatsoever you suffered in those times, his Majesty, who had much the greater loss, knowing that the memory of his glorious father will always be preserved, is the best judge how long the revenge ought to be pursued. But if indeed, out of your superlative care of his Majesty and your 'living', you are afraid of some new disturbance of the same nature, let me so far satisfy you as I am satisfied. The Nonconformists say, that they are bound in conscience to act as far as they can, and for the rest to suffer to the utmost. But because though they do mean honestly, 'tis so hard a chapter for one that thinks himself in the right to suffer extremities patiently, that some think it impossible; I say next, that it's very seldom seen that in the same age, a civil war, after such an interval, has been raised again upon the same pretences: but men are all so weary, that he would be knocked on the head that should raise the first disturbance of the same nature. A new war must have, like a book that would sell, a new title. I am ashamed, Mr Bays, that you put me on talking thus impertinently (for policy in us is so). Therefore, to be short, the King hath so indulged and obliged the Nonconformists by his late mercy, that if there were any such knave, there can be no such fool among them that would ever lift up an ill thought against him. And for you, Mr Bays, he is assured of your loyalty; so that I think you may enjoy your 'living' very peaceably, which I know is all your business. 'Twas well replied of the Englishman in Edward the fourth's time, to the Frenchman that asked him insulting, when they should see us there again? 'When your sins are greater than ours.' There are as many occasions of war as there are vices in a nation, and therefore it concerns a prince to be watchful on all hands. But should kings remember an injury as long as you implacable divines do, or should we take up arms upon your piques, because your *Ecclesiastical Policy* is answered, to revenge your quarrel, the world would never be at quiet. Therefore, Mr Bays, let all those things of former times alone, and mind your own business; for kings, believe me, as they have royal understandings, so have gentlemen's memories.

And now, Mr Bays, I think it is time to take my leave, having troubled you with so long a visit. Only, before I quit this matter, because I do not love to be accounted singular in my opinion, I will add the judgement of one author, and that as pertinent as I could pick out to our purpose. I have observed that not only other princes, but Queen Elizabeth too, hath the misfortune to be much out of your favour; but for what reason I cannot possibly imagine, for none ever deserved better as to the thing of uniformity, unless it be the ill-luck she had to pass that 'impertinent clause' in the Act 5° Elizabeth, of the *Jejunium Cecilianum*. You cannot, for her sake, endure the wit or learning of her times, but say, page 94 of your second book, 'though this trifling artifice of sprinkling little fragments of wit and poetry might have passed for wit and learning in the days of Queen Elizabeth, yet to men of learning, reading, and ingenuity, their vulgar use has sullied their lustre and abated their value.' This is indeed, Mr Bays, a very laboured period, and prepared by you, I believe, on purpose as a model of the wit and eloquence of your days. But not only so, but page 483 of the same book, I think you call her in derision, and most spitefully and unmanneredly, plain 'old Elizabeth'. And those that knew her humour, think you could not have disobliged her more than in styling her so; both as a woman, which sex never love to be thought old, and as a queen, who was jealous lest men should therefore talk of the succession. Besides the irreverent nickname you give her, that you might as well have presumed to call her Queen Bess, or Bold Beatrice. Now, to the end that that queen of famous memory may have a little female revenge upon you, and to give you a taste of the wit and learning even of her times, I will 'sprinkle' here one 'fragment', which not being a 'scholar-like saying of ancient poet or philosopher', but of a reverend divine, I hope, Mr Bays, may be less displeasing to you. The man is Parker; not Robert Parker° who wrote another treatise of Ecclesiastical Policy, and the book *de Cruce*,° for which, if they had catched him, he had possibly gone to the gallows, or at least the galleys; for he was one of those 'well-meaning zealots that are, of all villains, the most dangerous'; but it is the Archbishop of Canterbury, Parker° (for if I named him before without addition, 'twas what I learned of you speaking of Whitgift). He, in his book *de Antiquitatibus Ecclesiae Britannicae*,° page 47, speaking of the slaughter of the monks of Bangor, and so many Christians more, upon the instigation of Austin the monk, who stirred up Ethelbert, king of Kent, against them, because they would not receive the Romish ceremonies, useth these words: *Et sane illa prima de Romanis ritibus inducendis per Augustinum tunc excitata contentio, quae non nisi clade et sanguine innocentium Britannorum poterat extingui; ad nostra recentiora tempora, cum simili pernicie caedeque Christianorum pervenit. Cum enim illis gloriosis ceremoniis a pura primitivae ecclesiae simplicitate recesserunt, non de vitae sanctitate, de evangelii praedicatione, de spiritus sancti vi et consolatione multum laborabant; sed novas indies altercationes de novis ritibus per papas singulos additis, qui neminem tam excelso gradu dignum, qui aliquid, ceremoniosi non dicam, monstrosi, inauditi et inusitati ncn adjecisset, instituebant. Suggestaque et scholas fabulis rixisque suis implebant. Nam prima ecclesiae species simplicior et integro et interno Dei cultu, ab ipso verbo praescripto, nec vestibus splendidis, nec magnificis structuris decorata, nec auro, argento, gemmisque fulgens fuit: etsi liceat his exterioribus uti modo animum ab illo interiori et integro Dei cultu non abducant; curiosis et morosis ritibus ab illa primaeva et recta simplicitate evangelica degeneravit. Illa autem in Romana ecclesia rituum multitudo ad immensum illius magni Augustini Hipponensis episcopi temporibus creverat: ut questus sit Christianorum in ceremoniis et ritibus duriorem tunc fuisse conditionem quam Judaeorum, qui etiamsi tempus libertatis non agnoverint, legalibus tamen sarcinis non humanis praesumptionibus subjicieban-*

*tur; nam paucioribus in divino cultu quam Christiani ceremoniis utebantur. Qui si sensisset quantus deinde per singulos papas coacervatus cumulus accessit, modum Christianum credo ipse statuiset; qui hoc malum tunc in ecclesia vederat. Videmus enim ab illa ceremoniarum contentione nedum ecclesiam esse vacuam; quin homines, alioquin docti atque pii de vestibus et hujusmodi nugis adhuc rixoso magis et militari, quam aut philosophico aut Christiano more inter se digladiantur.* These words do run so direct against the genius of some men that contributed not a little to the late rebellion, and, though so long since writ, do so exactly describe that evil spirit with which some men are even in these times possessed, who seem desirous upon the same grounds to put all things in combustion, that I think them very well worth the labour of translating. 'And indeed, that first contention then raised by Augustine about the introducing of the Romish ceremonies, which could not be quenched but by the blood and slaughter of the innocent Britons, hath been continued e'en to our later times, with the like mischief and murder of Christians. For when once by those glorious° ceremonies they forsook the pure simplicity of the primitive Church, they did not much trouble themselves about holiness of life, the preaching of the Gospel, the efficacy and comfort of the Holy Spirit; but they fell every day into new squabbles about new-fangled ceremonies added by every Pope, who reckoned no man worthy of so high a degree but such as invented somewhat, I will not say ceremonious, but monstrous, unheard-of, and before unpractised; and they filled the schools and the pulpits with their fables and brawling of such matters. For the first beauty of the Church had more of simplicity and plainness, and was neither adorned with splendid vestments, nor magnificent structures, nor shined with gold, silver, and precious stones, but with the entire and inward worship of God, as it was by Christ himself prescribed. Although it may be lawful to use these external things, so they do not lead the mind astray from that more inward and entire worship of God, by those curious and crabbed rites it degenerated from that ancient and right evangelical simplicity. But that multitude of rites in the Romish Church had immeasurably increased in the times of that great Augustine, the Bishop of Hippo, in so much that he complained that the condition of Christians as to rites and ceremonies was then harder than that of the Jews; who, although they did not discern the time of their liberty, yet were only subjected to legal burdens, instituted first by God Himself, not to human presumptions. For they used fewer ceremonies in the worship of God than Christians. Who, if he could have foreseen how great a heap of them was afterwards piled up, and added to by the several popes, he himself doubtless would have restrained it within Christian measure, having already perceived this growing evil in the Church; for we see that even yet the Church is not free from that contention: but men, otherwise learned and pious, do still cut and slash about vestments and such kind of trifles, rather in a swash-buckler and hectoring way, than either like philosophers or like Christians.'

Now, Mr Bays, I doubt you must be put to the trouble of writing another Preface against this Archbishop; for nothing in your answerer's treatise of 'Evangelical Love' does so gird° or aim at you, for aught I can see, or at those whom you call the Church of England, as this passage. But the last period does so plainly delineate you to the life, that what St Austin did not presage, the Bishop seems to have foreseen most distinctly. 'Tis just your way of writing all along in this matter. You bring nothing sound or solid. Only you think you have got the 'great secret', or the 'philosopher's stone' of railing; and I believe it, you have so multiplied it in projection: and as they into gold, so you turn everything you meet with into railing. And yet the secret is not great, nor the process long or difficult, if a man would study it and make a trade on't. Every scold hath it naturally. It is but crying whore first, and having the last word, and whatsoever t'other says, cry, 'Oh, these are your Nonconformists' tricks; Oh, you have learned this of the Puritans in Grubstreet. Oh, you white-aproned gossip.' For indeed I never saw so provident a fetch:° you have taken in beforehand all the posts of railing, and so beset all the topics of just crimination,° foreseeing where you are feeble, that if this trick would pass, it were impossible to open one's mouth to find the least fault with you. For in your first chapter of your second book, beside what you do always in a hundred places where you are at a loss, you have spent almost a hundred pages upon 'a character of the fanatic deportment toward all adversaries'. And then, on the other side, you have so engrossed and bought up all the ammunition of railing, searched every corner in the Bible and *Don Quixote* for powder, that you thought, not unreasonably, that there was not one shot left for a fanatic. But truth, you see, cannot want words: and she will laugh too sometimes when she speaks, and rather than all fail too, be serious. But what will you say to that of the Archbishop's, 'than either like philosophers or like Christians'? For the excellency of your logic, philosophy, and Christianity in all your books, is either, as in conscience, to take away the subject of the question; or as in the magistrate, having gotten one absurdity, to raise a thousand more from it. So that, except the manufacture and labour of your periods, you have done no more than any school boy could have done on the same terms. And so, Mr Bays, good-night.

And now good-morrow, Mr Bays; for though it seems so little a time and that you are but now gone to bed, it hath been a whole live-long night, and you have tossed up and down in many a troublesome dream, and are but just now awaked at the title-page of your book: 'A preface showing what grounds there are of fears and jealousies of Popery'. It is something artificially couched, but looks as if it did allow that there are some grounds of fears and jealousies of that nature. But here he words it, 'a consideration what likelihood, or how much danger there is of the return of Popery into this nation.' Had he not come to this at last, I should have thought I had been all this while reading a chapter in Montaigne's *Essays*; where you find sometimes scarce one word in the discourse of the matter held forth in the title. But now indeed he takes up this argument and debates it to purpose. For I had before begun to show that he had writ not only his two former books, but especially too this Preface, with an evil eye and aim at his Majesty, and the measures he had taken of government. And whoever will take the pains to read here, will soon be of my mind. His Majesty had, I said, the 15th of March 1671, issued his Declaration of Indulgence to tender consciences. He, on the contrary, issues out thereupon, all in haste and as fast as he could write, this his remonstrance or manifesto against Indulgence to tender consciences; and to make his Majesty's proceedings more odious, stirs up this seditious matter, of what probability there is of Popery.

And this he discourses, to be sure, in his own imagination very cunningly. For he knows that there was an act of parliament in this king's reign with a greater penalty than that of 5° Eliz. of spreading false news, against reports of this nature. And therefore he resolves to handle it so warily, that he himself might escape, but might draw others that should answer him within the danger of that Act, and that he might lay the crime at their doors. But notwithstanding all his sleights and legerdemain, it doth enough detect his malice and ill-intention to his Majesty's government, that he should take this occasion, altogether foreign and unseasonable, to raise a public and solemn discourse through the whole nation, concerning a matter the most odious and dangerous that could be exposed. So that now no man can look at the wall, no man can pass by a bookseller's stall, but he must see *A Preface showing what grounds there are for fears and jealousies of Popery*.

It had been something a safer and more dutiful way of writing a Preface showing the CAUSELESSNESS of the Fears and Jealousies of Popery.

For I do not think it will excuse a witch, to say that she conjured up a spirit only that she might lay it; nor can there be a more dexterous and malicious way of calumny, than by making a needless apology for another in a criminal subject. As, suppose I should write a Preface showing what grounds there are of fears and jealousies of Bays's being an atheist. But this is exactly our author's method and way of contrivance; whereby, more effectually by far, than by any flying coffee-house tattle, he traduces the state, and by printing so pernicious a question fills all men's mouths and beats out all men's eyes with the probability of the return of Popery. Had he heard any that malignly and officiously talked to such a purpose, it had been the part of one so prudent as he is, not to have continued the discourse. Had he (as he hath a great gift that way) picked up out of any man's talk or writing, matter whereof to make an ill story, there was a better and more regular way of proceeding, had he meant honestly to his Majesty's government, to have prevented the evil, and to have brought the offender to punishment. He should have gone to one of the Secretaries of State, or to some other of his Majesty's Privy Council, and have given them information. But instead of that, I am afraid that in the survey of this business we shall find, that even some of them are either accused, or shrewdly° marked out with a character of our author's displeasure. Therefore, I will not come nearer to his matter in hand, although it concerns me to be careful of coming too near; nor shall I dwell too long upon so jealous and impertinent a subject.

'To consider what likelihood or how much danger there is of the return of Popery into this nation.' The very first word is, 'for my part, I know none'. Very well considered. Why then, Mr Bays, I must tell you, that if I had printed a book or Preface upon that argument, I should have thought myself at least a fool for my labour. The next considerer is mine enemy; I mean he is an enemy to the State, whoever shall foment such discourses without any likelihood or danger. Yet, Mr Bays, you know I have for a good while had no great opinion of your integrity; neither here. I doubt you prevaricate a little with somebody. For I suppose you cannot be ignorant that some of your superiors of your robe did, upon the publishing that Declaration, give the word, and deliver orders through their ecclesiastical camp, to beat up the pulpit drums against Popery. Nay, even so much that there was care taken too for arming the 'poor Readers, that though they came short of preachers in point of efficacy, yet they might be enabled to do something in point of common security'. So that, though for so many years some of your superiors had forgot there was any such thing in the nation as a Popish recusant; though 'polemical and controversial divinity' had for so long been hung up in the halls, like the rusty obsolete armour of our ancestors, for monuments of antiquity, and for derision rather than service; all on a sudden (as if the 15th of March had been the 5th of November°) happy was he that could climb up first to get down one of the old cuirasses or an habergeon° that had been worn in the days of Queen Elizabeth. Great variety there was and a heavy do. Some clapped it on all rusty as it was, others fell of oiling and furbishing their armour; some pissed in their barrels, others spat in their pans, to scour them. Here you might see one put on his helmet the wrong way; there one buckle on a back in place of a breast. Some by mistake catched up a Socinian or Arminian argument, and some a Popish to fight a Papist. Here a dwarf lost in the accoutrements of a giant: there a Don Quixote in an equipage of differing pieces and of several parishes. Never was there such incongruity and Nonconformity in their furniture. One ran to borrow a sword of Calvin; this man for a musket from Beza;° that for a bandoleer° even, from Keckerman.° But when they came to seek match, and bullet, and powder, there was none to be had. The fanatics had bought it all up, and made them pay for it most unconscionably and through the nose. And no less sport was it to see their leaders. Few could tell how to give the word of command, nor understood to drill a company: they were as unexpert as their soldiers awkward; and the whole was as pleasant a spectacle as the exercising of the trained bands in shire. But, Mr Bays (for I believe you do nothing but upon common advice), either this was all intended but for a false alarm, and was only for a pretence to take arms against the fanatics (which you might have done without raising all this din and obloquy against the state, and disquieting his Majesty's good subjects); or else you did really think (and who can help misapprehensions?) that you did know some likelihood or danger of the return of Popery. I crave you mercy, Mr Bays, I took you a little short. 'For my part, I know none', you say, 'but the Nonconformists' boisterous and unreasonable opposition to the Church of England.'

This, I confess, hath some weight in it. For truly before 'I knew none' too, I was of your opinion, Mr Bays, and believed that Popery could never return into England again, but by some very sinister accident. This expression of mine is something uncouth; and therefore, because I love to give you satisfaction in all things, Mr Bays, I will acquaint you with my reason of using it. Henry the Fourth of France, his Majesty's grandfather, lived (you know) in the days of Queen Elizabeth. Now the wit of France and England, as you may have observed, is much of the same mode, and hath at all times gone much after the same current rate and standard; only there hath been some little difference in the alloy, and advantage or disadvantage in the exchange, according to men's occasions. Now Henry the Fourth was (you know too) a prince, like Bishop Bramhall, 'of a brave and enterprising temper, and had a mind large and active enough to have managed the Roman Empire at its utmost extent'; and particularly 'as far as the prejudice of the age ("Old Elizabeth's" age) would permit him' he was very witty and facetious, and the courtiers strove to humour him always in it, and increase the mirth. So one night after supper he gave them a subject (which recreation did well enough in those times, but were now insipid), upon which like boys at Westminster, they should make a French verse extempore. The subject was, *Un accident sinistre*. Straight answers, I know not whether 'twas Bassompierre or Abigne:

> *Un sinistre accident et un accident sinistre,*
> *De voir un Père Capuchin chevaucher un Ministre.*

For when I said, to see Popery return here would be a very sinister accident, I was just thinking upon that story; the verses, to humour them in translation, being only this:

> O what a trick unlucky, and how unlucky a trick,
> To see friend Doctor Patrick° bestrid by Father Patrick!°

Which seemed to me, would be the most improbable and preposterous spectacle that ever was seen; and more ridiculous for a sight than the *Friendly Debate* is for a book. And yet if Popery come in, this must be, and worse.

But now I see there is some danger by the Nonconformists' opposition to the Church of England. And now your business is all fixed. The fanatics are ready at hand to bear the blame of all things. Many a good job have I seen done in my time upon pretence of the fanatics. I do not think Mr Bays ever breaks his shins but it is by stumbling upon a fanatic. And how shall they bring in Popery? Why thus, three ways. 'First, by creating disorders and disturbances in the State. Secondly, by the assistance of atheism and irreligion. Thirdly, by joining with crafty and sacrilegious statesmen in confederacy.' Now here I remark two things. One, that however you do not find that the fanatics are inclinable to Popery, only they may accommodate it by creating disturbances in the state. Another is, that I see these gentlemen, the fanatics, the atheists, and the sacrilegious statesmen are not yet acquainted; but you have appointed them a meeting (I believe it must be at your lodgings, or nowhere), and I hope you will treat them handsomely. But I think it was not so wisely done, nor very honestly, Mr Bays, to lay so dangerous a plot as this, and instruct men that are strangers yet to one another, how to contrive together such a consipiracy. But first to your first.

The 'fanatics', you say, 'may probably raise disturbance in the State.' For they 'are so little friends to the present government, that their enmity to that is one of the main grounds of their quarrel to the Church'. But now, though I must confess it is very much to your purpose, if you could persuade men so, I think you are clear out, and misrepresent here the whole matter. For I know of no enmity they have to the Church itself, but what it was in her power always to have remedied; and so it is still. But such as you it is that have always strove by your leasings° to keep up a strangeness and misunderstanding betwixt the King and his people; and all the mischief hath come on't does lie much at your door. Whereas they, as all the rest of mankind, are men for their own ends too; and no sooner hath the King shown them this late favour, but you, Mr Bays, and your partners, reproach them for being too much friends to the prerogative. And no less would they be to the Church, had they ever at any age in any time found her in a treatable temper. I know nothing they demand, but what is so far from doing you any harm, that it would only make you better. But that indeed is the harm, that is the thing you are afraid of. Here our author divides the discourse into a great eulogy of the Church of England, that if he were making her funeral sermon he could not say more in her commendation; and a contrary invective against the Nonconformists, upon whom (as if all he had said before had been nothing) he unloads his whole laystall,° and dresseth them up all in *sanbenitas*,° painted with all the flames and devils in hell, to be led to the place of execution, and there burned to ashes. Nevertheless, I find on either side only the natural effect of such hyperboles and oratory; that is, not to be believed. The Church of England (I mean as it is by law established, lest you should think I equivocate) hath such a stock of solid and deserved reputation, that it is more than you, Mr Bays, can spoil or deface by all the pedantry of your commendation. Only there is that party of the clergy, that I not long ago described, and who will always presume to be the only Church of England, who have been a perpetual eye-sore, that I may not say a cancer and gangrene, in so perfect a beauty. And as it joys my heart to hear any thing well said of her, so, I must confess, it stirs my choler when I hear those men pride and boast themselves under the mask of her authority. Neither did I therefore approve of an expression you here use: 'the power of princes would be a very precarious thing without the assistance of ecclesiastics, and all government does and must owe its quiet and continuance to the Church's patronage.' That is as much as to say, that but for the assistance of your *Ecclesiastical Policy*, princes might go a-begging; and that the Church, that is you, have the *jus patronatus*° of the kingdom, and may present whom you think fitting to the crown of England. This is indeed something like the return of Popery, and right

*Petra dedit Petro, Petrus diadema Rudolpho.*°

The crown were surely well helped up, if it were to be held at your convenience, and the Emperor must lead the Patriarch's ass all his lifetime. And little better do I like your 'We may rest satisfied in the present security of the Church of England, under the protection of a wise and gracious prince; especially when, besides the impregnable confidence that we have from his own inclination, it is so manifest that he never can forsake it either in honour or interest.' This is a pretty way of coaxing indeed, while you are all this while cutting the grass under his feet, and animating the people against the exercise of his ecclesiastical supremacy. Men are not so plain-hearted but they can see through this oblique rhetorication and sophistry. If there be no danger in his time of taking a 'pin out of the Church' (for that it is you intend), why do you then speak of it in his time, but that you mean mischief? But here you do not only mow the grass under his feet, but you take the pillow from under his head. 'But should it ever happen that any King of England should be prevailed with to deliver up the Church, he had as good at the same time resign up his crown.' This is pretty plain dealing, and you have doubtless secured hereby that prince's favour. I should have thought it better courtship in a divine to have said, 'O King, live for ever.' But I see, Mr Bays, that you and your partners are very necessary men, and it were dangerous disobliging you. But as in this imprudent and nauseous discourse you have all along appropriated or impropriated all the loyalty from the nobility, the gentry, and the commonalty, and dedicated it to the Church; so I doubt you are a little too immoderate against the body of the Nonconformists. You represent them to a man, to be all of them of republican principles, most pestilent, and *eo nomine*° enemies to monarchy, traitors and rebels; such miscreants as never was in the world before, and fit to be packed out of it with the first convenience. And I observe that all the argument of your books is but very frivolous and trivial; only the memory of the late war serves for demonstration, and the detestable sentence and execution of his late Majesty is represented again on the scaffold; and you having been, I suspect, better acquainted with Parliament Declarations formerly upon another account, do now apply and turn them all over to prove that the late war was wholly upon a fanatical cause, and the dissenting party do still go big with the same monster. I grew hereupon much displeased with my own ignorance of the occasion of those troubles so near our own times, and betook myself to get the best information concerning them, to the end that I might, if it appeared so, decline the dangerous acquaintance of the Nonconformists, some of whom I had taken for honest men, nor therefore avoided their company. But I took care, nevertheless, not to receive impressions from any of their party; but to gather my lights from the most impartial authorities that I could meet with. And I think I am now partly prepared to give you, Mr Bays, some better satisfaction in this matter. And because you are a dangerous person, I shall as little as possible, say any thing of my own, but speak too before good witnesses. First of all, therefore, I will, without further ceremony, fall upon you with the butt end of another Archbishop. 'Tis the Archbishop of Canterbury, Abbot,° in the narrative under his own hand, concerning his disgrace at Court in the time of his late Majesty. I shall only in the way demand excuse, if, contrary to my fashion, the names of some eminent persons in our Church, long since dead, be revived here under no very good character; and most particularly that of Archbishop Laud,° who, if for nothing else, yet for his learned book against Fisher deserved far another fate than he met with,° and ought not now to be mentioned without due honour. But those names having so many years since escaped the press, it is not in my power to conceal them; and I believe Archbishop Abbot did not write but upon good consideration.

This I have premised for my own satisfaction; and I will add one thing more, Mr Bays, for yours—that whereas the things now to be alleged relate much to some impositions of money° in the late King's time, that were carried on by the clergy, I know you will be ready to carp at that, as if the Nonconformists had, and would be always enemies to the King's supply. Whereas, Mr Bays, if I can do the Nonconformists no good, I am resolved I will do them no harm, nor desire that they should lie under any imputation on my account. For I write by mine own advice, and what I shall allege concerning the clergy's intermeddling with supplies, is upon a particular aversion that I have, upon good reason, against their disposing of our money. And, Mr Bays, I will acquaint you with the reason, which is this: 'Tis not very many years ago that I used to play at piquet; and there was a gentleman of your robe, a dignitary of Lincoln, very well known and remembered in the ordinaries,° but being not long since dead, I will save his name. Now, I used to play pieces,° and this gentleman would always go half a crown° with me; and so all the while he sat on my hand he very honestly 'gave the sign'; so that I was always sure to lose. I afterwards discovered it; but of all the money that ever I was cheated of in my life, none ever vexed me so as what I lost by his occasion. And ever since I have borne a great grudge against their fingering of anything that belongs to me. And I have been told, and showed the place where the man dwelt in the late King's time, near Hampton Court, that there was one that used to rob on the highway in the habit of a bishop, and all his fellows rode too in canonical coats. And I can but fancy how it madded those, that would have perhaps been content to relieve an honest gentleman in distress, or, however, would have been less grieved to be robbed by such a one, to see themselves so episcopally pillaged. Neither must it be less displeasing always to the gentry and commonalty of England, that the clergy (as you do, Mr Bays), should tell them that they are never *sui juris*,° not only as to their consciences, but even as to their purses; and you should pretend to have this 'power of the keys' too, where they lock their money. Nay, I dare almost aver, upon my best observation, that there never was, nor ever will be, a Parliament in England that could or can refuse the King supplies proportionable to his occasions, without any need of recourse to extraordinary ways, but for the pickthankness° of the clergy, who will always presume to have the thanks and honour of it, nay, and are ready always to obstruct the parliamentary aids, unless they may have their own little project pass too into the bargain, and they may be gratified with some new ecclesiastical power, or some new law against the fanatics. This is the naked truth of the matter. Whereas Englishmen always love to see how their money goes, and if there be any interest or profit to be got by it, to receive it themselves. Therefore, Mr Bays, I will go on with my business, not fearing all the mischief that you can make of it.

'There was,'° saith he, 'one Sibthorp, who not being so much as Bachelor of Arts, by the means of Doctor Pierce,° Vice-chancellor of Oxford, got to be conferred upon him the title of Doctor. This man was Vicar of Brackley in Northamptonshire, and hath another benefice. This man preaching at Northampton, had taught that princes had power to put poll-money upon their subjects' heads. He, being a man of a low fortune, conceived the putting his sermon in print might gain favour at Court, and raise his fortune higher.' It was at the same time that the business of the loan was on foot. In the same sermon 'he called that loan a tribute; taught that the King's duty is first to direct and make laws; that nothing may excuse the subject from active obedience, but what is against the law of God or nature, or impossible; that all antiquity was absolutely for absolute obedience in all civil and temporal things.' And the imposing of poll-money by princes he justified out of St Matthew:° and in the matter of the loan, 'What a speech is this!' saith the Bishop; 'he observes the forwardness of the Papists to offer double.' For this sermon was sent to the Bishop from Court, and he required to licence it, not under his chaplain, but his own hand. But he, not being satisfied of the doctrine delivered, sent back his reasons why he thought not fit to give his approbation; and unto these Bishop Laud, who was in this whole business, and a rising man at Court, undertook an answer. 'His life in Oxford,' saith Archbishop Abbot, 'was to pick quarrels in the lectures of Public Readers,° and to advertise them to the Bishop of Durham, that he might fill the ears of King James with discontent against the honest men that took pains in their places, and settled the truth (which he called puritanism) in their auditors. He made it his work to see what books were in the press, and to look over epistles dedicatory, and prefaces to the reader, to see what faults might be found. 'Twas an observation what a sweet man this was like to be, that the first observable act he did was the marrying of the Earl of D. to the Lady R.° when she had another husband, a nobleman, and divers children by him.' Here he tells how, for this very cause, King James would not a great while endure him, till he yielded at last to Bishop Williams's importunity, whom notwithstanding he straight strove to undermine, and did it at last to purpose: for, saith the Archbishop, 'Verily such is his undermining nature, that he will underwork any man in the world, so he may gain by it. He called in the Bishop of Durham, Rochester, and Oxford, tried men for such a purpose, to the answering of my reasons, and the whole style of the speech runs We, We. In my memory, Doctor Harsnet, then Bishop of Chichester, and now of Norwich (as he came afterward to be Archbishop of York), preached at Whitehall upon *Give unto Caesar the things that are Caesar's*; a sermon that was afterwards burned, teaching that goods and money were Caesar's, and so the King's: whereupon King James told the Lords and Commons, that he had failed in not adding, "according to the laws and customs of the country wherein they did live". But Sibthorp was for absolutely absolute. So that if the King had sent to me for all my money and goods, and so to the clergy, I must, by Sibthorp's proportion, send him all. If the King should send to the City of London to command all their wealth, they were bound to do it. I know the King is so gracious he will attempt no such matter; but if he do it not, the defect is not in these flattering divines.' Then he saith, reflecting again upon the loan, which Sibthorp called a tribute, 'I am sorry at heart the King's gracious Majesty should rest so great a building on so weak a foundation, the treatise being so slender, and without substance, but that proceeded from a hungry man.' Then he speaks of his own case as to the licensing this book, in parallel to the Earl of Essex's divorce; which to give it more authority, 'was to be ratified judicially by the archbishop'. He concludes, how finally he refused his approbation to this sermon, and saith, 'it was thereupon carried to the Bishop of London, who gave a great and stately allowance of it, the good man not being willing that anything should stick with him that came from Court, as appears by a book commonly called the Seven Sacraments, which was allowed by his lordship with all the errors, which have been since expunged.' And he adds a pretty story of one Doctor Worral,° the Bishop of London's chaplain, 'scholar good enough, but a free-fellow-like man, and of no very tender conscience', who before it was licensed by the Bishop, Sibthorp's sermon being brought to him, 'hand over head° approved it, and subscribed his name'. But afterwards hearing more of it, went to a counsel at the Temple,° who told him, that by that book 'there was no *meum* nor *tuum*° left in England, and if ever the tide turned, he might come to be hanged for it', and thereupon 'Worral scraped out his name again', and left it to his lord to license. Then the Archbishop takes notice of the instructions for that loan. 'Those that refused, to be sent for soldiers to the King of Denmark. Oaths to be administered with whom they had conference; and who dissuaded them, such persons be sent to prison, etc. He saith that he had complained thrice of Montague's Arminian book,° to no purpose: Cosins put out his book of "Seven Sacraments" (strange things), but I knew nothing of it; but as it pleased my Lord of Durham and the Bishop of Bath, so it went.' In conclusion, the good Archbishop, for refusing this license of Sibthorp's sermons, was, by the underworking of his adversaries, first commanded from Lambeth, and confined to his house in Kent, and afterwards sequestered, and a commission passed to exercise the Archiepiscopal Jurisdiction to the Bishops of London, Durham, Rochester, Oxford, and Bishop Laud (who from thence arose in time to be the Archbishop). If I had leisure, how easy a thing it were for me to extract out of this narrative a just parallel of our author, even almost upon all points: but I am now upon a more serious subject, and therefore shall leave the application to his own ingenuity and the good intelligence of the reader.

About the same time (for I am speaking within the circle of 2°, 3°, and 4° Caroli) that this book of Sibthorp's called *Apostolical Obedience* was printed, there came out another of the same stamp, entitled *Religion and Allegiance*, by one Doctor Manwaring.° It was the substance of two sermons preached by him at Whitehall, beside what of the same nature at his own parish of Saint Giles. Therein he delivered for truth, 'that the king is not bound to observe the laws of the realm concerning the subject's rights and liberties, but that his royal word and command, in imposing loans and taxes without common consent in Parliament, does oblige the subject's conscience upon pain of eternal damnation; that those who refused to pay this loan offended against the law of God and the king's supreme authority, and became guilty of impiety, disloyalty, and rebellion; that the authority of Parliament was not necesary for the raising of aids and subsidies, and the slow proccedings of such great assemblies were not fitted for the supply of the state's urgent necessities, but would rather produce sundry impediments to the just designs of princes.' And after he had been questioned for this doctrine, nevertheless he preached again, 'that the king had right to order all as to him should seem good, without any man's consent; that the king might, in time of necessity, demand aid, and if the subjects did not supply him, the king might justly avenge it; that the property of estate and goods was ordinarily in the subject, but extraordinarily in the king; that in case of the king's need he hath right to dispose them.' He had besides, entering into comparison, called the refusers of the loan 'temporal recusants', and said, 'the same disobedience that they (the Papists, as they then called them) practise in spirituals, that or worse, some of our side, if ours they be, dare to practise in temporals.' And he aggravated° further upon them under the resemblance of Turks, Jews, Corah, Dathan and Abiram;° 'which last,' said he, 'might as well liken themselves to the three children; or Theudas and Judas,° the two incendiaries in the days of Caesar's tribute, might as well pretend their cause to be like that of the Maccabees,° as what the refusers alleged in their own defence.'

I should not have been so large in these particulars, had they been only single and volatile Sermons; but because this was then the doctrine of those persons that pretended to be the Church of England. The whole choir sung that tune, and instead of the Common Law of England and the statutes of Parliament, that part of the clergy had invented these Ecclesiastical Laws, which, according to their predominancy, were sure to be put in execution. So that between their own revenue, which must be held *jure divino*,° as every thing else that belonged to them, and the prince's, that was *jure regio*,° they had not left an inch of property for the subject. It seemed that they had granted themselves 'letters of reprisal'° against the laity for the losses of the Church under Henry the Eighth, and that they would make a greater havoc upon their temporalities in retaliation. And indeed, having many times since pondered with my greatest and earnest impartiality, what could be the true reason of the spleen that they manifested in those days, on the one hand against the Puritans, and on the other against the gentry (for it was come, they tell me to 'Jack Gentleman'), I could not devise any cause, but that the Puritans had ever since the Reformation obstructed that laziness and splendour which they enjoyed under the Pope's supremacy, and the Gentry had (sacrilegiously) divided the abbey lands, and other fat morsels of the Church at the Dissolution, and now was the time to be revenged upon them.

While, therefore, the kingdom was turned into a prison upon occasion of this ecclesiastical loan, and many of the eminentest of the gentry of England were under restraint, they thought it seasonable to recover once again their ancient glory, and to magnificate° the Church with triumphal pomp and ceremony. The three ceremonies that have the countenance of Law would not suffice; but they were all upon new inventions; and happy was he that was endued with that capacity, for he was sure before all others to be preferred. There was a sccond service, the table set *altarwise*, and to be called the *altar*; 'candles, crucifixes, paintings, images, copes, bowing to the east, bowing to the altar', and so many several cringes and genuflexions, that a man unpractised stood in need to entertain both a dancing master and a remembrancer.° And though these things were very uncouth to English Protestants, who naturally affect a plainness of fashion, especially in sacred things; yet if those gentlemen could have contented themselves with their own formality, the innovation had been more excusable. But many of these additions, and to be sure, all that had any colour of law, were so imposed and pressed upon others, that a great part of the nation was e'en put to it as it were to fine and ransom upon this account. What censures, what excommunications, what deprivations, what imprisonments! I cannot represent the misery and desolation as it hath been represented to me. But wearied out at home, many thousands of his Majesty's subjects, to his and the nation's great loss, thought themselves constrained to seek another habitation; and every country, even though it were among savages and cannibals, appeared more hospitable to them than their own.

And although I have been told by those that have seen both, that our Church did even then exceed the Romish in ceremonies and decorations—and indeed, several of our Church did thereby frequently mistake their way, and from a Popish kind of worship fell into the Romish religion—yet I cannot upon my best judgement believe that that Party had generally a design to alter the religion so far, but rather to set up a new kind of Papacy of their own here in England. And it seemed they had, to that purpose, provided themselves of a new religion in Holland. It was Arminianism, which, though it were the republican opinion there, and so odious to King James that it helped on the death of Barnevelt,° yet now they undertook to accommodate it to monarchy and Episcopacy. And the choice seemed not imprudent. For, on the one hand, it was removed at so moderate a distance from Popery that they should not disoblige the Papists more than formerly, neither yet could the Puritans, with justice, reproach these men as Romish Catholics; and yet, on the other hand, they knew it was so contrary to the ancient reformed doctrine of the Church of England, that the Puritans would never embrace it, and so they should gain this pretence further to keep up that convenient and necessary quarrel against Nonconformity. And accordingly it happened; so that here again was a new shibboleth. And the Calvinists were all studiously discountenanced, and none but an Arminian was judged capable and qualified for employment in the Church. And though the king did declare, as I have before mentioned, that Montague's Arminian book had been the occasion of the schisms in the Church, yet care was immediately taken, by those of the same robe and party, that he should be the more rewarded and advanced. As also it was in Manwaring's case; who, though by censure in Parliament made incapable of any ecclesiastical preferment, was straight made rector of Stamford-Rivers in Essex, with a dispensation to hold too his living in St Giles's. And all dexterity was practised to propagate the same opinions, and to suppress all writings or discourses to the contrary.

So that those who were of understanding in those days tell me that a man would wonder to have heard their kind of preachings. How, instead of the practical doctrine which tends to the reforming of men's lives and manners, all their sermons were a very mash of Arminian subtleties, of ceremonies and decency, and of Manwaring and Sibthorpianism brewed together; besides that in their conversation they thought fit to take some more license, the better to *dis-Ghibelline* themselves from the Puritans. And though there needed nothing more to make them unacceptable to the sober part of the Nation, yet moreover they were so exceeding pragmatical, so intolerably ambitious, and so desperately proud, that scarce any gentleman might come near the tail of their mules. And many things I perceive of that nature do even yet stick upon the stomachs of the old gentlemen of those times. For the English have been always very tender of their religion, their liberty, their property, and (I was going to say) no less of their reputation. Neither yet do I speak of these things with passion, considering at more distance how natural it is for men to desire to be in office, and no less natural to grow proud and intractable in office; and the less a clergyman is so, the more he deserves to be commended. But these things before mentioned grew yet higher, after that Bishop Laud was once not only exalted to the See of Canterbury, but to be chief minister. Happy had it been for the King, happy for the nation, and happy for himself, had he never climbed that pinnacle. For whether it be or no, that the clergy are not so well fitted by education as others for political affairs, I know not, though I should rather think they have advantage above others, and even, if they would but keep to their Bibles, might make the best ministers of State in the world; yet it is generally observed that things miscarry under their government. If there be any counsel more precipitate, more violent, more rigorous, more extreme than the other, that is theirs. Truly, I think the reason that God does not bless them in affairs of state is, because He never intended them for that employment. Or if government and the preaching of the Gospel may well concur in the same person, God therefore frustrates him, because, though knowing better, he seeks and manages his greatness by the lesser and meaner maxims. I am confident the Bishop studied to do both God and his Majesty good service; but, alas, how utterly was he mistaken. Though so learned, so pious, so wise a man, he seemed to know nothing beyond Ceremonies, Arminianism, and Manwaring. With that he began, and with that ended; and thereby deformed the whole reign of the best prince that ever wielded the English sceptre.

For his late Majesty being a prince truly pious and religious, was thereby the more inclined to esteem and favour the clergy. And thence, though himself of a most exquisite understanding, yet thought he could not trust it better than in their keeping. Whereas every man is best in his own post, and so the preacher in the pulpit. But he that will do the clergy's drudgery must look for his reward in another world. For they having gained this ascendancy upon him, resolved, whatever became on't, to make their best of him; and having made the whole business of State their Arminian jangles and the persecution for ceremonies, did for recompense assign him that imaginary absolute government, upon which rock we all ruined.

For now was come the last part of the Archbishop's indiscretion; who, having strained those strings so high here, and all at the same time, which no wise man ever did, he moreover had a mind to try the same dangerous experiment in Scotland, and sent thither the Book of the English Liturgy to be imposed upon them. What followed thereupon is yet within the compass of most men's memories. And how the war broke out, and then to be sure hell's broke loose. Whether it be a war of religion or of liberty, is not worth the labour to inquire. Whichsoever was at the top, the other was at the bottom; but upon considering all, I think the cause was too good to have been fought for. Men ought to have trusted God; they ought and might have trusted the King with that whole matter. The 'arms of the Church are prayers and tears'; the arms of the subjects are patience and petitions. The King himself, being of so accurate and piercing a judgement, would soon have felt where it stuck. For men may spare their pains where nature is at work, and the world will not go the faster for our driving. Even as his present Majesty's happy Restoration did itself, so all things else happen in their best and proper time, without any need of our officiousness.

But after all the fatal consequences of that rebellion, which can only serve as sea-marks unto wise princes to avoid the causes, shall this sort of men still vindicate themselves as the most zealous assertors of the rights of princes? They are but at the best 'well-meaning zealots'. Shall, to decline so pernicious counsels, and to provide better for the quiet of government, be traduced as the author does here, under these odious terms of 'forsaking the Church, and delivering up the Church'? Shall these men always presume to usurp to themselves that venerable style of the 'Church of England'? God forbid! The Independents at that rate would not have so many distinct congregations as they. There would be Sibthorp's church, and Manwaring's church, and Montague's church, and a whole bead-roll° more, whom for decency's sake I abstain from naming. And every man that could invent a new opinion, or a new ceremony, or a new tax, should be a new Church of England.

Neither, as far as I can discern, have this sort of the clergy, since his Majesty's return, given him better encouragement to steer by their compass. I am told that preparatory to that, they had frequent meetings in the city, I know not whether in Grubstreet, with the divines of the other Party, and that there, in their Feasts of Love, they promised to forget all former offences, to lay by all animosities; that there should be a new heaven and a new earth, all meekness, charity, and condescension. His Majesty, I am sure, sent over his gracious Declaration° of liberty to tender consciences, and, upon his coming over, seconded it with his commission under the broad seal for a Conference° betwixt the two parties, to prepare things for an Accommodation, that he might confirm it by his royal authority. Hereupon what do they? Notwithstanding this happy conjuncture of his Majesty's Restoration, which had put all men into so good a humour, that, upon a little moderation and temper of things, the Nonconformists could not have stuck out; some of these men so contrived it, that there should not be the least abatement to bring them off with conscience and (which insinuates into all men) some little reputation. But, to the contrary, several unnecessary additions were made, only because they knew they would be more ungrateful and stigmatical to the Nonconformsits. I remember one in the Litany, where to 'false doctrine and heresy' they added 'schism', though it were to spoil the music and cadence of the period. But these things were the best. To show that they were men like others, even cunning men, revengeful men, they drilled things on, till they might procure a law, wherein, besides all the Conformity that had been of former times enacted, there might be some new conditions imposed on those that should have or hold any church-livings, such as they assured themselves that, rather than swallow, the Nonconformists would disgorge all their benefices. And accordingly it succeeded; several thousands of those ministers being upon one memorable day° outed of their subsistence. His Majesty in the mean time, although they had thus far prevailed to frustrate his royal intentions, had reinstated the Church in all its former revenues, dignities, and advantages; so far from the author's mischievous aspersion of ever thinking of converting them to his own use, that he restored them free from what was due to him by law upon their first admission. So careful was he, 'because all government must owe its quiet and continuance to the Church's patronage', to pay them even what they ought. But I have observed, that if a man be in the Church's debt once, 'tis very hard to get an acquittance: and these men never think they have their full rights, unless they reign. What would they have had more? They rolled on a flood of wealth, and yet in matter of a lease would make no difference betwixt a Nonconformist and one of their own fellow-sufferers, who had ventured his life and spent his estate for the king's service. They were restored to Parliament, and to take their places with the king and the nobility. They had a new liturgy to their own hearts' desire; and to cumulate all this happiness, they had this new law against the Fanatics. All they had that could be devised in the world to make a clergyman goodnatured.

Nevertheless, after all their former sufferings, and after all these new enjoyments and acquisitions, they have proceeded still in the same track. The matter of ceremonies, to be sure, hath not only exercised their ancient rigour and severity, but hath been a main ingredient of their public discourses, of their sermons, of their writings. I could not (though I do not make it my work, after a great example, to look over Epistles Dedicatory) but observe by chance the title-page of a book t'other day, as an emblem how much some of them do neglect the scripture in respect to their darling ceremonies. *A Rationale upon the Book of Common-Prayer of the Church of England*, by A. Sparrow, D.D. Bishop of Exeter. With the *Form of Consecration of a Church or Chapel, and of the place of Christian Burial.* By Lancelot Andrews, late Lord Bishop of Winchester. Sold by Robert Pawlet at the sign of the Bible in Chancery Lane. These surely are worthy cares for the Fathers of the Church.

But, to let these things alone, how have they of late years demeaned themselves to his Majesty, although our author urges their immediate dependence on the King to be a great obligation he hath upon their loyalty and fidelity? I have heard that some of them, when a great minister of state grew burdensome to his Majesty and the Nation, stood almost in defiance of his Majesty's good pleasure, and fought it out to the uttermost in his defence. I have been told that some of them in a matter of divorce, wherein his Majesty desired that justice might be done to the party aggrieved opposed him vigorously, though they made bold too with a point of conscience in the case, and went against the judgement of the best divines of all parties. It hath been observed, that whensoever his Majesty hath had the most urgent occasions for supply, others of them have made it their business to trinkle° with the members of Parliament for obstructing it, unless the king would buy it with a new law against the fanatics. And hence it is that the wisdom of his Majesty and the Parliament must be exposed to after-ages for such a superfoetation° of Acts in his reign about the same business. And no sooner can his Majesty upon his own best reasons try to obviate this inconvenience, but our author, who had before outshot Sibthorp and Manwaring in their own bows, is now for retrenching his authority, and moreover calumniates the State with a likelihood, and the reasons thereof, 'of the Return of Popery into this Nation'. And this hath been his first method by the 'fanatics raising disturbance': whereupon if I have raked further into things than I would have done, the author's indiscretion will I hope excuse me, and gather all the blame for reviving those things which were to be buried in oblivion. But, by what appears, I cannot see that there is any probability of disturbance in the state but by men of his spirit and principles.

The second way whereby the fanatic party, he saith, may at last work the ruin of the Church, is 'by combining with the atheists; for their union is like the mixture of nitre and charcoal, it carries all before it without mercy or resistance'. So, it seems, when you have made gunpowder of the atheists and fanatics, we are like to be blown up with Popery. And so will the larks too. But his zeal spends itself most against the atheists, because they use to 'jeer the parsons'. That they may do, and no atheists neither; for really, while clergymen will, having so serious an office, play the drolls and the boon-companions, and make merry with the scriptures, not only among themselves, but in gentlemen's company, 'tis impossible but that they should meet with at least an unlucky repartee sometimes, and grow by degrees to be a tail and contempt to the people. Nay, even that which our author always magnifies, the reputation, the interest, the secular grandeur of the Church, is indeed the very thing which renders them ridiculous to many, and looks as improper and buffoonish, as to have seen the porter lately in the good Doctor's cassock and girdle. For, so they tell me, that there are nowhere more atheists than at Rome, because men seeing that princely garb and pomp of the clergy, and observing the life and manners, think therefore the meaner of religion. For certainly the reputation and interest of the clergy was first gained by abstracting themselves from the world, attending their callings, humility, strictness of doctrine, and the same strictness in conversation: and things are best preserved by the same means they were at first attained. But if our author had been as concerned against atheism as he is against their disrespect of his function, he should have been content that the fanatic preachers might have spent some of their pulpit-sweat upon the atheists, and made a noise in their ears about 'faith, communion with God, attendance upon ordinances,' which he himself jeers at so pleasantly. Neither do I like upon the same reasons his manner of discourse with the atheists, where he complains that ours are not like those good atheists of former times, who never did thrust themselves into public cares and concerns, 'minding nothing but love, wine, and poetry'. Nor, in another place, 'put the case the clergy were cheats and jugglers, yet it must be allowed they are necessary instruments of state to awe the common people into fear and obedience, because nothing else can so effectually enslave them' ('tis this, it seems our author would be at) 'as the fear of invisible powers, and the dismal apprehensions of the world to come: and for this very reason, though there were no other, it is fit they should be allowed the same honour and respect as would be acknowledged their due, if they were sincere and honest men.' No atheist could have said better. How mendicant a cause has he here made of it! They will say, they see where the shoe wrings him, and that though this be some ingenuity in him, yet it is but little policy. Nay, perhaps they will say, that they are no atheists neither, but only, I know not by what fate, every day one or other of the clergy does or saith some so ridiculous and foolish thing, or some so pretty accident befalls them, that in our author's words, 'a man must be very splenetic that can refrain from laughter'. I would have quoted the page here, but that the author has, I think for evasion sake, omitted to number them in his whole Preface. But whether there be any atheists or no, which I question more than witches, I do not, for all this, take our author to be one, though some would conclude it out of his principles, others out of his expressions. Yet really, I think he hath done that sort of men so much service in his books by his ill-handling, and while he personates one party, making all religion ridiculous, that they will never be able to requite him but with the same manner. He hath opened them a whole treasury of words and sentences, universally applicable, where they may rifle or choose things, which their pitiful wit, as he calls it, would never have been able to invent and flourish. But truly, as the simple Parliament 5° Eliz. never imagined what consequence that clause in the Wednesday fast would have to Puritanism, neither did he what his periods would have to atheism; and yet, though he is so more excusable, I hope I may have the same leave on him, as he on that Parliament, to censure his impertinence. To close this: I know a lady that chid her master of the horse, for correcting the page that had sworn a great oath. 'For,' saith she, 'the boy did therein show only the generosity of his courage, and his acknowledgement of a deity.' And indeed he hath approved his religion, and justified himself from atheism, much after the same manner.

The third way and last (which I, being tired, am very glad of) by which the fanatics may raise disturbances, and so introduce Popery, is by joining crafty and sacrilegious statesmen into the confederacy. But really here he doth speak concerning king and counsellors at such a rate, and describe and characterize some men so, whomsoever he intends, that though I know there are no such, I dare not touch; it is too hazardous. 'Tis true he passes his compliment ill-favouredly enough. 'The Church has at present an impregnable affiance in the wisdom etc. of so gracious a prince, that is not capable of such counsels, should they be suggested to him: though certainly no man that is worthy to be admitted to his Majesty's favour or privacy can be supposed so foolhardy or presumptuous as to offer such weak and dishonourable advice to so wise and able a prince; yet princes are mortal, and if ever hereafter (and some time or other it must happen) the crown should chance to settle upon a young and inexperienced head, this is usually the first thing in which such princes are abused by their keepers and guardians, etc.' But this compliment is no better, at best, than if, discoursing with a man of another, I should take him by the beard. Upon such occasions in company, we use to ask, 'Sir, whom do you mean?' I am sure our author takes it always for granted that his answerer intends him upon more indefinite and less direct provocations. But our author does even personate some men as speaking at present against the Church: 'They will entangle your affairs, endanger your safety, hazard your crown. All the reward you shall have to compensate your misfortunes by following Church counsels, shall be that a few churchmen, or such-like people, shall cry you up for a saint or a martyr.' Still 'your, your', as if it were a close° discourse unto his Majesty himself. Though if this were the worst that they said, or that the author fathers upon them, I wish the King might never have better counsellors about him. But if the author be secure, for the present, in his Majesty's reign, fears not Popery, not forsaking the Church, not assuming the Church revenues, why is he so provident? why put things in men's heads they never thought of? why stir such an odious, seditious, impertinent, unseasonable discourse? why take this very minute of time, but that he hath mischief, to say no worse, in his heart? He had no such remote conceit (for all his talk) of an infant coming to the crown. He is not so weak but he knows too much, and is too well instructed, to speak to so little purpose. That would have been like a set of Elizabeth players, that in the country having worn out and over acted all the plays they brought with them from London, laid their wits together to make a new one of their own. No less man than Julius Caesar was the argument; and one of the chief parts was Moses persuading Julius Caesar not to make war against his own country, nor pass the Rubicon. If our author did not speak of our present times (to do which nevertheless had been sufficiently false and absurd), but writ all this merely out of his providence for after-ages, I shall no more call him Bays, for he is just such a 'second Moses'. I ask pardon if I have said too much; but I shall deserve none, if I meddle any further with so improbable and dangerous a business.

To conclude, the author gives us one ground more, and perhaps more seditiously insinuated than any of the former; that is, if 'it should so prove', that is, if the 'fanatics, by their wanton and unreasonable opposition to the ingenious and moderate discipline of the Church of England, shall give their governors too much reason to suspect that they are never to be kept in order by a milder and more gentle government than that of the Church of Rome, and force them at last to scourge them into better manners, with the briars and thorns of their discipline.' It seems, then, that the discipline contended about is worth such an alteration. It seems that he knows something more than I did believe of the design in the late times before the war. Whom doth he mean by 'our governors'? the King? No, for he is a single person. The Parliament, or the Bishops.

I have now done; after I have (which is, I think, due) given the reader and the author a short account how I came to write this book, and in this manner. First of all, I was offended at the presumption and arrogance of his style; whereas there is nothing either of wit or eloquence in all his books worthy of a reader's, and more unfit for his own, taking notice of. Then, his infinite tautology was burdensome, which seemed like marching a company round a hall upon a pay-day so often, till, if the muster-master were not attentive, they might receive the pay of a regiment. All the variety of his treat is *pork*° (he knows the story), but so little disguised by good cookery, that it discovers the miserableness, or rather the penury, of the host. When I observed how he inveighs against the 'trading part' of the Nation, I thought he deserved to be within the 'five-mile Act',° and not to come within that distance of any corporation. I could not patiently see how irreverently he treated kings and princes, as if they had been no better than King Phys and King Ush° of Brentford. I thought his profanation of the scripture intolerable; for though he alleges that 'tis only in order to shew how it was misapplied by the fanatics, he might have done that too, and yet preserved the dignity and reverence of those Sacred Writings; which he hath not done, but on the contrary, he hath, in what is properly his own, taken the most of all his ornaments and embellishments thence, in a scurrilous and sacrilegious style; insomuch that, were it honest, I will undertake out of him to make a better, that is a more ridiculous and profaner book, than all the *Friendly Debates* bound up together. Methought I never saw a more bold and wicked attempt than that of reducing grace, and making it a mere *Fable*, of which he gives us the *Moral*. I was sorry to see that even prayer could not be admitted to be a virtue, having thought hitherto it had been a grace, and a peculiar gift of the Spirit; but I considered that that prayer ought to be discouraged, in order to prefer the Liturgy. He seemed to speak so little like a divine in all those matters, that the poet° might as well have pretended to be the Bishop Davenant,° and that description of the poet's of prayer and praise was better than our author's on the same subject. Canto the 6th, where he likens prayer to the Ocean:

For prayer the Ocean is, where diversly
Men steer their course each to a several coast;

> Where all our interests so discordant lye,
> That half beg winds, by which the rest are lost.°

And praise he compares to the union of fanatics and atheists, etc., that is, gunpowder; 'praise is devotion fit for mighty minds,° etc.'

> Its utmost force, like powder, is unknown.
> And though weak kings excess of praise may fear,
> Yet when 'tis here, like powder, dangerous grown,
> Heaven's vault receives what would the palace tear.°

Indeed all Astragon° appeared to me the better 'scheme of religion'. But it is unnecessary here to recapitulate all, one by one, what I have in the former Discourse taken notice of. I shall only add what gave, if not the greatest, yet the last impulse to my writing. I had observed in his first book, page 57, that he had said 'some pert and pragmatical divines had filled the world with a buzz and noise of the Divine Spirit'; which seemed to me so horribly irreverent, as if he had taken similitude from the hum and buzz of the humble-bee in the *Rehearsal*.°

In the same book, I have before mentioned, that most unsafe passage of our 'Saviour being not only in a hot fit of zeal, but in a seeming fury and transport of passion'. And striving to unhook himself hence, page 152 of his second book, swallows it deeper, saying, 'Our Blessed Saviour did in that action take upon Him the person and privilege of a Jewish Zealot.' Take upon Him the person, that is, *personam induere*. And what part did He play? of a Jewish zealot.

The second Person of the Trinity (may I repeat these things without offence?) to take upon Him the person of a Jewish zealot, that is, of a notorious rogue and cut-throat!

This seemed to proceed from too slight an apprehension and knowledge of the duty we owe to our Saviour. And last of all, in this Preface, as before quoted, he saith, the 'Nonconformist preachers do spend most of their pulpit-sweat in making a noise about communion with God'. So that there is not one Person of the Trinity that he hath not done despite to; and lest he should have distinct communion with the Father, the Son, and the Holy Ghost, for which he mocks his answerer, he hath spoken evil distinctly of the Father, distinctly of the Son, and distinctly of the Holy Ghost. That only remained behind, wherein our Author might surpass the character given to Aretino,° a famous man of his faculty.

> *Qui giace il Aretino,*
> *Chi de tutti mal disse fuor d' Iddio:*
> *Ma di questo si scusa, perche no 'l conobbe.*

> Here lies Aretine,
> Who spoke evil of all, except God only;
> But of this he begs excuse, because he did not know Him.

And now I have done. And shall think myself largely recompensed for this trouble, if any one that hath been formerly of another mind shall learn by this example, that it is not impossible to be merry and angry as long time as I have been writing, without profaning and violating those things which are and ought to be most sacred.

# THE KING'S SPEECH

## 13 APRIL 1675

My Lords and Gentlemen,

I told you last meeting, the winter was the only time for business, and truly I thought so till my Lord Treasurer° assured me that the spring is the fittest season for salads and subsidies. I hope therefore this April will not prove so unnatural as not to afford plenty of both. Some of you may perhaps think it dangerous to make me too rich, but do not fear it, for I promise you faithfully, whatsoever you will give me, I will always take care to want; for the truth of which you may rely upon the word of a King.

My Lords and Gentlemen,

I can bear my own straits with patience, but my Lord Treasurer protests that the revenue, as it now stands, is too little for us both; one of us must pinch for it, if you do not help us out. I must speak freely to you. I am under encumbrances, for besides my harlots in service my reformado° ones lie hard upon me. I have a pretty good estate I confess, but, Godsfish, I have a great charge upon it. Here is my Lord Treasurer can tell you that all the money designed for the Summer Guards must of necessity be employed to the next year's cradles and swaddling-clothes. What then shall we do for ships? I only hint this to you—it is your business, not mine. I know by experience I can live without them—I lived ten years without them abroad and was never in better health in my life. But how will you live without them you had best say; and therefore I do not intend to insist upon it. There is another which I must press more earnestly, which is, it seems a good part of my revenue will fail in two or three years, except you will be pleased to continue it. Now I have this to say for it. Pray why did you give me as much except you resolved to go on? The nation hates you already, for giving me so much, and I will hate you now if you do not give me more. So that now your interest obliges you to stick to me or you will not have a friend left in England. On the other side, if you will continue the revenue as desired, I shall be enabled to perform those great things for your religion° and liberty which I have had long in my thoughts, but cannot effect them without this establishment. Therefore look to it, if you do not make me rich to undo you, it shall lie at your doors. For my part, I can with a clear conscience say, 'I have done my best and I shall leave the rest to my successor.' That I may gain your good opinion, the best way is to acquaint you with what I have done to deserve it, out of my royal care for your religion and your property.

For the first, my late proclamation° is a true picture of my mind: he that can't as in a glass see my zeal for the Church of England, doth not deserve any further satisfaction, for I declare him wilful, abominable and not good. Some may perhaps be startled, and cry, 'How comes this sudden change?' To which I answer, 'I am a changeling.' I think that is a full answer. But to convince men that I mean as I say there are three arguments. First, I tell you so, and you know I never broke my word. Secondly, my Lord Treasurer says so, and he never told a lie. Thirdly, my Lord Lauderdale° will undertake for me, and I shall be loath by any act of mine to forfeit the credit he hath with you.

If you desire more instances of my zeal, I have them for you. For example, I have converted all my natural sons from Popery, and may say without vanity, it was my own work, and so much the more peculiar than the begetting of them. It would do your heart good to hear how prettily little George° can read the Psalter.

They are all fine children, (God bless 'em), and so like me in their understandings. But as I was saying, I have, to please you, given a pension to your favourite, the Lord Lauderdale, not so much that I thought he wanted it, as I thought you would take it kindly. I have made Portsmouth° a Duchess, and have married her sister to my Lord of Pembroke.° I have at my brother's request sent my Lord of Inchiquin° to settle Protestant religion at Tangiers and my Lord Vaughan° to propagate Christian religion at Jamaica. I have made the reverend and learned Crew° Bishop of Durham and at the first word of my Lady Portsmouth, I have preferred Brideoak° to be Bishop of Chichester.

Now for your properties: my behaviour to the bankers, and letting the customers to my Lord St John° and his partners take for public instances; and my proceeding between Mistress Hyde and Emerton° for a private one, as such convincing evidences, that it will be needless to say more. I must now acquaint you, that by my Lord Treasurer's advice I have made a considerable retrenchment upon my expenses of candles and charcoal, and do not intend to stop there, but will, with your help, look into the late embezzlements of my kitchen staff, for which, by the way on my conscience [neither] the Lord Treasurer nor my Lord Lauderdale are guilty. I speak my opinion, but if you shall find them daubing in that business I tell you plainly, I leave them to you, for I would have the world know, I am not a man to be cheated.

My Lords and Gentlemen,

I desire you to believe of me as you have ever found, and I do solemnly promise that whatsoever you give me shall be managed with the same conduct, thrift, and prudence, that I have ever practised since my happy Restoration.

# NOTES

ABBREVIATIONS

| | |
|---|---|
| 1681 | *Miscellaneous Poems*. By Andrew Marvell. 1681. (The British Library copy, C.59.i.8.) |
| Bodleian MS | Bodleian MS. Eng. poet. d. 49. (A copy of *1681* with manuscript corrections and additions.) |
| Cooke | *The Works of Andrew Marvell*, ed. Thomas Cooke (2 vols., 1726). |
| Donno | *Andrew Marvell: The Complete Poems*, ed. Elizabeth Story Donno (1972). |
| Grosart | *The Complete Works of Andrew Marvell*, ed. A. B. Grosart (4 vols., 1872–5). |
| Kermode | *The Selected Poetry of Marvell*, ed. Frank Kermode (New York, 1967). |
| Lord | *Andrew Marvell: Complete Poetry*, ed. George de F. Lord (New York, 1968; London, 1984). |
| Margoliouth | *The Poems and Letters of Andrew Marvell*, ed. H. M. Margoliouth (2 vols., Oxford, 1927); 3rd edn. revised by Pierre Legouis and E. E. Duncan-Jones (2 vols., Oxford, 1971). |
| Smith | *'The Rehearsal Transpros'd' and 'The Rehearsal Transpros'd The Second Part'*, ed. D. I. B. Smith (Oxford, 1971). |
| Thompson | *The Works of Andrew Marvell*, ed. Edward Thompson (3 vols., 1776). |
| Wilcher | *Andrew Marvell: Selected Poetry and Prose*, ed. Robert Wilcher (1986). |

Other works abbreviated will be found under 'Further Reading' (p. 359), or in the headnotes to individual poems.

1 *An Elegy upon the Death of my Lord Francis Villiers*. Surviving in an apparently unique quarto from the library of Sir George Clarke (1661–1736), now in the library of Worcester College, Oxford. The poem is ascribed to Marvell by Clarke (for details see Margoliouth). The attribution has not gone unchallenged. Hilton Kelliher thinks that the 'command of classical myth and literary convention and his [lack of] familiarity with contemporary poetic conventions', and the verse on 'heavy Cromwell' and 'long-deceived Fairfax' ('which contrast so flagrantly with his [later] admiration of these men'), do not tell against Marvell's authorship so much as 'the concluding lines which are the utterance of a bigoted royalist'. On the other hand, Elsie Duncan-Jones thinks 'the reminiscences of Ecclesiasticus, Virgil, and Du Bartas all favour Marvell's authorship'.

Francis Villiers (b. 1629), second and posthumous son of George, first Duke of Buckingham, was killed 7 July 1648, at a skirmish near Kingston. (See S. R. Gardiner, *History of the Great Civil War*, 4, 160.)

l. 18. *bloody bays*. Rosemary dipped in blood was placed upon the coffins of soldiers (Duncan-Jones).

l. 30. *his princess*. Lady Katherine Manners, Duchess of Buckingham.

ll. 31–2. *As the wise Chinese . . . entomb*. Cf. 'The First Anniversary', ll. 19–20.

2 l. 40. *acceptably*. Stressed on the first and third syllables.

l. 41. *'Tis truth . . . dispraise*. Cf. Ecclesiasticus 11: 2: 'Commend not a man for his beauty' (Duncan-Jones).

l. 61. *Fair Richmond*. Mary Villiers married James Stuart, first Duke of Richmond.

l. 69. *Clora*. Probably Mary Kirke, daughter of the poet Aurelian Townsend.

3 l. 81. *modest plant*. *Mimosa pudica* or sensitive plant.

l. 84. *feign*. Dissemble or conceal.

ll. 98–100. *The Sad Iliads . . . would spare*. Cf. *Aeneid*, iv. 186–7.

4 l. 123. *own*. Due.

*To his Noble Friend Mr Richard Lovelace, upon his Poems*. A commendatory poem prefixed to Lovelace's *Lucasta* (1649).

l. 2. *with*. Proposed by R. G. Howarth, *Notes and Queries*, 198 (1953), 380; 'which' *1649*.

l. 5. *could tell*. Knew.

l. 12. *civic crown*. An oak-leaf garland bestowed on one who saves a citizen's life in battle.

l. 20. *Of wit . . . sons*. Born ill-shaped of the corruption of wit, as insects of rotten matter.

ll. 21–2. *censurers . . . book*. In June 1643 Parliament issued an Ordinance against the printing of unlicensed books and this remained in force (despite *Areopagitica*, 1644, and other protests). *Lucasta* was licensed in 1648. *consistory*. Court of presbyters.

5 l. 24. *young presbytery*. Established in 1643.

l. 28. *wronged*. Abused. *the House's privilege*. Parliamentary privilege includes immunity from action in the courts for what is said in the House. In *Lucasta*, that is, Lovelace assumed a freedom of speech granted only to Members in the House.

l. 29. *sequestration*. Lovelace's estate was ordered to be confiscated on 28 November 1648.

l. 30. *Because . . . war*. Referring to Lovelace's song 'To Lucasta going to the wars'.

5 ll. 31–2. *Kent . . . sent.* Lovelace was sent to prison for presenting to the House of Commons (April 1642) a Kentish petition asking for control of the militia and the use of the Book of Common Prayer.

*Upon the death of Lord Hastings.* First published in *Lachrymae Musarum* (1649), a collection of funeral elegies for Henry Lord Hastings, who died of smallpox on 24 June. The funeral elegy was a genre requiring a high conceited style. Contributors to the collection included Herrick, Denham, and the 18-year-old Dryden.

6 l. 11. *phlegmatic.* Accented on the first syllable.

l. 12. *remora.* The sucking-fish, believed by the ancients to have the power of staying the course of any ship to which it attached itself.

ll. 13–16. *What man . . . already old.* Some live to attain a second childhood; others develop so fast that while still young they are treated as if they were old, and die prematurely.

l. 18. *But . . . geometric year.* The contrast may be between arithmetical numbering and geometrical proportion, the former sequential, the latter accounting not for numbers but ratios (so that maturity, early or late, qualifies for death). E. E. Duncan-Jones finds an explanation in J. Wilkins, *Mathematical Magic* (1648), which refers to an artificial motion geometrically contrived to be swifter than the revolution of the heavens. A 'geometric year' would be one that passes more quickly.

l. 23. *that state.* Heaven.

l. 24. *more than one.* Both the trees.

ll. 25–6. *Therefore . . . ostracize.* The astrological conjunction which brought about his death acted like democracy—it excluded someone of worth (i.e. an aristocrat).

l. 34. *carousels.* Knightly tournaments.

l. 43. *Hymeneus.* The god of marriage, traditionally clothed in a saffron robe and carrying a torch, now wears purple for mourning.

l. 46. *Reversed.* On the analogy of reversed arms at state funerals.

l. 47. *Aesculapius.* God of medicine.

l. 48. *Mayern.* Physician to the king, and Hastings' prospective father-in-law.

l. 50. *leap.* Explode *(OED*, s.v. 5b).

7 ll. 53–4. *laurels . . . balsam.* Poetry and medicine.

l. 60. *art . . . short.* Attributed to Hippocrates, Greek physician.

## POEMS PUBLISHED IN *MISCELLANEOUS POEMS*, 1681

9 *A Dialogue, between the Resolved Soul and Created Pleasure.* Mary Palmer or her friends set this poem at the head of *1681*. It could well have been written as a brief cantata. The scheme of the poem—the rejection of each sense, progressing upwards, and then of the temptations of sex, wealth,

and forbidden knowledge—is the scheme used in the temptation of Christ in *Paradise Regained* (1671), book 4.

*Resolved*. Resolute, free from doubt and uncertainty; perhaps with a pun on the musical sense—(having) passed from discord to concord.

ll. 2–4. *shield . . . helmet . . . sword*. Ephesians 6: 11, 13, 16, 17: 'Put on the whole armour of God, that ye may be able to stand against the wiles of the devil . . . Wherefore take unto you the whole armour of God, that ye may be able to withstand in the evil day, and having done all, to stand . . . Above all, taking the shield of faith . . . and take the helmet of salvation, and the sword of the Spirit, which is the word of God.' The traditional equipment of the *miles Christi*.

ll. 5–6. *army . . . banners*. Song of Solomon 6: 4, 10.

ll. 15–16. *Where . . . yours*. Where the essences of fruits and flowers will be ready to stimulate your (lower or sensitive) soul.

l. 18. *bait*. Take refreshment during a pause on a journey.

l. 21. *plain*. Flat or evenly.

l. 22. *Lest one leaf . . . strain*. This line refers to the legend of the Sybarite king Mindyrides whose comfort was impaired by resting on crumpled rose-petals. Seneca, *de Ira*, II. XXV. 2.

10 l. 34. *crystal*. Looking-glass.

l. 36. *but*. Only.

l. 39. *posting*. Hastening.

l. 44. *chordage*. Puns on 'cordage' and 'chord' in its sense of 'concord', perhaps remembering a famous passage in which St Augustine wrote that the pleasures of the ear had more firmly bound and subdued him than those of the lower senses (*voluptates aurium tenacius me implicaverant et subiugaverant*). William Empson notes that the pun is 'exquisitely pointed, especially in that most cords are weaker than *chains*, so that the statement is paradox, and these *chords* are impalpable, so that it is hyperbole'. See *Seven Types of Ambiguity* (1930), ch. 3.

ll. 45–8. *Earth . . . delight*. Cf. Seneca, *Ecce spectaculum dignum ad quod respiciat Deus . . . vir fortis cum mala fortuna compositus* (L. N. Wall, *Notes and Queries* (1961), 185–6). *fence*. Ward off.

l. 53. *within one beauty*. The painter Zeuxis, believing that in nature all possible beauties could not be found in one body (*uno in corpore*: Cicero, *de Inventione*, ii), used many models for his portrait of Helen. The theme was often moralized. Cf. Cowley, 'The Soul'.

11 l. 69. *each hidden cause*. The origins of natural phenomena.

l. 71. *centre*. Of the earth.

l. 73. *degree*. (1) scale; (2) ladder (with pun on academic sense).

l. 76. *not . . . more*. Cf. Luke 4: 13, when Jesus has overcome *all* the temptations.

12 *On a Drop of Dew*. Marvell wrote a Latin version entitled 'Ros' ('dew'). It is not known which came first.

l. 1. *orient*. Sparkling as a pearl. ('Orient' pearls came from the Indian ocean and had more lustre than the European variety.)

l. 3. *blowing*. Blossoming.

l. 5. *For*. Because of.

l. 6. *encloses*. Closes in (intransitive), or '[it] encloses in itself . . .'.

l. 8. *as*. So far as. *native element*. The heavens.

l. 19. *So the soul*. The association between manna, dew, and grace is ancient, as is the notion that each soul is a small part of the divine.

l. 23. *swart*. Dark (conjectured by Sparrow); 'sweat' *1681*, Margoliouth.

l. 24. *recollecting*. Collecting again; remembering.

l. 26. *heaven less*. A lesser heaven.

l. 27. *coy*. Modest.

l. 29. *the world excluding round*. Thus shutting out the world on every side.

l. 34. *girt*. Prepared for action.

13 l. 39. *dissolving*. Exodus 16: 21.

*The Coronet*

l. 7. *towers*. Women's tall head-dresses.

l. 11. *chaplet*. Coronet.

l. 14. *twining in*. Entwining.

l. 16. *wreaths*. Coils.

l. 19. *thou*. Christ.

l. 22. *curious*. Elaborately wrought. *frame*. Structure (the chaplet).

l. 26. *feet*. When the serpent's head is bruised by Christ, he will, in the process, destroy the carefully made, but worthless, poems offered him by the penitent; thus what could not serve to crown his head will, accidentally, crown his feet.

14 *Eyes and Tears*. There was a contemporary vogue for conceited poems about tears, especially those of Mary Magdalene; cf. Crashaw, 'The Weeper'. Cf. also Shakespeare, *Venus and Adonis*, ll. 962–3.

l. 3. *That . . . vain*. Cf. Ecclesiastes 1: 14.

ll. 5–6. *the . . . height*. The sight, being liable to self-induced error, makes wrong estimates of height by misjudging angles.

ll. 21–4. *So . . . pours*. The sun, like an alchemist, extracts each day the essence of the world, but it turns out to be moisture, which is then poured back.

15 ll. 29–32. *So Magdalen . . . feet*. *1681* prints a Latin version of this stanza.

15 l. 29. *So Magdalen . . . wise.* See Luke 7: 37–8.

l. 35. *Cynthia teeming.* Full moon.

16 l. 55. *each . . . bears.* The characteristics of each are transferred to the other.

*Bermudas.* John Oxenbridge, in whose house at Eton Marvell lived in 1653, had twice been to the Bermudas, first in 1635 as a result of the ecclesiastical persecutions. The islands, thus associated with Puritan exile, had for long been celebrated as a kind of earthly paradise, and the traditional attributes of that place associated with them. They were sometimes known as the Somers Islands, in memory of the captain whose wreck there gave rise to pamphlets on which Shakespeare drew in *The Tempest.* These already suggest that the islands were a paradise, with continual spring and summer together, rich in natural resources and reserved for the godly English. All the traditional lore of the earthly paradise became attached to the Indies, especially Bermuda; see, for instance, Ralegh's *History of the World*, H. R. Patch, *The Otherworld according to Descriptions in Medieval Literature* (Cambridge, Mass., 1950), Arthur O. Lovejoy *et al.*, *Primitivism and Related Ideas in Antiquity* (Baltimore, 1935), and J. W. Bennett, 'Britain among the Fortunate Isles', *Studies in Philology*, 53 (1956). Leishman discusses the well-known parallels with Waller's poem 'The Battle of the Summer Islands', published in 1645, and see also Rosalie L. Colie, 'Marvell's "Bermudas" and the Puritan Paradise', *Renaissance News*, 10 (1957).

To summarize, the conventional descriptions of the earthly paradise, with a few specific details added, had been very generally applied to the Bermudas from the time of John Smith's *General History of Virginia the Summer Isles, and New England* (1624), if not from that of the pamphlets used by Shakespeare.

l. 1. *ride.* Like a ship at anchor.

l. 7. *so long unknown.* The islands were discovered by Juan Bermudez in 1515.

ll. 9–16. *Where he . . . the air.* As in the *Benedicite*, the whales precede the fowls (Craze, 269).

l. 9. *sea-monsters.* Waller had written of a battle between Bermudans and stranded whales.

l. 15. *fowls.* As the ravens were sent to succour Elijah (1 Kings 17: 6), William Strachey, in his *True Reportory of the Wreck* (1610), published in *Purchas's Pilgrims* (1625), describes at length how one easily caught a kind of bird something like a plover and 'fat and full as a partridge'.

l. 20. *Ormus.* Hormuz on the Gulf of Iran.

ll. 21–2. *He makes . . . our feet.* Cf. 'The Garden', stanza 5.

l. 23. *apples plants.* i.e. plants apples.

l. 28. *Proclaim.* Make manifest. *ambergris.* Fragrant substance excreted by sperm whales.

17 ll. 35–6. *Which thence . . . Mexique Bay*. Cf. 'First Anniversary', ll. 345–72 and l. 381. *beyond . . . Bay*. In order, presumably, to convert the heathen and the papist in South America.

ll. 37–40. *Thus sung . . . time*. The 'story of Richard More's party landing in Bermuda in 1612 contains the description of men rowing and the singing of a Psalm of thankfulness' (Margoliouth, quoting Douglas Bush).

*Clorinda and Damon*

l. 3. *scutcheon*. Heraldic arms.

l. 4. *Flora*. Goddess of flowers.

l. 7. *Grass . . . fade*. Isaiah 40: 8.

l. 8. *vade*. Pass away, decay.

l. 9. *cave*. Perhaps recalling the cave (*spelunca*) which Dido entered with Aeneas (*Aeneid*, iv. 124).

l. 10. *den*. Stanyhurst (1582) translates *spelunca* by 'den'.

l. 20. *Pan*. Christ, as often in pastoral poetry.

18 l. 23. *oat*. Flute.

*A Dialogue between the Soul and Body*. A note in the annotated Bodleian copy (Bodleian MS), '*Desunt multa*' (much is missing), suggests this poem is incomplete. The second speech of the Body has four extra lines, which may have survived from a third rejoinder to the Soul. (In the Bodleian MS the lines are scored through.)

The debate between soul and body is a form with a long history, and seems at this time to have had a minor revival. The conflicts and paradoxes of the debate are inherent in Galatians 5: 17: 'For the flesh lusteth against the spirit, and the spirit against the flesh: and these are contrary the one to the other.' In a Latin poem surviving in a thirteenth-century manuscript in the library of Corpus Christi College, Cambridge, Caro (the flesh) argues for sensual pleasure as natural and desirable, while Spiritus stresses the long and unpleasant consequences of this course. See F. J. E. Raby, *Secular Latin Poetry* (Oxford, 1934). Leishman gives several examples from English seventeeth-century poetry, none of them very closely resembling Marvell's strong paradoxical manner. K. S. Datta, 'New Light on Marvell's "A Dialogue . . . " ', *Renaissance Quarterly*, 22 (1969), stresses the relation between Marvell's poem and an emblem in Hermann Hugo's *Pia Desideria* (1624), citing relevant passages in Plutarch's *Moralia*. She mentions as a notable predecessor Francis Davison's 'A Dialogue between the Soul and Body' in *Poetical Rhapsody* (1602–21). See further Rosalie Osmond, 'Body and Soul Dialogues in the Seventeenth Century', *English Literary Renaissance*, 4 (1974).

Rosalie Colie observes that Marvell is, in this poem as elsewhere, expanding and exhausting a genre; he breaks with precedent in presenting soul and body not in opposition, but as reluctantly interdependent ('*My Ecchoing Song*', 1970, 57).

18 ll. 1–10. *O, who shall . . . double heart.* Borrowed from Hermann Hugo's emblem book *Pia Desideria* (1624), as in *'Pes compes, manicaeque manus, nervique catenae/Ossaque cancellis nexa catasta suis'* ('feet fetters, hands manacles, nerves chains, bones a cage for showing off a slave in the market, bound together with its own lattice-work of bars').

ll. 3–4. *bolts . . . hands.* The violent antithesis between the function of the various organs from the point of view of soul and body is expressed by an assonance, alliterations delayed and given point by the line-ending, and finally by an exploitation of the Latin root (*manus:* hand) of the verb, which represents hand as fetters. This prepares for the flatter paradox concerning the eye, and for the further paradox (reinforced by onomatopoeia) concerning the ear.

l. 10. *vain . . . heart.* So far, the organs have been literally treated, as the physical agents of the soul's imprisonment. A characteristic swerve of wit now uses *head* and *heart* ambiguously; they are still parts of the anatomy for the soul's prison, but carry also the figurative senses of egotism and treachery.

ll. 11-12. *O, who shall . . . tyrannic soul.* Cf. Romans 7: 24.

l. 13. *impales.* The Body is also a prisoner, held erect and penetrated by the animal function of the Soul, given a life that is a mere preparation for death, and as if possessed by a demon.

l. 14. *precipice.* Cf. 'after he [Marvell's antagonist Parker] was stretched to such a height in his own fancy, that he could not look down from top to toe but that he eyes dazzled to the precipice of his stature', *Rehearsal Transprosed*, i (1672), 64. Marvell enjoyed the word; see 'Appleton House', l. 375.

l. 15. *needless.* Unnecessary; or, having not need of it.

l. 16. *A fever.* Perhaps a reminiscence of Donne's 'A Fever'; the disease stimulating the corrupt flow of animal spirits.

19 l. 21. *magic.* Referring to the common magical torment of enclosing familiar spirits in trees.

ll. 23–4. *whatsoever . . . pain.* 'It' is the body. Whatever complaint *it* has, *I* experience the pain of, although (except in so far as I animate a body) I am immaterial and so impervious to pain.

ll. 25–6. *all . . . destroys.* I am obliged to devote myself to the preservation of a body whose health is directly opposed to my interests.

l. 29. *the port.* Death.

l. 30. *shipwrecked.* The paradox of the happy shipwreck is employed in Shakespeare's *The Tempest* and its sources; also by Crashaw in a Latin epigram *'Ad Bethesdae piscinam positus'* which Marvell probably knew. There is a French seventeenth-century ballet called *Ballet du naufrage heureux* by Claude de l'Estoille.

l. 32. *maladies.* Here spiritual ills are rendered in physical terms, as befits the body.

19 ll. 43–4. *So . . . grew.* The emphasis is on the perversity of the soul in altering what was natural (indifferent to sin) into something which could, like a building, be occupied by it.

*The Nymph complaining for the Death of her Fawn.* Poems on the death of a pet have a long history (see e.g. Ovid's poem on Corinna's parrot, *Amores*, ii. 6, and Catullus's on Lesbia's sparrow, 2), and were common in European poetry of this period including English poetry. A fawn is lamented in *Aeneid*, VII. 475 ff.—the lament starts a war—and a hind in the *Punica* of Silius Italicus (13), where the wanton killing again gives rise to conflict. A stag is mourned in Ovid, *Metamorphoses*, X. 106.

l. 1. *troopers.* The word came into use about 1640, and was first applied to soldiers in the Covenanting army, but soon became associated with Cromwell's men.

20 l. 13. *so.* Forgotten, unavenged by heaven.

l. 17. *deodands.* Any chattel that caused the death of a man was forfeit under law to the king for pious uses; the nymph is saying that if men do not kill beasts with justice, this rule ought to apply equally to them. They would thus be deodands.

ll. 23–4. *There is . . . sin.* The troopers have killed the only thing that could have redeemed the sin they committed in doing so.

ll. 32–6. dear . . . *heart.* These very old pastoral puns are here used as a contribution to the sophisticated naïveté of the tone.

21 l. 70. *four.* Disyllabic, as in 'Appleton House', l. 323.

22 ll. 71–92: *I have a garden . . . roses within.* Cf. The Song of Solomon 2: 8–9, 16–17; 5: 1, 10, 13, 16; 6: 2–3; 3: 1. Like the Song of Solomon, 'The Nymph complaining' lacks continuous or systematic allegory, but lends itself to shifting meanings in an allegorical field.

l. 97. *balsam.* Applied both to the balsam-tree and its resin.

ll. 99–100. *Heliades . . . tears.* The three daughters of Helios, sisters of Phaeton, after mourning their brother's death, were turned into amber-dropping trees. Ovid, *Metamorphoses*, ii. 364–5.

l. 104. *Diana's shrine.* Diana was goddess of both chastity and hunting.

l. 106. *turtles.* Doves.

l. 116. *stone.* The nymph imagines her fate as parallel to that of Niobe who, lamenting the death of all her children, was turned to stone. *Metamorphoses*, VI. 146–317.

l. 119. *There. 1681*; 'Then' conjectured by E. E. Duncan-Jones.

23 *Young Love*

l. 6. *beguiled.* (1) charmed; and (2) deceived.

23 l. 9. *stay*. Wait (till they are fifteen).

l. 11. *green*. Immature.

l. 21. *of*. Over.

l. 23. *antedate*. Anticipate.

24 *To his coy Mistress*. This poem draws upon ancient lyric traditions—the persuasion of a young woman to 'seize the day', the *blason* or catalogue of charms, etc., though the irony and force of the lover's arguments are highly original. The structure of the poem has often been compared to a syllogism: for a comment on its logic see R. I. V. Hodge, *Foreshortened Time* (Cambridge, 1978), 22–4.

l. 8. *ten years before the flood*. i.e. quite near the beginning of time.

l. 10. *conversion of the Jews*. This was expected during the Last Days, near the end of time. There was contemporary interest in the expectation and Cromwell allowed the Jews back into England after centuries of exclusion to allow it to happen.

ll. 13–18. *A hundred . . . heart*. The hyperbole derives from the old *blason*, or catalogue of a mistress's beauties, and is anticipated by Cowley in 'My Diet', *The Mistress* (1647), stanza 3:

> On a sigh of pity I a year can live,
> One tear will keep me twenty at least,
> Fifty a gentle look will give,
> A hundred years on one kind word I'll feast:
> A thousand more will added be,
> If you an inclination have for me;
> And all beyond is vast eternity.

l. 22. *Time's . . . chariot*. Time is often represented as winged, and often has a chariot; but the conflation appears to be Marvell's own.

ll. 25–9. *Thy beauty . . . dust*. The theme is as old as *The Greek Anthology*: 'You would keep your virginity? What will it profit you? It is among the living that we taste the joys of Venus. You will find no lover in Hades, girl. In Acheron, child, we shall only be bones and dust.' Asklepiades, tr. Forrest Reid, *Poems from the Greek Anthology* (1943), 23.

25 l. 29. *quaint honour*. While this is perfectly intelligible as a figurative expression, abstract for concrete, Marvell is punning on other than the more common senses of the words, each of which is, in the English of the time, used concretely to mean the female pudenda.

ll. 33–4. *youthful hue . . . morning dew*. Margoliouth; 'youthful hue . . . morning glue (glew)' *1681*; 'youthful glue (glew) . . . morning dew' Bodleian MS. The equivalent couplet in Bodleian MS Don. b. 8 reads:

> Now then whil'st ye youthfull Glue
> Stickes on your Cheeke, like Morning Dew.

The reading 'glue' or 'glew' is attractive, however. As Wilcher says, from its presence in the three earliest texts (the Haward MS [Bodleian Don.

b. 8], *1681*, and the Bodleian MS [Eng. Poet. d. 49]), it looks as if the word 'glue' figured in Marvell's poem at some stage of its composition. In Shakespearian English, 'hue' means not only 'colour' but 'appearance' and 'complexion'. *OED* gives an example from 1653.

25 ll. 35–6. *willing . . . fires*. Despite her professed coyness, her amorous spirit shows in her flushed face; it breathes through every pore. Donne uses the same idea in a different context when he says of Elizabeth Drury that 'her pure and eloquent blood/Spoke in her cheeks' ('The Second Anniversary', ll. 244–5).

l. 40. *slow-chapped power*. The power of his slowly devouring jaws (*sub . . . lentis maxillis*, Suetonius, *Tiberius*, 21).

ll. 41–6. *Let . . . run*. There has been much discussion of the significance of these images. They are certainly discontinuous, and perhaps derive urgency from that. Margoliouth thinks that in l. 42 Marvell is thinking of a pomander, and in l. 44, 'where the sexual strife is waged', the 'gates of life' suggest the narrow reach known as the Iron Gate, which separates the upper from the lower Danube. Or: Alexander was supposed to have built vast gates to hold back the Scythians—the tribes of Gog and Magog—behind the Urals; at the coming of Antichrist these tribes would break through and this would herald the end of history.

Whether or not any of this is plausible, the lines certainly refer to the act of defloration. The final couplet, as Christopher Hill first suggested (*Puritanism and Revolution*, 1958, 347, n. 1), recommends that the lovers should not imitate Joshua's sun, which stood still, but David's, which came forth like a bridegroom to run his race. Cf. 'The First Anniversary', ll. 7–8:

> Cromwell alone with greater vigour runs,
> (Sun-like) the stages of succeeding suns . . .

Cromwell resembles the lover in other ways: he could 'ruin the great work time' ('Horatian Ode', l. 34) and it is also he who 'the force of scattered time contracts, /And in one year the work of ages acts' ('First Anniversary', ll. 13–14).

l. 44. *Thorough*. Through. *gates*. *1681*; 'grates' Bodleian MS; 'grates' is preferred by some modern editors, but as Wilcher points out, 'gates of life' is an unexpected, and Marvellian, variant of 'gates of death'.

*The unfortunate Lover*. The imagery, which is an elaborate distortion of familiar Petrarchan conceits (sighs=gales, tears=seas, etc.), bears some resemblance to a series of emblems depicting a lover's suffering in Otto van Veen, *Amorum Emblemata* (1608). R. L. Colie comments on the emblematic elements in *'My Ecchoing Song'*, 109-13. See also Annabel M. Patterson, *Marvell and the Civic Crown*, 20 ff. Margarita Stocker argues at length that the poem is a historical allegory concerning Charles I. See also P. R. K. Davidson and A. K. Jones, 'New Light on Marvell's "The Unfortunate Lover"?', *Notes and Queries* (June 1985), 170–2, another political reading. Elsie Duncan-Jones gives an unpolitical reading in Reuben Brower *et al.*, *I. A. Richards: Essays in his Honor* (New York, 1973).

Cf. also Lovelace's 'Against the Love of Great Ones'.

25 ll. 6–8. *meteors . . . time.* Meteors were thought to be exhalations of vapours from the interior of the earth; they ascended to the sphere of fire and burned out. If they could pass the sphere of fire and the moon they would reach a region of incorruptibility and timelessness.

l. 9. *shipwreck.* The lover's birth, by Caesarian section, is represented as a shipwreck.

26 l. 16. *section.* Three syllables.

ll. 17–19. *tears . . . winds.* The attributes of the Petrarchan lover are explained here by the circumstances of his birth.

l. 22. *forked.* Two syllables.

l. 26. *masque.* Show, representation.

l. 36. *bill.* Peck.

l. 38. *consumed.* Three syllables.

l. 39. *languished.* Three syllables.

l. 40. *amphibium.* A being that lives equally well in water and on land. (The stress is on the third syllable.)

l. 44. *At sharp.* With unbated weapons.

27 l. 48. *Ajax.* Son of Oileus, shipwrecked, stranded, and destroyed by the gods (*Aeneid*, i. 41–5).

ll. 55–6. *And all . . . relish best.* 'Some commentators take the last line and a half to be the words spoken by the Lover. Donno glosses "says" as a shortened form of "assays". Less violence is done to the text by taking "all he says" as the object of "relish"; the couplet can then be paraphrased: "A lover covered in his own blood can best appreciate everything that this unfortunate lover says" ' (Wilcher).

l. 57. *banneret.* One created knight on the battlefield. Probably from Lovelace, 'Dialogue—Lucasta, Alexis', in *Lucasta* (1649), to which Marvell contributed commendatory verses (see above, pp. 4–5):

> Love near his standard when his host he sets,
> Creates alone fresh-bleeding bannerets.

After the battle of Edgehill (1642) the word 'must have taken on new life' (Duncan-Jones).

l. 60. *Forced.* Disyllabic.

ll. 61–2. *Yet dying . . . ear.* Cf. Ecclesiasticus 49: 1: 'The remembrance of Josias is like the composition of the perfume that is made by the art of the apothecary . . . and as music at a banquet of wine' (Duncan-Jones).

l. 63. *And he . . . rules.* 'This may mean either "he is supreme in the world of fiction [the romances of chivalry]" or "it is only in the world of fiction that he rules". However, "story" might mean "a painting . . . representing a historical subject. Hence any work of pictorial . . . art containing figures" (*OED*; and cf. "The loyal Scot", l. 171). This would agree better with the next line, and confirm the emblematic character of the poem' (Legouis).

27 l. 64. *In . . . gules.* Heraldic terms: a red lover in a black field. 'The same contrast of heraldic colours on a blood-stained warrior is found in *Hamlet*, II. ii. 437–42: Pyrrhus' arms are sable before the massacre; after it he is "total gules" ' (Duncan-Jones). Patterson cites a similar figure in Cleveland's *The Character of a London Diurnal* (1647).

*The Gallery.* Poems describing real or imaginary galleries occur among Marino and his followers, Italian and French, but Marvell uses the trope in a wholly individual way. L. N. Wall (*Notes and Queries*, 202, 170) suggests a debt to Lovelace's 'Amyntor's Grove'. The relation of Marvell's 'paintings' to actual Baroque pictures is considered in Charles H. Hinnant, 'Marvell's Gallery of Art', *Renaissance Quarterly*, 24 (1971).

ll. 7–8. *for . . . mind.* You will find that the mental gallery contains only my pictures of you (all other 'furniture' having been put away).

l. 11. *Examining.* Testing.

28 l. 12. *shop.* Business equipment, tools, instruments; '[work]-shopful' (Legouis).

l. 13. *Engines.* Instruments (of torture).

l. 14. *cabinet.* Picture gallery.

l. 24. *perfecting.* Stressed on first syllable.

l. 27. *light obscure.* 'A witty glance at the technical term (*chiaroscuro*) for the disposition of brighter and darker masses in a picture' (Wilcher).

l. 35. *halcyons.* These birds were thought to nest on the surface of the water and were thus regarded as ensuring calm at their nesting season.

l. 38. *ambergris.* See 'Bermudas', l. 28.

l. 40. *smell.* Sense of smell.

29 l. 42. *does.* Grierson; 'dost' *1681*; 'do' Bodleian MS.

l. 48. *Whitehall's . . . were.* Charles I had made a great collection of paintings, partly by purchasing those of Vincenzo Gonzaga, Duke of Mantua. The collection was ordered to be sold after the king's death by an act of Parliament (1650). It is argued that Marvell must have written the poem or altered this and the preceding line after 1650. But 'were' may be subjunctive; and in any case the alteration ('are' to 'were') is an easy one.

*The Fair Singer.* A common theme of the period, occurring in Góngora, the Marinisti, the French poet Voiture ('Sur une belle voix'), and in Carew, Lovelace, Cowley, Stanley, Waller, and Milton.

l. 9. *curled.* Disyllabic.

30 l. 18. *gained.* Disyllabic.

*Mourning*

l. 1. *You . . . the fate.* i.e. astrologers.

l. 3. *infants.* Tears; but some editors think 'babies', in the common sense of the image as reflected in the mistress's eyes.

30 l. 9. *moulding of.* 'Taking their shape from the moist spheres (of her eyes)' (Wilcher).

l. 20. *Danae.* She was visited by Jupiter in a shower of gold.

31 l. 27. *donatives.* Largesses 'at the installing of a new' emperor.

l. 29. *wide.* Inaccurately.

*Daphnis and Chloe.* This poem seems to be related to some lines from Suckling's play *Aglaura* (III. i.) where Aglaura asks Thersames not to consummate their marriage at a critical moment in his life:

> Gather not roses in a wet and frowning hour,
> They'll lose their sweets then, trust me they will, sir,
> What pleasure can love take to play his game out,
> When death must keep the stakes?

(E. E. Duncan-Jones, reported in Leishman, p. 121.) The title is that of the pastoral romance of Longus. The stanza is that of Shakespeare's 'The Phoenix and Turtle', and of Carew's 'Separation of Lovers'.

32 l. 12. *comprised.* Included as a condition.

l. 13. *does use.* Is accustomed.

l. 15. *separate.* Withdraw from conjugal cohabitation.

l. 27. *But . . . more.* But at this point they were mere legacies.

33 l. 42. *loved.* Disyllabic.

l. 44. *resolved.* Trisyllabic.

l. 53. *condemned.* Trisyllabic.

l. 61. *alone.* By itself.

34 l. 65. *enrich my fate.* By having had her.

l. 78. *gourmand Hebrew.* Numbers 11: 33, where Jehovah having provided quails and manna strikes the Israelites 'with a very great plague'—presumably as a punishment for eating greedily. (While still wandering in the desert he is due to die.)

l. 79. *with.* Cooke; 'he' *1681*.

l. 80. *He.* Cooke; 'And' *1681*.

l. 83. *seed.* Ferns have no seed; it was once thought that they had, but that the seed was invisible and could confer invisibility on anyone who contrived to get hold of it. ('We have the recipe of fernseed, we walk invisible', *1 Henry IV*, II. i. 96.)

35 *The Definition of Love.* The title suggests that 'The Definition of Love' belongs to the genre so named, but it is very unlike the other poems which belong to that genre (the most accessible is probably Ralegh's 'Now what is Love') and seems to resemble another kind of poem, which develops the topic 'in love despair is nobler than hope'. Stampa, a follower of Marino, has a poem called 'Amante che si pregia di non avere alcuna speranze': 'a noble heart thinks its excellence diminished if hope intrudes its flattering

foot to reduce the ardent flames—base comfort of common minds, depart! He who asks relief values little his gentle torments.' The *Chanson* in which Desportes writes a somewhat similar argument also represents the nobly hopeless love as spendidly contrary to nature, and is that much closer to Marvell. In Marvell's poem, all the dispositions of fate, including the structure of the world, must be altered before the lovers may be joined. John Carey points out that the Latin *definire* means not only 'define', but also 'restrict' or 'limit' (in Patrides).

36 l. 10. *extended soul*. His soul resides in his mistress, not in him.

l. 24. *planisphere*. Astrolabe, in which the poles are 'clapt flat together' in the example of 1594 in *OED*. There is a possibility (D. M. Schmitter 'The Cartography of "The Definition of Love" ', *Review of English Studies*, 12 (1961), 49–51, and P. Legouis's partly dissenting comment, 51–4) that the figure is terrestrial rather than celestial. Thus the poles would be terrestrial, the planisphere a crushed globe, with the lines of latitude parallel, and those of longitude meeting at the poles. Ann E. Berthoff (*RES*, 17 (1966), 21–5) argues strongly for the celestial interpretation.

ll. 31–2. *conjunction . . . opposition*. Astrological terms, the first borrowed for the spiritual union of the lovers, the second also at a remove (the opposition is not between the stars themselves but between the stars and the lovers).

37 *The Picture of Little T. C. in a Prospect of Flowers*. For an elaborate study of this poem, see Joseph H. Summers, 'Marvell's "Nature" ', *English Literary History*, 20 (1953), 121–35. Margoliouth suggested that T. C. was Theophila Cornewall, born 1644. A year earlier, the same parents had a child, also christened Theophila, who died two days old. If, as Maren-Sofie Røsvig suggests (*Huntington Library Quarterly*, 18 (1954–5), 13 ff.), Marvell borrows from Benlowes' *Theophila*, and if E. E. Duncan-Jones is right in her subsequent conjecture (*Huntington Library Quarterly*, 20 (1955–6), 183–4) that in this title he takes over the literal sense of Benlowes (see note to l. 10), this poem is presumably later than 1652, the date of publication of Benlowes' book. Leishman, at considerable length (pp. 165-89), traces the tradition of sub-amorous addresses to the prepubescent from the Greek Anthology through Homer to the poetry of Marvell's century, and to Prior and Phillips. Cf. 'Young Love' (above, p. 23).

l. 5. *gives them names*. A task traditionally attributed to Eve in Eden.

l. 10. *darling of the gods*. Theophila means 'dear to the gods'.

l. 17. *in time*. In good time.

l. 22. *but*. Only.

38 *The Match*. [Title]. 'The word *match* . . . afforded multiple meanings in seventeenth-century idiom: antagonist, counterpart, equal, contest, pairing, alliance, and, aptly for the second half of the poem, the wick used to ignite gunpowder . . . ' (Donno).

39 l. 19. *magazine*. Arsenal.

l. 29. *vicinity*. Propinquity.

40 *The Mower against Gardens*. The theme of complaint against gardens as wanton human perversion of nature is ancient. There is a rhetorical exercise reported by Seneca the Elder which complains that great houses include streams and woods—*mentita nemora*, fake or 'enforced' (cf. l. 31) groves—within the buildings; that their owners prefer imitations to the real thing and hate what is natural. The art of gardening (grafting, budding, etc.) could be represented as encouraging a sort of botanical adultery; Pliny said so of a fabulous tree of many fruits. The argument as to whether interference with nature is benign was a set piece, a famous version being the debate between Polixenes and Perdita in *A Winter's Tale*, IV. The opposite view of the matter is taken in Thomas Randolph's 'Upon Love fondly refused for Conscience' Sake', a poem in the same rather unusual 'epode' manner—alternating decasyllabic and octosyllabic lines—which uses horticultural 'inoculation' as an argument in favour of fornication. The horticulture of the poem, especially as relating to ll. 27–30, has been much discussed, e.g. by Bradbrook and Thomas, Nicholas A. Salerno, *Études Anglaises*, 21 (1968), R. Wilcher, *Études Anglaises*, 23 (1970), and by John Carey in Patrides (ed.) *Approaches to Marvell*.

l. 1. *Luxurious*. Sinful, lecherous. *bring . . . use*. To reap interest on his vice, to make it profitable (Kermode); spread his vice to other creatures and make it the universal custom (Margoliouth).

l. 6. *standing pool of air*. The phrase is used in books by Henry Wotton (1624) and James Howell (1642).

l. 7. *luscious*. (1) cloying; (2) voluptuous.

l. 8. *stupefied*. (1) astounded; (2) benumbed.

l. 15. *onion root*. Bulb. *did hold*. Was valued.

l. 16. *one . . . sold*. Tulip bulbs were sold by weight in Holland during the 1630s; one cost 5,500 florins, or as much as 550 sheep (Margoliouth). See Simon Schama, *The Embarrassment of Riches* (1987), 350 ff.

l. 18. marvel of Peru. A tropical American plant, *mirabilis jalapa*, called by the botanist Parkinson *mirabilis peruviana* 'the marvel of Peru'.

l. 21. *dealt between*. Tilley records the expression as proverbial for interference between man and wife. It can also mean 'pandered for'.

l. 22. *Forbidden mixtures*. 'Thou shalt not sow thy vineyard with divers seeds: lest the fruit of thy seed, which thou hast sown, and the fruit of thy vineyard, be defiled' (Deuteronomy 22: 9).

l. 24. *tame*. Cultivated.

l. 25. *uncertain . . . fruit. Pirus invito stipite mala tulit* ('the pear bore apples from its unwilling stock'), Propertius, IV. ii. 18.

l. 30. *to procreate without a sex*. There is a long-running argument about the sense here (see references in headnote). But it seems that the cherries are made eunuchs, deprived of their stones. William Harrison (1581) credits gardeners with the power of 'bereaving . . . some . . . fruits of their kernels', and Evelyn, in Marvell's time, says something similar. Carey quotes a passage from Gervase Markham's *The Country Farm* (1616), which explains that stoneless cherries are the result of a graft uniting a young cherry with a barren cherry. Of course this would not be so; but the evidence for the interpretation 'stoneless', though still contested, is strong.

41 *Damon the Mower*. The archetypes of such rustic love-complaints are Theocritus, *Idyll* xi, and Virgil, *Eclogue* ii. The mower was socially the lowest of agricultural workers, and it was unusual to substitute him for the shepherd.

l. 2. *Juliana*. i.e. Gillian.

ll. 3–4. *paint . . . scene*. A theatrical figure.

l. 12. *hamstringed*. Lamed (by the heat).

ll. 19–28. *It from . . . bend*. This description of hot weather is developed from Virgil, *Eclogue* ii.

l. 21. *mads*. Bodleian MS; 'made' *1681*, Margoliouth.

l. 22. *Phaeton*. The charioteer of the sun, who was unable to control it.

l. 28. *gelid*. Frozen.

42 ll. 35–40. *To . . . brought*. The pastoral tradition of such gifts as inducements of love begins, as Leishman points out, with those of Polyphemus to Galatea in *Idyll* xi of Theocritus, elaborated by Ovid in his Polyphemus-Galatea passage, *Metamorphoses*, xiii. 798 ff., and best known from Marlowe's 'Passionate Shepherd to his Love'. There is a particularly charming gift-catalogue in Richard Barnfield's 'Second Day's Lamentation of the Affectionate Shepherd' (1594).

l. 48. *cowslip-water*. Used by women to cleanse the skin.

ll. 49–56. *What . . . hay*. The Mower is something of a novelty in pastoral, but the rivalry between rustics of different professions is not.

l. 53. *golden*. Unlike the sheep's, but resembling Jason's.

l. 54. *closes*. Enclosed fields.

l. 57. *Nor . . . sight*. Almost literally translated from Virgil's *Eclogue* ii. 25: *nec sum adeo informis*. Leishman rightly remarks that what is so characteristic of Marvell is to transform this by having the speaker look not into a calm sea (as Virgil's Corydon does) but into the curved and polished blade of his scythe.

43 l. 64. *ring*. The 'fairy ring', which is actually caused by the mycelium of certain fungi.

l. 83. *shepherd's purse. Capsella bursa-pastoris*, a weed supposed to check bleeding. *clown's-all-heal. Stachys palustris*, said to heal wounds.

*The Mower to the Glow-worms.*

43 l. 1. *lamps.* 'You will make me believe that glow-worms are lanterns'. This proverb (Tilley G 143) is cited by Kitty Scoular, *Natural Magic* (1965), 106.

44 ll. 7–8. *Shining . . . fall.* 'Glowbards never appear before hay is ripe upon the ground, nor yet after it is cut down' (Pliny, *Natural History*, XI. 28, tr. Philemon Holland; quoted by Leishman, 125). Kitty Scoular (107) points to an exact parallel in Remy Belleau, for whom the insect is also the countryman's prophet:

> *Qui au laboureur prophétise*
> *Qu'il faut pour faucher, qu'il aguise*
> *Sa faulx, et face les moissons.*

l. 9. *officious.* Attentive.

l. 12. *foolish fires. Ignis Fatuus*, will-o'-the-wisp; marsh-gas (methane) spontaneously ignited.

*The Mower's Song*

l. 1. *survey.* A coloured estate map (Craze).

ll. 3–4. *And in . . . glass.* Green was traditionally the colour of hope.

45 *Ametas and Thestylis making Hay-ropes.* Probably meant to be sung, like the other pastoral dialogues. Colie (54) has a good comment on this poem.

46 l. 10. *yourselves.* Bodleian MS.; 'yourself (your selve)' *1681*, Margoliouth.

*Music's Empire.* For the relation of this poem to the *laus musicae* tradition see James Hutton, 'Some English Poems in Praise of Music', *English Miscellany*, 2 (1951), and John Hollander, *The Untuning of the Sky: Ideas of Music in English Poetry, 1500–1700* (Princeton, 1961).

l. 5. *Jubal.* 'The father of all such as handle the harp and organ' (Genesis 4: 21).

l. 6. *jubilee.* (Jewish) year of emancipation and restoration; more generally, a season of joyful celebration. Did Marvell associate the ritual Jubilee with Jubal?

l. 7. *sullen.* Gloomy.

ll. 9–12. *Each sought . . . withdrew.* These lines may reflect Davenant's *Gondibert* (1651), II. vi. 80, which has 'virgin trebles' and 'manly voice' (Craze).

l. 9. *consort.* (1) mate; (2) harmony.

47 l. 16. *choir.* This word may be a verb (cf. *Merchant of Venice*, V. i. 62), as Craze argues. (Then, of course, we should read, 'heavens choir'.)

l. 17. *mosaic.* Presenting a unity made up of diverse sound.

l. 22. *conqueror.* Margoliouth suggests Fairfax, comparing the following line with 'Upon the Hill and Grove', ll. 75–6. Hollander argues for Cromwell.

*The Garden.* Like 'The Mower against Gardens', this poem is always confronting a silent set of antithetical attitudes, and from this derives its wit.

47 Like Adam, the poet is placed *in paradiso deliciarum*, in a paradise of delights, and like him he has a duty to contemplate them. The resemblance to Adam, we are told, is rather to the man alone, to the period before Eve's creation; the poet echoes St Ambrose's misogyny and Joseph Hall's 'I do not find that man, thus framed, found the want of a helper. His fruition of God gave him fulness of contentment.' This situation does not require to be explained in the language of hermeticism or alchemy; there is no hermaphrodite, no conflict between sense and mind, no code. The points are made within the large context of garden topics. The poet makes the green of the garden stand for solitude against crowds, retirement against action, sensual delight free of sexual pursuit, the satisfaction of the sense against that of the mind; it is not the green of hope, the *benedicta viriditas* of alchemy, the green of the hermetic emblem, but the poet's green. Within this shade, the poet galls the horsemen of the opposition. Perhaps Marvell was thinking of Randolph—the opening stanza of that poet's 'Pastoral Courtship', at any rate, introduces an antithetical garden, in which all trees combine to form, not the garland of repose, but a bower for lovemaking:

> Behold these woods, and mark, my sweet,
> How all the boughs together meet.
> The cedar his fair arms displays,
> And mixes branches with the bays.
> The lofty pine deigns to descend,
> And sturdy oaks do gentle bend,
> One with another subtly weaves
> Into one loom the various leaves,
> As all ambitious were to be
> Mine and my Phyllis' canopy!

Marvell's tone is always light. The second stanza plays wittily on the plants as virtues, a theme familiar in biblical commentary, and on the old paradox that solitude is more pleasantly companionable than company. By the same token, it is more amorous than love, a paradox stated in terms of the emblematic colours of solitude and love. Ever since Ovid's Oenone pastoral, lovers have cut the name of the desired person on trees; what we should do, it seems to follow, is to cut on trees their own names. Gods chase girls not for sex but to turn them into trees. And so the paradoxes continue. The figures of natural abundance are as old as Hesiod; like Adam, the solitary has easily what others must labour for; and unlike Adam, he may fall without being greatly upset. However, neither for him nor Adam is sensual repletion all; the garden provides no more than a mirror of creation, since it also enables the mind to withdraw from sensibilia and produce its own fantasies, establishing worlds other than the visible. Thus begins a formal garden ecstasy; but there is still an element of antithesis carried over from the earlier wit: this ascending love is traditionally, in the familiar Platonic formula, contrasted with that which descends to mere sensual contact. The soul, ascended, is as it were between the worlds, like a bird on a bough; the figure was used by Spenser. It is poised between the white light of eternity, and the varieties of colour that light assumes in the creation. This, we learn, was the position of Adam before Eve and the Fall; there is a newly witty reprise of the earlier antitheses on love. After this exercise comes a quiet close, but it is still constructed as an antithesis. Other sundials boast that they count only sunny hours, depending on the unmediated light of direct sunshine; this one, which is a new kind of sundial (he is calling the garden a sundial), reckons hours much more sweet and wholesome; its light is filtered through the greenness of trees and so is 'milder'. Its face consists not of figures cut in stone, but flowers, which yield to us, as to the bee, sweetness and light.

The tone and content of few poems in the English language have been so misunderstood; the first ignored and the second fantasticated in a manner more consonant with scholarly clumsiness than with poetic wit. It is a learned poem, in its way, but it has been packed with learned lumber; we have gained a heap of studious conjectures and almost lost a good poem.

The Latin version, 'Hortus', is not throughout a translation, nor yet an original, but partly an exercise in a related mode (see G. Williamson, 'Marvell's "Hortus" and "Garden" ' in *Milton and Others*, 1965).

l. 1. *vainly*. (1) futilely; (2) arrogantly (Donno).

l. 2. *the palm . . . bays*. Rewards for achievement in war, statesmanship, poetry.

l. 6. *upbraid*. (1) censure; (2) braid up (Donno).

l. 7. *all . . . all*. As opposed to the 'single' of l. 4; retirement offers greater rewards. *close*. Unite.

l. 14. *plants*. i.e. the plants symbolizing them, as palm, oak, and bays symbolize the various activities of l. 2.

ll. 15–16. *Society . . . solitude*. A paradox: ordinarily it is society that is thought of as 'polished'. The paradox goes back to Scipio's *nunquam minus solus quam cum solus*.

48 l. 17. *white . . . red*. Emblematic of female beauty.

l. 18. *amorous*. Beautiful, worthy of love. *green*. By association, emblematic of rural and solitary retirement.

ll. 19–20. *Fond lovers . . . name*. Ralph Austen, *Observations on Sir Francis Bacon's Natural History*, 1658, says one can inscribe young trees so that later 'the letters or figures will be more plain' (Colie, 159).

l. 24. *No . . . own*. If lovers behave thus to celebrate woman's beauty, it is logical—since the trees are more beautiful than the women—to carve only the tree's name on the tree. *your*. Cooke; 'you' *1681*.

l. 25. *heat*. (1) ardour; (2) race.

l. 26. *retreat*. (1) a military and (2) a religious figure.

l. 28. *Still*. Always.

ll. 29–32. *Apollo . . . reed*. In the myth, Apollo is thwarted when Daphne turns into a laurel, and Pan when Syrinx turns into a reed; Marvell inverts the myths to establish the 'amorous' superiority of trees to women.

48 ll. 33–40. *What . . . grass.* In this paradise, as in that of Adam, one is exposed to all sensual delight, but here one can be ensnared and fall without serious consequences. The catalogue of readily available fruit is a commonplace with a long history.

l. 41. *Meanwhile . . . less.* Experiencing less pleasure in nature than the delighted senses do (and so turning upwards). According to Aristotle, *de Anima*, I. iv, 'the mind is less subject to passion'. *pleasure. 1681*; 'pleasures' Bodleian MS.

l. 43. *that ocean.* Alluding to the belief that there is in the sea a parallel creation for everything on land. The implied theory of knowledge is that we can know the world because of the pre-existence of related forms in our minds.

l. 44. *straight.* At once.

ll. 45–6. *Yet . . . seas.* But the mind does more than merely provide such correspondences; the fancy or imagination can create forms which have no equivalent in reality.

ll. 47–8. *Annihilating . . . shade.* Making the created world seem as nothing compared with what can be imagined by the retired contemplative.

49 l. 49. *Here . . . foot.* Cf. 'Upon Appleton House', l. 645.

l. 54. *whets.* Preens.

l. 55. *And . . . flight.* Resting, as it were, between the created and the intelligible worlds in the process of its Platonic ascent; the same figure in a similar context is used by Spenser, 'Hymn of Heavenly Beauty', ll. 22–8.

l. 56. *various light.* The neo-Platonic image (familiar to all from Shelley's 'Adonais') was common enough in the Renaissance: see, e.g., Chapman, 'Ovid's Banquet of Sense', stanza 55.

ll. 57–8. *Such . . . mate.* The Garden of Eden was like this before the introduction of Eve; a point earlier made by St Ambrose (*Epistles*, i. 49).

l. 60. *What other . . . meet.* 'And the Lord God said, It is not good that the man should be alone; I will make a help meet for him' (Genesis, 2: 18). The words were later conflated giving 'helpmate'.

l. 66. *dial.* Sundial. We suspect that here the whole garden is meant; it is metaphorically a sundial; hence 'new'—a new version of the sundial.

ll. 67–8. *Where from . . . zodiac run.* The sun is made milder because in this 'floral zodiac' the sunlight is filtered through the trees, on to plants.

l. 70. *time.* A pun. The clock enables us to tell the time and enables the bee to take nectar from thyme. Marvell gets this point into his Latin version also.

*The second Chorus from Seneca's Tragedy 'Thyestes'.* Compare Cowley's translation of the same passage from *Several Discourses by Way of Essays, in Verse and Prose*, iii, 'Of Obscurity' (1668):

Upon the slippery tops of human state,
The guilded pinnacles of fate,

Let others proudly stand, and for a while
The giddy danger to beguile,
With joy, and with disdain look down on all,
Till their heads turn, and down they fall.
Me, O ye gods, on earth, or else so near
That I no fall to earth may fear,
And, O ye gods, at a good distance seat
From the long ruins of the great.
Here wrapped in the arms of quiet let me lie;
Quiet, companion of obscurity.
Here let my life, with as much silence slide
As time that measures it does glide.
Nor let my homily death embroidered be
With scutcheon or with elegy.
An old plebeian let me die,
Alas, all then are such as well as I.
To him, alas, to him, I fear,
The face of death will terrible appear:
Who in his life flattered his senseless pride
By being known to all the world beside,
Does not himself, when he is dying, know,
Nor what he is nor whither he's to go.

Sir Matthew Hale also has a version, in *Contemplations, Moral and Divine* (1676).

*Title*. Donno; 'Senec. Traged. ex Thyeste Chor. 2.' *1681*.

l. 2. *Tottering favour's pinnacle. 1681*; 'Giddy favour's slippery hill' Bodleian MS.

50 *An Epitaph upon Frances Jones*. The information on Frances Jones's epithet in the crypt of St Martin-in-the-Fields shows that Frances Jones was the daughter of Arthur Jones and Katherine Boyle, Lady Ranelagh, daughter of the Earl of Cork. Frances was born in 1633, and died in 1672. (See Hugh Brogan, 'Marvell's *Epitaph on —* ', *Renaissance Quarterly*, 32, 1979, 197–8.) Lady Ranelagh was a frequent visitor to Milton when he was living in Petty France, as, too, was Marvell.

*Title*. This edn.; 'An Epitaph on — ' *1681*.

l. 14. *came and went*. i.e. in prayer.

*Upon the Hill and Grove at Bilbrough*. For similar poems on mountains, see Kitty Scoular, *Natural Magic*, pp. 154 ff. In part an exercise in a manner much more fully developed in 'Upon Appleton House'. Bilbrough was a house of Fairfax's near Nun Appleton. Thomas Fairfax (1612–71), the third baron, had been commander-in-chief of the Parliamentary army. He refused to condone the execution of the king, and resigned in June 1650 because he disapproved of the proposed campaign against the Scots, which Cromwell undertook. Fairfax then retired to his Yorkshire properties, where he led the life of the great landowner who was also a scholar and a poet. He had married Ann Vere, who came of a distinguished military family, in 1637; she appears to have been an imposing woman of strong Presbyterian faith. In 1651 they appointed Marvell tutor to their daughter Mary, and he seems to have remained with them at Nun Appleton and Bilbrough for two years. Related to this poem is the Latin 'Epigramma in duos montes' printed before it in *1681*.

51 l. 5. *pencil*. Paintbrush. *draw*. Paint.

l. 7. *model*. i.e. in its perfect circularity.

l. 13. *For*. On account of.

l. 14. *new centre*. Because the irregularity of the mountains has made the earth imperfectly spherical.

l. 28. *Tenerife*. Volcanic peak in Canary Islands (12,192 ft). Bilbrough is 145 ft.

l. 29. *seamen*. They used the hill as a landmark for entering the Humber.

l. 34. *plume*. Cooke; it means 'crest' here; 'plum' *1681*. 'plump' Margoliouth.

l. 36. *sacred*. Clumps of trees so placed were often thought of as sacred groves.

52 l. 38. *Of the great . . . terror*. The authority of Fairfax.

l. 43. *Vera*. Ann Vere, Lady Fairfax.

l. 56. *this*. i.e. this lord.

53 l. 73. *ye*. Bodleian MS; 'the' *1681*.

l. 74. *oak*. Referring to the sacred oak at Dodona, from the rustlings of whose leaves the will of Zeus could be discovered.

*Upon Appleton House*. Marvell seems to have spent some two years (1650–2) as tutor to Maria Fairfax at Nun Appleton, the seat of Thomas, Lord Fairfax. The received idea of the house was until 1972 based on the account of it in C. R. Markham's *Life of the Great Lord Fairfax* (1870), 365, but it now seems likely that the building he describes was put up later, and that in Marvell's time the family lived in what remained of the nunnery. The new house was large, modern, and grand; its predecessor much humbler, which makes the opening stanzas of the poem more appropriate. (See John Newman, letter in *TLS*, 28 Jan. 1972, and further comment by A. A. Tait, *TLS*, 11 Feb. 1972.) The original house belonged to a Cistercian priory, and was acquired by the Fairfax family at the dissolution in 1542. Some nunnery ruins remain. Marvell's poem is a somewhat anomalous member of the genre of poems about country houses.

The poem, which certainly has allusions to the recent civil wars, has lately been treated as a political or political-apocalyptic allegory, with much questioning as to whether Marvell was criticizing Fairfax's retirement. See e.g. Margarita Stocker's exaggeratedly eschatological *Apocalyptic Marvell* (1986), 46–66, and Ernest B. Gilman, *The Curious Perspective* (1978), 204–31, an interesting account of Marvell's non-committed style, which denies the reader a fixed point of view. The best study of the political setting is Michael Wilding, *Dragons Teeth* (1987), 138–72.

*To my Lord Fairfax*. For Fairfax, see note on 'Upon the Hill and Grove', above (p. 300).

ll. 1–8. *Within . . . gazed*. This somewhat chauvinistic praise for the modesty of Nun Appleton is probably intended to be a criticism of poems praising the architectural grandeur of a patron's house; specifically, perhaps, Saint-Amant's praise of the Duc de Retz's hunting lodge in his *Palais de la Volupté*:

> *L'invention en est nouvelle,*
> *Et ne vient que d'une cervelle*
> *Qui fait tout avec tant de poids,*
> *Et prend de tout si bien le chois*
> *Qu'elle met en claire évidence*
> *Que sa grandeur et sa prudence*
> *Sont aussi dignes, sans mentir,*
> *De régner comme de batir.*

ll. 5–6. *Who . . . gazed*. Who, in agony to bring his great design to birth, employed his skull as a model for the vault.

l. 8. *arch*. Architectural pun.

l. 12. *equal*. Appropriate (to their size).

54 l. 22. *mote*. Bodleian MS; 'mose' *1681*; 'mole' Cooke.

l. 24. *the first builders*. Of the tower of Babel (Genesis 11: 1–9).

l. 30. *loop*. Opening.

l. 31. *door so strait*. Matthew 7: 13.

l. 36. *Vere*. Anne Vere, Fairfax's wife.

l. 40. As *Romulus's . . . cell*. The thatched hut in which Romulus, founder of Rome, had lived, here is compared to a beehive.

l. 45. *immure*. Enclose.

l. 46. *The circle . . . quadrature*. A reference to the old problem of squaring the circle. For an interesting account, see R. L. Colie, 'Some Paradoxes in the Language of Things' in *Reason and the Imagination*, ed. J. A. Mazzeo (1962), 121–3. The problem 'became the standard trope for time-wasting intellectual activity'. But to live by the square of human constancy, and respect the circle of heaven (the circle a symbol for God), is 'holy mathematics'.

55 l. 56. *That*. Its humility.

l. 64. *invent*. Find out.

l. 65. *frontispiece of poor*. The door is conceived as the frontispiece of a book, and the poor, confidently expecting the alms of Fairfax, are its decoration.

l. 71. *inn*. A reference to some lines of Fairfax, preserved in the Bodleian Library, called 'Upon the newbuilt house at Appleton':

Think not, O man that dwells herein
This house's a stay but as an inn
Which for convenience fittly stands
In way to one not made with hands
But if time here thou take rest
Yet think eternity's the best.

l. 73. *Bishop's Hill.* Fairfax's house in York. *Denton.* Another Fairfax estate thirty miles from Nun Appleton.

l. 90. *Thwaites.* Isabella Thwaites married an ancestor of Fairfax. Shc had been left in the charge of the Prioress of Nun Appleton, who tried, by shutting her up, to prevent her marriage to Sir William Fairfax; but her authority was overridden, and the marriage took place in 1518.

l. 105. *white.* The colour of the Cistercian habit.

ll. 107–8. *And our . . . dim.* A reference to the parable of the wise and foolish virgins (Matthew 25: 1–13).

57 l. 121. *prayed.* Two syllables.

l. 122. *legend.* A saint's life.

l. 141. *crown.* Of lilies.

58 l. 152. devoto. Religious devotee.

l. 169. *nice.* Fastidious.

l. 172. *perfecting.* Stressed on the first syllable.

59 l. 180. *sea-born amber.* Ambergris (see 'Bermudas', l. 28), used to perfume linen in storage.

l. 181. *grieved.* Hurt, wounded. *pastes.* Pastries.

l. 182. *baits.* Refreshments.

l. 184. *These . . . confess.* i.e. unless they needed a priest as confessor.

ll. 197–9. *Now . . . begin.* 'i.e. Now claim her plighted word, from which religion (which she henceforward doth begin) has released her' (Margoliouth).

60 l. 216. *And vice . . . wall.* So that the stone, even though laid by a just hand, would not fall on the girl's seducer, simply because it has been infected by the vice of the inmates of the nunnery.

l. 221. *state.* Estate.

l. 232. *First from a judge . . . soldier.* Sir William Fairfax's father was a judge; his mother the daughter of Lord Roos, a distinguished soldier.

l. 233. *in the storm.* In taking her from the nunnery by force.

61 ll. 241–4. *Is . . . Germany.* Sir Thomas, son of this pair, fought in Germany, and his son in France; a son of the next generation fought in Germany and was Marvell's patron in France.

l. 244. *either Germany.* i.e. high (Germany) and the low countries.

61 l. 245. *one.* Probably an allusion to the present Lord Fairfax, a general of the victorious Parliamentary army.

l. 248. *would intercept.* Wished to interrupt (the family succession).

l. 253. *disjointed.* Distracted.

l. 268. *had.* Bodleian MS; 'hath' *1681*, Margoliouth.

62 l. 274. *escheat.* A legal term: if the tenant died without an heir, the estate reverted to the lord.

ll. 281–2. *From that blest bed . . . fame.* It is uncertain whether 'hero' means Sir Thomas Fairfax, son of this marriage, or Lord Fairfax, Marvell's employer.

l. 288. *every sense.* There were gardens so laid out in sixteenth-century France.

l. 292. dian. Reveille.

l. 295. *pan.* The part of the musket lock holding the priming.

l. 296. *flask.* Powder-flask (the flowers represented as infantrymen).

l. 301. *virgin nymph.* Mary Fairfax.

l. 303. *think.* Imperative (addressed to the flowers). *not compare.* Do not invite comparison with.

63 l. 305. *firemen.* Soldiers using firearms (as distinct, for example, from bowmen).

l. 320. *or.* *1681*; 'nor' Cooke; 'or' must mean 'ere': 'before asking the password'.

l. 322. *garden of the world.* For this familiar topic (England as garden of the world) see J. W. Bennett, 'Britain among the Fortunate Isles', *Studies in Philology* 53 (1956), 114 ff., and Leishman, 283 ff.

l. 323. *four.* Disyllabic.

l. 326. *flaming sword.* Genesis 3: 24.

l. 328. *thee.* Bodleian MS; 'The' *1681*, Margoliouth.

l. 330. *militia.* Four syllables.

l. 336. *Switzers.* Referring to the yellow and red stripes of the papal Swiss guards.

64 l. 341. *stoves.* Hot houses.

l. 345. *there.* Bodleian MS; 'their' *1681*, Margoliouth.

l. 349. *Cinque Ports.* A group of ancient ports on the south-east coast of England. The wardenship was an important military appointment. Fairfax was warden 1650–1.

l. 351. *half-dry.* Because of the gradual extension of the land, which progressively cut off some of these ports from the sea. Fairfax was not nominally warden, but exercised the power of the warden by virtue of his being a member of the Council.

64 l. 356. *earthy*. Bodleian MS; 'earthly' *1681*, Margoliouth. *want*. Lack.

l. 358. *As that . . . shrinks*. The sensitive plant.

l. 363. *Cawood Castle*. Until 1642, a seat of the Archbishop of York, about two miles from Nun Appleton.

l. 365. *quarrelled*. Found fault with.

l. 368. *gaze*. *1681*, Margoliouth; 'graze' Bodleian MS.

l. 372. *giants*. 'And there we saw the giants . . . and we were in our own sight as grasshoppers, and so we were in their sight' (Numbers 13: 33).

l. 380. *Whether . . . go*. Whether he is falling or walking.

l. 382. *the ground*. Mud or sand from the seabed.

l. 385. *No scene . . . strange*. Referring to the elaborate machinery of the Renaissance theatre. For details see L. B. Campbell, *Scenes and Machines on the English Stage* (Cambridge, 1923), and Richard Southern, *Changeable Scenery* (1952).

ll. 389–90. *Who seem . . . green sea*. E. E. Duncan-Jones cites Wisdom of Solomon 19: 7: 'where water stood before, dry land appeared; and out of the Red sea a way without impediment: and out of the violent stream a green field'.

l. 392. *And crowd . . . side*. Crowd to either side to form a lane.

l. 395. *rail*. Corncrake or landrail. Cf. Virgil, *Georgics*, iv. 510–12, where a ploughman kills a baby nightingale.

l. 399. *untimely mowed*. Cf. 'Damon the Mower', l. 88.

66 l. 401. *Thestylis*. The name comes from Virgil, *Eclogues*, ii. 10–11; there she cooks for the reapers.

l. 402. *mowing camp*. As the mowers are represented as soldiers, Thestylis is a camp follower. *cates*. Food.

l. 406. *He*. The poet. This extremely unusual effect—a character in the poem jokes with the poet as if she had read the preceding verses and can develop the scriptural allusion—is well described by Frank J. Warnke as 'cracking the frame of fiction' (in Patrides, ed.).

l. 408. *Rails rain . . . dew*. Exodus 16: 13–14.

l. 416. *sourdine*. Mute on trumpet producing hoarse effect.

l. 417. *Or*. Either.

l. 419. *traverse*. A stage curtain on which a scene was depicted, but here used metaphorically for a track across a field.

l. 426. *hay*. (1) country dance; (2) mown grass.

l. 428. *Alexander's sweat*. Had a 'passing delightful savour' (Plutarch, *Life of Alexander*).

l. 430. *fairy circles*. The effect of mycelium. See 'Damon the Mower', l. 64.

67 l. 437. *Memphis*. Egyptian city near the Pyramids.

67 l. 439. *Roman camps*. Tumuli, now known to be of British origin.

l. 441. *This scene . . . withdrawing brings*. Continuing the theatrical figure of l. 385.

l. 442. *face*. Aspect.

l. 444. *cloths*. Canvases. *Lely*. Sir Peter Lely, celebrated Dutch portrait painter, who came to England in 1641.

l. 446. *table rase*. *Tabula rasa* (blank tablet).

l. 447. toril. A reminiscence of Marvell's visit to Spain, but by 'toril' he means 'bull-ring' and not the bull's enclosure, which is the modern sense.

l. 448. *Madril*. Madrid.

l. 450. *Levellers*. Egalitarian political movement of the period, favouring levelling out differences in rank, parliamentary representation, etc. Fairfax had suppressed a Leveller movement at Burford in 1649. *take pattern at*. Use as a model.

l. 451. *in common*. Not only is the meadow level, it is also a common for grazing, which strengthens its use as a model to levellers.

l. 453. *increased*. Grew.

l. 454. *beast*. Bodleian MS; 'breast' *1681*.

l. 456. *Davenant . . . herd*. Sir William Davenant, a contemporary of Marvell's, describes in his admired experimental epic, the unfinished *Gondibert*, a painting of the six days of creation. On the sixth day a 'universal herd' appears (II. vi. 60). The reference carries on the comparison with the new-created world begun in ll. 445–6.

l. 458. *A landskip . . . looking-glass*. A landscape shown in a painting as reflected in a looking glass and thus reduced in size.

ll. 461–2. *Such fleas . . . lie*. So do fleas appear on the glass before one looks at them through the microscope. Leishman cites James Howell's *Epistolae Ho-Elianae*: 'such glasses as anatomists use in the dissection of bodies, which can make a flea look like a cow'; and John Carey says Marvell is making fun of Howell: 'If fleas, when magnified, really look like cows . . . then it follows that unmagnified fleas can't look like fleas at all but like very, very tiny cows . . . what Howell had at the end of his "Multiplying Glasses" was a midget dairy herd, and that is why Marvell, seeing a midget dairy herd, likens it not just to fleas, but to fleas waiting in Howell's multiplying glasses to be magnified and identified by him as cows' (in Patrides, ed.).

68 l. 466. *Denton*. See note to l. 73.

l. 472. *And isles . . . round*. Makes an island around.

l. 476. *leeches*. Refers to the superstition that horsehair in water turned into eels or leeches.

ll. 477–80. *How boats . . . pound*. Cf. Ovid, *Metamorphoses*, i. 295-6.

l. 485. *the first carpenter*. Noah.

68 l. 486. *pressed.* Commandeered.

l. 490. *union.* The two woods are joined at one point, just as the Vere and Fairfax pedigrees are joined.

l. 491. *pedigrees.* Genealogical trees.

l. 493. *in war.* They were cut down to meet a wartime demand for timber.

ll. 495–6. *And . . . expect.* Cf. Saint-Amant, 'La Solitude', ll. 6–10:

> . . . ces bois, qui se trouverent
> A la nativite du temps,
> Et que tous les siecles reverent,
> Estre encore aussi beaux et vers,
> Qu'aux premiers jours de l'univers.

69 l. 499. *neighbourhood.* Proximity.

l. 502. *fifth.* Of a different substance from the existing four (earth, air, fire, and water).

ll. 505–12. *Dark . . . fires.* This is part of Marvell's considerable debt to Benlowes. See M.S. Røstvig, *The Happy Man* (Oslo, 1954), 247–8.

l. 508. *Corinthian.* The most ornate of the Greek architectural orders.

l. 526. *Sad pair . . . moan.* Imitated from Virgil's *nec gemere aeria cessabit turtur ab ulmo* ('and the turtle-dove shall not cease moaning from the high elm'), *Eclogues*, i. 58.

70 l. 535. *stork-like.* The stork was held to leave behind one of its young as a tribute to the owner; the heron is imagined as dropping one young bird in similar tribute.

l. 537. *hewel's.* Green woodpecker's.

l. 538. *holt-felster's.* Woodcutter's.

71 l. 568. *inverted tree.* 'Man is like an inverted tree' is a commonplace explored in historical depth by A. B. Chambers (*Studies in the Renaissance*, 8 (1961), 291–9), where it is traced back to Aristotle and even to Plato (*Timaeus*, 90A—the same work, 91E, may be the original of ll. 565–6). A famous example is Swift, *Meditation on a Broomstick.*

l. 577. *Sybil's leaves.* Palm leaves from which the Romans foretold the future.

l. 580. *Like Mexique . . . plumes.* Pictures made by sticking feathers together.

l. 582. *mosaic.* An image assembled from various (natural) materials; but also with a reference to the 'Mosaic' books. Of the two books of God—the Bible and Nature—Nature is the 'lighter'.

l. 586. *mask.* An allegorical garb, as for some masque-like entertainment.

l. 591. *antic cope.* Cf. Milton's *Apology for Smectymnuus (Complete Prose Works*, New Haven, 1953, i. 930); 'cope' means 'ecclesiastical outer garment'.

l. 592. *prelate.* Also developed from the Latin of Benlowes.

72 l. 599. *shed*. Separate, part.

l. 610. *gadding vines*. A reminiscence of Milton's 'Lycidas', l. 40.

73 ll. 629–30. *No . . . Nile*. Referring to the belief that the Nile floods begat serpents and crocodiles from the mud (cf. *Antony and Cleopatra*, II. vii. 29–31).

l. 631. *itself*. The river.

l. 636. *slick*. Sleek.

l. 639. *shade*. Reflected image.

l. 640. *Narcissus-like*. Narcissus was in love with his own reflection in a pool.

l. 649. *quills*. Floats.

l. 650. *angles*. Fishing tackle. *utensils*. Accent on first syllable.

l. 651. *Maria*. Mary Fairfax, to whom Marvell was tutor.

74 l. 659. *whisht*. Hushed.

l. 660. bonne mine. Good appearance; puts on its best behaviour (*bonne* is disyllabic).

l. 668. *ebon shuts*. Ebony (black) shutters.

l. 669. *halcyon*. This bird produces absolute calm. See 'The Gallery', l. 35, though here the bird is the kingfisher. Virgil (*Georgics*, iii. 335 ff.) describes its appearance in the calm of evening. Hodges remarks that Pliny (*History of the World*, tr. Philemon Holland, 1634, I. x. 32) offers two alternatives: the halcyon either *makes* the calm, or it knows when the calm will occur. Marvell chooses the first, slightly more magical explanation.

l. 671. *horror*. Awe, reverence (Latinism).

l. 673. *she*. The halcyon.

l. 677. *stupid*. Stupefied.

l. 679. *assist*. Attend; are present at.

l. 680. *sapphire-winged mist*. The kingfisher in flight.

ll. 681–2. *Maria . . . evening hush*. The influence of a lady over a landscape, especially an evening landscape, is a poetic theme discussed by Leishman, 81 ff., and Scoular, 147 (with parallel from Théophile de Viau) and 172 ff.

l. 684. *star new-slain*. A meteor, thought to be 'exhaled' from the earth (l. 686).

l. 688. *vitrified*. Turned to glass, as in the incorruptible crystalline sphere of the fixed stars; or, like the 'sea of glass' described in Revelation 4: 6.

75 l. 708. *all the languages*. Mary Fairfax, under Marvell's tuition, was proving a capable linguist.

l. 714. *trains*. (1) plots, stratagems; (2) trains of artillery.

76 l. 724. *the starry Vere*. Perhaps a reference to the Vere coat of arms.

76 l. 733. *grin*. Grimace.

l. 734. *black-bag*. Mask.

l. 738. *line*. Lineage.

l. 744. *choice*. Mary Fairfax later married the profligate Duke of Buckingham.

77 l. 753. *Thessalian Tempe*. The vale of Tempe in Thessaly was a famous ancient 'paradise'.

l. 755. *Aranjuez*. Spanish royal gardens on the Tagus, south of Madrid.

l. 756. *Bel-Retiro*. Buen Retiro, another royal residence near Madrid.

l. 757. *Idalian Grove*. Cyprus, the garden of Venus.

l. 761. *'Tis . . . world*. Referring to the disorder of a fallen world, to be compared with the order and balance of the microcosm, or little world, of Nun Appleton, which reflects the state of the world before paradise was lost.

l. 771. *Antipodes in shoes*. Those who dwell directly opposite us on the globe (their feet pointing to our feet).

l. 772. *Have shod . . . canoes*. Coracles ('leathern boats') were so carried. There was a Cambridge student song about lawyers, which had a similar point (Craze, 252).

l. 774. *rational amphibii*. Amphibii are animals at home in and out of water; the salmon-fishers, men being 'rational animals', are rational amphibii. Cf. Sir Thomas Browne, *Religio Medici*, i. 34: 'man that great and true amphibium'.

l. 776. *Does now . . . appear*. In being covered by a hemisphere of darkness, like the boat on the salmon-fisher's head.

*Flecknoe, an English Priest at Rome*. Richard Flecknoe, Roman Catholic priest and minor poet, was in Rome during 1645 and 1646, when Marvell may have met him. Flecknoe was later satirized as the reigning prince of Dulness in Dryden's *Mac Flecknoe*. Marvell's encounter with him is somewhat in the manner of Donne's *Satires*. The poem may have been written soon after the meeting but remained unpublished until *1681*.

l. 3. *Melchizedek*. King of Salem and priest of the most high God (Genesis 14: 18); also a prophetic type of Christ.

l. 4. *Lord Brooke*. Fulke Greville, first Lord Brooke (1554–1628). Flecknoe published verses complimenting Brooke's *Works* (published in 1670).

l. 6. *The Sad Pelican*. The Pelican was a common inn sign. *subject divine*. The pelican was thought to feed its young with its own flesh and blood; it became a symbol for Christ. See ll. 127–8.

78 l. 8. *triple property*. Another reference to the joke about Melchizedek.

l. 12. *ceiling*. (1) wall-hanging; (2) wainscot. *sheet*. (1) bedsheet; (2) winding sheet.

l. 14. *show*. Appear.

78 l. 18. *stanzas*. Rooms (It.).    appartement. Suite of rooms (Fr.) (four syllables).

l. 19. *information*. Five syllables.

l. 20. *and*. *1681*, Margoliouth; 'in' Bodleian MS.

l. 26. *prepared*. Three syllables.

l. 28. *The last . . . brain*. A reminiscence of Milton's 'Lycidas', l. 70–1:

> Fame is the spur that the clear spirit doth raise
> (That last infirmity of noble mind)

l.43. *straitness*. Tension.

79 l. 53. *he was sick*. And thereby exempt from the Ordinance.

l. 54. *the ordinance*. Against eating meat in Lent.

l. 55. *him scant*. G. A. Aitken, *Satires*; 'him: Scant' *1681*.    *scant*. Fasting.

l. 57. *our dinner*. Bodleian MS; 'dinner' *1681*.

l. 63. basso relievo. Low relief (It.).

ll. 64–5 *Who as a camel . . . stitch*. Matthew 19: 24.    *stitch*. (1) grimace of pain; (2) shred of clothing; (3) in sewing (Donno).

l. 66. *rich*. To be rich is what makes it as difficult to enter heaven as a camel to go through the eye of a needle (Mark 10: 25; Luke 18: 25).

l. 69. *circumscribes*. Wraps himself in writing.

l. 74. *sottana*. Cassock (It.).

l. 75. *antic*. (1) old (antique); (2) grotesque; (3) crazy.

l. 76. *first Council of Antioch*. AD 264.

l. 78. *tradition*. The authority of unwritten tradition, held by Roman Catholics to be equal to that of the Scriptures, was an important difference between Catholic and Protestant; it underlies this joke.

l. 83. *disfurnish*. i.e of its occupants (the only furniture).

80 l. 90. *thorough*. Through.

l. 92. *the palace*. E. E. Duncan-Jones suggests the Casa Barberini, the Cardinal of that name being Protector of the English in Rome.

l. 98. *Delightful*. Delighted.

l. 99. *penetration*. The occupation of the same space by two bodies at the same time (cf. 'Horatian Ode', l. 42).

l. 100. *Two . . . three*. 'When two or three are gathered together in thy name . . . ' (Book of Common Prayer).

l. 101. *in one substance*. Like the Trinity.

l. 126. *Nero's poem*. Suetonius tells us that 'no one was allowed to leave the theatre during his recital', however pressing the need (*Nero*, 23).

81 l. 130. *foul copies*. Rough drafts (and the more obvious sense).

81 l. 136. *ordinaries*. Inns.

l. 137. *chancres and poulains*. Syphilitic sores.

l. 152. *Perillus*. The contriver of the Brazen Bull of Phalaris, and its first victim.

l. 156. *is no lie*. Does not give you the lie (require you to challenge him).

l. 170. *for a vow*. As an *ex voto* offering.

82 *An Horatian Ode upon Cromwell's Return from Ireland*. This poem was cancelled from all extant copies of *1681* except for two, one in the British Library and one in the Huntington Library, California.

Cromwell returned from his ferocious reconquest of Ireland in May 1650, and in the following month undertook the preventive campaign against Scotland, Fairfax having resigned as commander-in-chief because he thought the Scots should not be compelled to war. Cromwell entered Scotland on 22 July 1650; the poem, presumably, was written between his return from Ireland and that date.

Though the tone and some of the details derive from Horace, as the title suggests, Marvell also remembered and imitated what Lucan had written, in his epic *Pharsalia*, about Julius Caesar and Pompey. On the Horatian antecedents see John Coolidge, 'Marvell and Horace', *Modern Philology*, 43 (1965), 111–20. Annabel Patterson takes the mode of the poem to be that of 'conditional praise' (61). A strong and direct indebtedness to *Il Principe* is demonstrated by Brian Vickers, 'Machiavelli and Marvell's "Horatian Ode" ', *Notes and Queries* (March 1989), 32–8.

The historical circumstances of the moment are fully described by Blair Worden, *The Historical Journal*, 27 (1984), 525–47; and see also Michael Wilding, *Dragons Teeth*, ch. 5.

l. 2. *now*. 'In times like these'. The opening lines (1–8) are adapted from Lucan, *Pharsalia*, i. 239–43.

l. 9. *restless*. A trait of Lucan's Caesar. The passage ll. 9–24 imitates *Pharsalia*, i. 143–55. *cease*. Rest.

l. 15. *thorough*. (Bodleian MS); Through (*1681* has 'through'). *side*. (1) party; (2) the lightning is conceived as tearing through the side of its own body the cloud.

ll. 19–20. *And with . . . oppose*. To pen him in will produce an even more violent reaction than to fight against him.

l. 20. *more*. Worse; a Latinism; cf. *Pharsalia*. l. 1: '*Bella . . . plus quam civilia*' (Craze).

l. 23. *Caesar*. Charles I, beheaded in 1649.

l. 24. *laurels*. Thought to be proof against lightning and worn by Roman emperors.

l. 32. *bergamot*. A variety of pear.

l. 35. *kingdom*. *1681*; 'kingdoms' Bodleian MS, Thompson.

83 l. 38. *ancient rights*. See 'Tom May's Death', l. 69.

83 l. 42. *penetration*. See note on 'Flecknoe', l. 99.

l. 46. *Where his . . . scars*. i.e. the scars he gave.

ll. 47–52. *And . . . case*. In 1648 the king, noting the increased hostility of the Army Council, took flight from his palace at Hampton Court to Carisbrooke on the Isle of Wight. He did not receive the expected welcome; the governor treated him as a prisoner. Thus his flight was in part responsible for what happened later, but the contemporary rumour that Cromwell engineered it—which is what Marvell here has in mind—appears to be without foundation.

l. 49. *subtile*. Finely woven.

l. 52. *case*. Cage.

l. 53. *actor*. This theatrical figure is sustained in l. 54 ('tragic scaffold'—stages for the acting of tragedies), l. 56 (clapping), and l. 58 (scene). John Carswell (*TLS*, 1 Aug. 1952, 501) suggests that ll. 57–64 constitute a faintly ironic criticism of an actor's performance. One contemporary account of the king's trial was called *Tragicum Theatrum Actorum*. The fact that the execution took place on a scaffold outside the Banqueting House (in which Charles had formerly acted in masques) gives additional force to the figure.

l. 56. *clap*. Some said that the soldiers around the scaffold were ordered to clap, with the object of rendering the king's words inaudible.

l. 59. *keener*. Keener than the axe's edge. The Latin *acies* means both 'eyesight' and 'blade'. 'It is recorded by one [eye-witness] that he had never seen the king's eyes brighter than in his last moment, and by another he more than once inquired about the sharpness of the axe' (C. V. Wedgwood, *Poetry and Politics under the Stuarts* (1961), 101–2).

l. 60. *try*. Test.

ll. 63–4. *But bowed . . . bed*. 'The Venetian ambassador reported that the executioners had prepared for resistance on the part of the king by arranging to drag his head down by force; but he told them this was unnecessary, and voluntarily placed his head on the block' (Christopher Hill, *Puritanism and Revolution* (1958), 360 n.).

l. 66. *forced*. Two syllables; gained by force.

ll. 67–72. *So . . . fate*. The story is told by Pliny, *Natural History*, xxviii. 4. The excavator of the foundations of the temple of Jupiter Capitolium found 'a man's head, face and all, whole and sound: which sight . . . plainly foretold that [Rome] should be chief castle of the empire and the capital place of the whole world' (Livy, *Annals*, i. 55.6, tr. Philemon Holland) (Donno). Hodge remarks that in Livy's version of the story the head is discovered by Tarquin, later banished as a tyrant (126).

l. 74. *in one year*. Cromwell's Irish campaign lasted from August 1649 to May 1650.

l. 76. *act and know*. This commends Cromwell for contemplative and active virtue, resuming the theme of ll. 29–37.

84 ll. 77–90. *They . . . skirt.* Reminiscent of Lucan, ix. 192–200 (magnanimity of Pompey).

l. 77. *They.* The Irish. Irish testimony would be hard to find at the time (or, one might add, later).

l. 81. *yet.* Either 'nevertheless' or 'up to now'.

l. 82. *still.* Either 'always' or 'up to now'.

ll. 85. *Commons'.* Bodleian MS, Thompson; 'Common' *1681.*

l. 87. *what he may.* So far as he can.

l. 95. *lure.* A lure was made of feathers. During training the hawk could expect to find food in it, and on active service return to it when the falconer calls.

l. 100. *crown. 1681*; 'crowns' Bodleian MS, Thompson.

l. 104. *climacteric.* Critical, marking an epoch (stressed on the first and third syllables).

l. 105. *Pict.* The name of a Celtic tribe inhabiting Scotland; chosen rather than 'Scot' for the sake of the pun in the next line.

l. 106. *parti-coloured.* Variously coloured (Latin *pingere, pictum*, to paint). The Scots were generally regarded as fickle and treacherous. (There is also a pun on 'party'.)

l. 107. *sad.* Severe, or steadfast.

l. 110. *mistake.* Because of his coloured camouflage.

l. 116. *thy. 1681*; 'the' Thompson.

85 ll. 117–18. *Besides . . . night.* Usually interpreted as referring to the cross-hilt of the sword, but this would imply that the sword is held hilt upmost, and this sword is 'erect'. Its power against the forces of darkness may then derive, as John M. Wallace suggests, from 'a sun-like glitter on the blade' (*PMLA* 77, 1961, 44). Elsie Duncan-Jones explains the lines as alluding to the pagan belief (Odyssey, ix. 48, *Aeneid*, vi. 260) that spirits of the underworld fear cold iron.

ll. 119–20. *The same arts . . . maintain.* A commonplace of political theory; for references see John M. Wallace, 43 and notes, 48-54.

*Tom May's Death.* Thomas May (1599–1650), historian, court poet, and translator of Lucan in a version known to Marvell, had been close to Ben Jonson, who wrote commendatory verse for the Lucan translation. He transferred his allegiance to the Parliamentary cause during the Civil War and wrote a history of the Long Parliament. To reconcile this attack on May, presumably written soon after May's death on 13 November 1650, and so quite soon after the Horatian Ode, has been a puzzle to commentators, not made easier by the use of the expression 'ancient rights' (l. 69), since those rights were what Charles I had pleaded for in vain, not only in the poem but at his trial, for he asked how any citizen could call his life and possessions his own 'if power without right daily make new, and abrogate the fundemental laws of the land' (quoted in *The Letters and Speeches of*

*Oliver Cromwell*, ed. W. C. Abbott (1937), i. 738). No entirely convincing solution has been offered. See, for example, Warren Chernaik, *The Poet's Time* (1983), 174 ff. ('from being incompatible, [the two poems] are in many ways companion pieces'; the disapproval of rebellion in the poem 'need not be taken as reflecting the poet's persuasion'). Gerard Reedy in ' "An Horatian Ode" and "Tom May's Death" ', *Studies in English Literature*, 20 (1980), argues that Marvell's quarrel was based on disagreements between rival positions on the Parliamentary side, and Marvell may have disliked Tom May for his venality or for some personal reason. Christopher Hill (in Patrides, ed.) suggests that the poem was revised after the Restoration, when May's body was exhumed from Westminster Abbey in 1661, ll. 85–90 being a comment on this event.

'Tom May's Death' was omitted from the Bodleian MS. The fact has led some to question its authenticity.

l. 1. *drunk*. May died in his sleep, according to Aubrey, 'after drinking with his chin tied with his cap (being fat); suffocated'.

l. 6. *Stephen's Alley*. May lived here in Westminster. It was a well-known street of many taverns. *grass?* Cooke; 'grass' *1681*.

l. 7. *The Pope's Head . . . Mitre*. Common inn signs.

l. 8. *still*. Always.

l. 10. *Ares*. Perhaps an innkeeper.

l. 13. *Ben*. Ben Jonson, famous for his 'mountain belly'. He contributed laudatory verse to May's translation of Lucan's *Pharsalia*.

l. 18. *Brutus and Cassius*. Heroes to republicans, but the darkest villains to royalists or imperialists. Dante places them with Judas in the mouths of Satan (*Inferno*, xxiv. 64–9).

l. 21. *Emathian*. Cooke; 'Emilthian' *1681*.

ll. 21–4. *'Cups . . . health'*. A parody of the opening lines of May's translation of Lucan:

> Wars more than civil on Emathian plains
> We sing; rage licensed; where great Rome distains
> In her own bowels her victorious swords . . .

l. 26. *translated*. By death and by Jonson.

l. 27. *stumbling*. May is said to have had a stammer, but of course he is here represented as drunk.

l. 29. *friend*. Jonson had addressed May as his 'chosen friend' in *Underwoods*, 21.

86 l. 38. *Like Pembroke . . . masque*. Lord Pembroke, the Lord Chamberlain, broke his staff on May at a court masque in 1634, 'not knowing who he was'. May was compensated.

l. 41. *Polydore . . . Goth*. Polydore Vergil, an Italian historian who came to the English court early in the sixteenth century and wrote a flattering *Historia Anglia* for Henry VII; and three barbarian tribes.

86 ll. 44–8. *On them . . . Cicero.* May was fond of making parallels between Parliamentary and Roman history.

l. 50. *As Bethlem's . . . Loreto walk.* The house of the Virgin was said to have been miraculously transported to Loreto (from Nazareth, not from Bethlehem).

l. 54. *Those . . . May.* May wrote a *Continuation of Lucan's Historical Poem till the death of Julius Caesar.*

l. 57. *more worthy.* Sir William Davenant became Poet Laureate in 1637, when May thought he would be appointed.

l. 60. *gazette-writer.* Hack journalist ('gazette' stressed on the first syllable).

l. 62. *basket.* The *borsa* in which the warring Florentine sects of Guelphs and Ghibellines cast their votes.

l. 68. *world's.* Cooke; 'world' *1681.*

l. 69. *ancient rights.* See 'Horatian Ode', l. 38 (p. 83).

87 l. 74. *chronicler to Spartacus.* Historian of the revolutionary Parliament. Spartacus was the leader of the slaves' revolt against Rome in 73–71 BC.

l. 75. *equal.* Just.

l. 82. *Who thy . . . pay.* The Council of State voted £100 for May's burial in Westminster Abbey.

l. 88. *As th' eagle's . . . divide.* Eagle feathers were supposed to rot those of other birds.

l. 90. *Phlegethon.* A river of Hades.

l. 91. *Cerebus.* Three-headed watchdog of Hades.

l. 92. *Megaera.* Serpent-haired Fury.

l. 93. *Ixion's wheel.* One of the torments of Hades.

l. 94. *the perpetual vulture feel.* Like Prometheus.

*To his worthy Friend Doctor Witty upon his Translation of the 'Popular Errors'.* Robert Witty was a schoolmaster and later a physician in Hull. He translated a Latin work by James Primrose, another Hull doctor, on common errors and myths about medicine. Witty's translation was published in 1651 and Marvell's commendatory verses were printed therein. They were republished in 1681. Our text is mainly from *1681*, which does not differ in substance from *1651*.

l. 4. *cypress.* Of fine linen.

88 ll. 4–16. *Take . . . spots.* Margoliouth notices these lines as 'one of the few scraps of Marvell's literary criticism'.

l. 17. *Celia.* Perhaps Mary Fairfax; see 'Upon Appleton House', ll. 707–8 (p. 75). Marvell may have written these lines at Nun Appleton in the winter of 1650/1.

l. 30. *caudles.* Gruels. *almond-milk.* Preparation of blanched almonds and water, used as an emollient.

88 *The Character of Holland*. The first hundred lines, together with an eight-line conclusion not by Marvell, were published in 1665. The whole work appeared in *1681*. Marvell probably wrote it in 1653, after the English victory over the Dutch fleet off Portugal in February, and before the engagement of 3 June, in which Deane was killed.

89 l. 5. *alluvion*. Matter deposited by flood or inundation.

l. 18. *Thorough*. Through.

ll. 19–20. *And . . . ground*. The figure is from bear-baiting.

l. 24. *leap-frog*. 'Frogs' was an occasional nickname for the Dutch.

l. 26. Mare Liberum. The title of a book by the Dutch lawyer and scholar Grotius claiming the freedom of the seas. The Commonwealth government claimed the Channel as English and required foreign ships to salute the English flag.

l. 28. *level-coil*. A rough game, one player unseating the other (*lever le cul*).

l. 32. *cabillau*. Codfish.

l. 34. Heeren. Gentlemen.

l. 36. *duck and drake*. Skimming flat stones across water.

90 l. 45. *Leak*. Leaky.

l. 49. *dyke-grave*. Officer in charge of sea-walls.

l. 53. *Half-anders*. Literally, 'half-different'. Here, not Hollanders (Whole-anders).

l. 62. *Poor-John*. Dried hake.

l. 65. *Margaret*. A legendary Dutch countess who had 365 children at a birth.

l. 66. Hans-in-Kelder. A child in the womb. *Hans-town*. A member of the Hanseatic league of cities.

l. 78. *village*. The Hague, which was denied the status of a town till the Napoleonic wars.

l. 80. Hogs. *Hoog-mogenden* ('high and mighty') was the style in which the States-General were addressed. Bores. Boors (with a pun on 'boars' and 'Boers').

l. 82. *Civilis*. The chief of the Batavi in the fight against Rome, AD 69.

91 l. 86. *chafing-dish*. The Dutch carried stoves to church.

l. 88. *a wooden*. *1665*; 'wooden' *1681*.

1. 90. *western end*. Buttocks.

l. 94. *butter-coloss*. Butter-box (nickname for Dutchman).

l. 95. *towns of* Beer. Place-names beginning with Beer- or Bier-.

l. 96. *snick and sneer*. Cut and thrust.

91 l. 98. *Cut . . . to a man.* Deinocrates the sculptor wanted to cut Mount Athos into the shape of Alexander. They want to cut each other's formless bulk into human shape.

l. 107. *vail.* Salute by lowering colours; here, surrender.

l. 113. Jus Belli et Pacis. Grotius wrote *De Jure Belli et Pacis* (On the Law of War and Peace), 1625.

l. 114. *burgomaster of the sea.* Admiral Van Tromp.

l. 115. *gun power.* Fortified spirits. *brand wine.* Brandy.

l. 116. *linstock.* Forked staff holding a lighted match.

l. 118. *sore.* Genesis 34: 25.

l. 120. *case-butter.* Cannister shot using butter (i.e. in this instance consisting of nothing more dangerous than butter). *bullet-cheese.* Bullets made of cheese.

l. 123. *kindly.* According to her nature.

l. 124. *A wholesome danger . . . ports.* Blake took refuge in port after an engagement with Van Tromp in November 1652.

92 l. 127. *careen.* Be heeled over on their sides for repair.

l. 130. *halcyon.* See 'Appleton House', l. 669, and 'The Gallery', l. 135. The halcyon was sometimes said to nest during a period of calm at the winter solstice.

l. 134. *corruptible.* Accented on the third syllable. Several officers were discharged after an inquiry into the action of November 1652.

l. 135. Bucentore. The Doge's galley, centre of the ceremony of Venice wedding the sea.

l. 137. *now.* Bodleian MS; 'how' *1681.* *the.* *1681* (most copies); 'their' Bodleian MS, *1681* (Huntington copy). *seven provinces.* The United Provinces of the Netherlands.

l. 138. *our infant Hercules.* The original Hercules strangled serpents as a baby; but ours—the new Commonwealth—was like the mature hero, for he strangled the Hydra.

l. 139. *wants.* We have chopped it off (unlike the Hydra, the Dutch have only one neck).

ll. 141–4. *Or . . . refuse.* 'Unless, if the Dutch sued for peace, declined the approach lest the English youth lost the habit of war.' (The difficulty arises from the parenthesis; either, though the youth would surely not do so, or surely the government wouldn't behave so.)

l. 150. *Deane, Monck, and Blake.* Deane and Monck were colonels, appointed generals of the fleet in association with Blake. Deane was killed 3 June 1653.

*The First Anniversary of the Government under His Highness the Lord Protector.* Published as an anonymous quarto in 1655 in an edition apparently authorized by the Parliamentary government, and reprinted, but cancelled, in most copies of *1681* except the British Library's and the Huntington's.

Cromwell became Protector in December 1653. When Marvell wrote the poem he was at Eton, tutor to Cromwell's protégé William Dutton.

J. A. Mazzeo has argued that Marvell represents Cromwell here as a Davidic king; the concept of the *regnum Davidicum* had long been associated with European kingship, for example in coronation rituals (*Reason and the Imagination: Studies in the history of ideas 1660–1800*, ed. J. A. Mazzeo, 1962). John M. Wallace treats the poem as a 'deliberative' one, about Cromwell instituting a new dynasty, Marvell holding that the choice of king was human, but that anointment ought to follow; after election, the grace of God. The succession must be assured against accidents and Fifth Monarchy machinations (*Destiny his Choice*, 1968). See also, on the politics of the poem, Steven Zwicker, 'Models of Governance in Marvell's "The First Anniversary" ', *Criticism*, 16 (1974), 1–12 (Marvell is one who rejects the idea of Cromwell as king, but regards him as a divine instrument sent to heal England, herald of a millennium when Christ will rule). Also, on the chiliastic element, see Stocker, 13–27, Chernaik, 43–59 and 68 ff.

The poem has seven principal sections: (1) 1–48, celebration of Cromwell's monarchic vigour and integrity; (2) 49–116, his building of a harmonious state; (3) 117–58, delay in arriving of the millennium because of human folly; (4) 159–220, Cromwell's coach accident in Hyde Park, September 1654; (5) 221–92, defence of Cromwell against the charge that he exercised arbitrary power; (6) 293–324, attack on the Fifth Monarchy men; (7) 325–402, envy and admiration of foreign princes, and conclusion.

*Title. 1655*, Bodleian MS; 'The First Anniversary of the Government under O.C.' *1681*. (*1681*'s title is presumably a censored version.)

l. 1. *curlings*. Undulations.

93 l. 12. *And shines the jewel . . . ring*. Like the sun in the zodiac.

ll. 13–14. *'Tis he the force . . . acts*. Compare the different version in the 'Horatian Ode', ll. 34–6.

l. 16. *Saturn*. Which had the longest orbit of the known planets.

l. 17. *Platonic years*. A Platonic year is the period required for all the planets to return to their original relative positions; estimated at 36,000 years.

l. 20. *China clay*. It was believed that the Chinese made porcelain by burying clay in the ground.

l. 23. *some*. Some king.

ll. 24–6. *Took in by proxy . . . lost*. Various more or less contemporary instances of this not unusual practice of delegating the fighting and taking the credit have been adduced. There is an early example in 2 Samuel 11, where David leaves the work to Joab while he stays at home and seduces Bathsheba; in chapter 12 he takes the credit.

l. 27. *wrong*. A verb.

l. 30. *common enemy*. The subjects.

93 l. 33. *temple.* David was commanded to leave the building of the Temple to Solomon (1 Chronicles 28).

l. 36. *perfect.* Stressed on the first syllable.

l. 41. *image-like.* Like clock-figures striking the hour on a bell.

94 ll. 47–8. *Learning . . . sphere.* A reference to a familiar body of ideas about *musica humana* and its relation to the harmony of the spheres (see, for full treatment, J. Hollander, *The Untuning of the Sky*, 1961).

l. 49. *Amphion.* He built the wall at Thebes by making stones move to music; a stock figure in the formal *laudes musicae.*

l. 50. *the god.* Hermes.

l. 60. *joining. 1655:* 'joyng' *1681.*

l. 66. *seven.* Appropriately, since there were 'six notes of music' and the seventh provided what we would call an octave.

l. 68. *Instrument.* A pun; Cromwell's Protectorate was instituted by an Instrument of Government in 1653.

l. 69. *hack.* (1) muddle; (2) to break a note in music.

ll. 69–70 *While tedious statesmen . . . back.* Earlier attempts to frame a new constitution had succeeded only in reducing liberty.

l. 81. *bends.* Consents.

l. 85. *attractive.* In the literal sense: he draws them as Amphion did the stones.

95 l. 90. *contignation.* A joining together of boards.

l. 95. *opposed.* Three syllables.

l. 99. *a place.* Archimedes said that given a fulcrum he could move the earth.

l. 101. *aspects.* An astrological term. Planets in varying positions have good or ill aspects toward the earth.

l. 104. *influence.* Also astrological.

ll. 105–8. *O would they rather . . . lead.* Psalm 2: 10–12.

l. 110. *latter days.* Daniel 2: 28, 10: 14. These are apocalyptic lines. Many believed in the approaching Fifth Monarchy of the Saints (Daniel 7: 18), after the destruction of the Great Whore (Rome) (Revelation 17: 5) and the Beast (Revelation 17: 3).

l. 115. *Indians.* As one of the nations, they should be brought in, not subdued.

l. 116. *Jew.* The Conversion of the Jews would precede the Latter Days. Cromwell, in 1656, allowed the readmission of the Jews to England, expressly to facilitate this development.

l. 123. *prevents the east.* Anticipates the dawn.

l. 125. *hollo.* Huntsman's cry.

l. 128. *thorough.* Through.

96 l. 140. *latest day*. Day of Judgement.

ll. 151–2. *And stars . . . flail*. Revelation 12: 3–4, and Milton's 'Ode on the Morning of Christ's Nativity', ll. 168–72.

l. 152. *volumes*. Coils.

l. 153. *suspend*. At the Flood.

l. 157. *landing nature*. About to reach port; cf. 'A Dialogue between the Soul and Body', ll. 29–30.

l. 161. *saint-like mother*. Elizabeth Cromwell died 16 November 1654, aged 93.

ll. 171–2. *Thy breast . . . prophecies*. Alluding to various plots against Cromwell by Levellers and Fifth Monarchists, some gaining the impetus from interpretations of biblical prophecies.

l. 175. *How near*. By what a narrow margin.

97 ll. 177–8. *Our brutish fury . . . hurried thee*. On 29 September 1654, Cromwell's coach, drawn by six horses, overturned in Hyde Park.

l. 182. *yearly*. Celebrating the events of the year.

l. 184. *purling*. Embroidery.

l. 197. *Nor through . . . wanton air*. Cf. Virgil, *Georgics*, i. 375–6.

l. 203. *panic*. Marvell derives from the Greek for 'all' *(pan)*.

ll. 205–6. *centre*. Cf. 'Resolved Soul', l. 71, and ll. 363–4 below. *sphere*. The fourth sphere of the pre-Copernicus astronomy was the sun's.

98 ll. 215–20. *But thee . . . mantle rent*. Cf. 2 Kings 2: 11–13, and Plato, *Republic*, viii. 566.

ll. 221–32. *For all delight . . . body strong*. Cf. 'Horatian Ode', ll. 27–32.

ll. 233–8. *Till at the seventh time . . . king*. Cf. l Kings 18: 44–6.

l. 238. *though forewarned*. This refers to the king.

l. 239. *since*. Since 1649.

99 ll. 249–56. *When Gideon . . . son*. Judges 8: 1–23. Cromwell's achievements are aligned with Gideon's.

ll. 257–64. *Thou with . . . didst awe*. Judges 9: 7–15, a parable about the olive, the fig, the vine, and the bramble, which accepts the crown the others refuse; Jotham gives it a political application.

l. 262. *Had quickly levelled . . . top*. An allusion to the Levellers.

ll. 265–78. *So have I seen . . . prevent*. The ship of state was safer in the open sea—the plots and plans of various parties being represented as the rocks and shallows some mistook for land.

l. 269. *Tritons*. Sea gods.

l. 270. *corposants*. Balls of fire seen on the masts and rigging of ships.

l. 281. *bounders*. Limits.

99 ll. 293–310. *Yet such a* . . . Alcoraned. Referring to the many religious sects of the times, especially the Fifth Monarchy Men (l. 297).

l. 293. *Chammish.* Like Ham (Vulgate, Cham), Genesis 9: 24–5.

100 l. 300. *Might muster* . . . *ten.* i.e. if one heresy equalled ten men.

ll. 301–2. *What thy misfortune* . . . *fall.* 'Spirit' alludes to the cant of the sects; those possessed by it behave like epileptics (l. 302).

l. 303. *Mahomet.* His revelation was supposedly accompanied by something like an epileptic fit.

l. 305. *Feake and Simpson.* Christopher Feake and John Simpson, imprisoned in 1654 for preaching against Cromwell.

l. 307 *rant.* Alluding to the sect of Ranters, fanatically antinomian.

l. 308. *tulipant.* Turban.

l. 310. Alcoraned. Turned into a holy book like the Koran.

l. 311. *locusts.* Revelation 9: 1–11; a passage generally associated with heretics.

l. 313. *Munser's rest.* Munster, in Westphalia, was taken in 1534 by the Anabaptists. This means: Anabaptists, etc., left over from Munster. Or possibly the reference is to Thomas Munzer, founder of the sect—'Munzer's leavings'.

l. 315. *deface.* Here with the sense 'efface'. They condemned the use of Scripture, having the Holy Spirit within them, and called all laws oppressive; thus these libertarians treated the Bible and law with the same contempt that they showed to personal adornment.

l. 319. *act the Adam and Eve.* Adamite sects went naked.

101 ll. 325–42. *So when* . . . *side.* This tale was first mentioned, mockingly, in Lucretius, *De Rerum Natura,* v. 973–6. It was again mentioned by Manilius, *Astronomica,* i. 66–70, and Statius, *Thebaid,* iv. 282–4. The most obvious resemblance in Marvell's poem is to the lines of Statius (Duncan-Jones). This ancient tale presumably refers here to Cromwell's accident; he was at first reported dead.

l. 350. *both wars.* The Civil War and the Dutch War.

l. 352. *their.* Conjectured by Margoliouth; 'our' *1655, 1681.*

l. 356. *shedding leaves.* A kind of goose was thought to be bred from leaves fallen into water.

l. 362. *brazen hurricanes.* Bronze cannon.

l. 363. *That through* . . . *side.* Cf. l. 205.

l. 366. *leaguers.* Besieging forces.

102 l. 381. *enchased.* Worked in together.

l. 384. *our knots.* Alluding to the Gordian knot: whoever untied it would be master of the world. Alexander cut it.

102 ll. 401–2. *And as the angel . . . heal.* In John 5: 4 an angel troubles the water and it cures the sick. Cromwell did the same by accepting the Protectorate in December 1653, and was about to do it again by dissolving Parliament in January 1655.

*On the Victory obtained by Blake over the Spaniards.* Presumably written June–early August 1657. News of Blake's victory reached England in late May; Blake died in early August. Sir Robert Blake (1599–1657) was one of Cromwell's admirals, a brilliant and aggressive commander. At the time of this action against Spanish treasure ships his health was much enfeebled by a wound received in an earlier battle.

The poem was first published anonymously in *A New Collection of Poems and Songs* (1674). In this version the allusions to Cromwell, to whom the poem is implicitly addressed, are excised, and there are other omissions. Our text is from *1681*. The poem is not in the Bodleian MS.

*Title. 1681*; 'On the victory over the Spaniards in the Bay of Santa Cruz, in the Island of Tenerife.' *1674*.

l. 4. *guilt.* With pun on 'gilt'.

103 l. 20. *streamers.* Pennons.

l. 25. *One of which.* Tenerife, with its tall peak.

l. 28. *Trees . . . supply.* The idea may be that dew descending from the trees on the top does the office of rain.

l. 39. *Your.* i.e. Cromwell's.

l. 40. *kings.* Notable because Marvell had thought it right for Cromwell to refuse the crown. This lends some force to Lord's rejection of the poem. (It is not in the Bodleian MS.)

l. 46. *this one peace.* England and Spain had been at peace since 1630.

104 l. 54. *the fancied drink.* Nectar.

l. 66. *your present.* Your present conquests.

105 l. 91. *sconces.* Small forts.

l. 117. *Stayner.* Sir Richard Stayner, one of Blake's captains.

l. 118. *To give him laurel . . . plate.* This victory gained him honours (he was knighted for his deed) as the last (he intercepted a Spanish treasure fleet) gave him prize money.

106 l. 132. *its.* The fire's.

l. 139. *or . . . or.* Either . . . or.

107 *Two Songs at the Marriage of the Lord Fauconberg and the Lady Mary Cromwell.* Mary Cromwell, third daughter of the Protector, married Lord Fauconberg on 19 November 1657. These songs probably belong to a musical entertainment devised for the wedding.

l. 1. *eyes.* As well as his stars.

108 l. 27. *thorough.* Through.

l. 30 *Anchises.* Lover of Venus, but here Robert Rich, who a week earlier had married Cromwell's fourth daughter, Frances.

l. 35. *Latmos' top.* The place where Cynthia wooed Endymion.

109 *Second song*

l. 3. *northern shepherd's.* Fauconberg was from Yorkshire and a kinsman of Fairfax.

l. 4. *Menalca's daughter.* Menalca is, like the other names here, familiar in pastoral poetry. In this case he is Cromwell.

110 l. 26. *silly.* Innocent.

l. 34. *Marina's.* Marina is another pastoral name, like Damon: perhaps arbitrarily chosen, it seems to have no marine connotation.

l. 43. *beauty's hire.* Obscure: presumably her beauty is the reward of his virtue?

*A Poem upon the Death of his late Highness the Lord Protector.* Together with the 'Horatian Ode' and 'The First Anniversary', this poem was cancelled from *1681*, though not from the British Library's copy. However, even in that copy 140 lines are wanting (185–324). These were provided by Thompson, but occur also in the Bodleian MS, which is probably the book Thompson describes as having come into his hands when he had finished preparing his edition.

Cromwell died on 3 September 1658. Marvell's poem is a long funeral elegy on the general pattern of such things (see Wallerstein, *Seventeenth-Century Poetic*, 1950), but it also reveals some of the political tensions experienced by the poet at this time.

Text: ll. 1-184 from *1681*; ll. 185–324 from Bodleian MS.

*title*: Bodleian MS; 'A Poem upon the Death of O.C.' *1681*.

l. 2. *every hair.* Matthew 10: 30.

111 l. 4. *seen the period.* Foreseen the completion.

l. 14. *or clemency that would. His* clemency was such that no one wanted to hurt him.(?)

l. 16. *angry heaven.* Cf. 'Horatian Ode', l. 26.

l. 21. *signed.* Assigned.

l. 22. *Those nobler weaknesses . . . mind.* Cf. Milton's 'Lycidas', l. 71, and 'Flecknoe', ll. 27–8. *mind. 1681*; 'kind' Thompson, Bodleian MS.

l. 30. *Eliza.* Cromwell's second daughter Eliza died on 6 August 1658.

ll. 39–40. *When with . . . fairer mind.* Cf. Donne, 'The Second Anniversary', ll. 244–5, and 'A Funeral Elegy' (following 'The First Anniversary'), ll. 59–61.

112 l. 45. *not knowing.* i.e. not by knowing.

112 l. 54. *And him . . . racks*. The father having within him the image of the daughter must suffer the pain of her fever (the image compared to a wax doll melting in the heat).

l. 62. *feigns*. Dissimulates.

ll. 67–8. *And now . . . had worn*. Alluding to Ovid's story of Scylla, who cut off from her father's head the purple lock on which his life depended.

113 l. 87. *immortal tried*. Tested in other ways, he proved immortal.

l. 108. *celebrates*. Bodleian MS, Margoliouth; 'celebrate' *1681*.

l. 113. *First the great thunder . . . sent*. There was a great storm on 2 September.

l. 121. *lead*. Grosart, Margoliouth; 'dead' *1681*, Bodleian MS.

114 l. 128. *not to see*. In order not to see. There had been an epidemic of fever earlier in the year.

l. 131. *air*. i.e. one of the elements.

l. 137. *The stars . . . power*. Judges 5: 20: 'They fought from heaven; the stars in their courses fought against Sisera.'

l. 139. *cast*. Calculate (as in astronomy).

l. 144. *Twice*. At Dunbar, 3 September 1650, and at Worcester, 3 September 1651.

l. 146. *He marched . . . ending war*. At the battle of Worcester, Cromwell bridged the Severn and fell on the flank of the Scots.

l. 154. *Gave chase . . . coast*. In September 1658 a Spanish force under the Prince de Ligne was defeated in Flanders by a French army with an English contingent.

l. 162. *Than those of Moses . . . eyes*. Deuteronomy 34: 6.

115 ll. 173–4. *Who planted . . . Indian ore*. The capture of Dunkirk from the Spaniards (1658), and of Jamaica (1655).

l. 176. *worthies*. King Arthur was made one of the nine worthies.

l. 178. *Confessor*. Edward.

l. 180. *manned*. 'Made a man of'.

l. 181. *inward mail*. Ephesians 6: 11.

l. 187. *Preston's field*. Cromwell defeated the Scots under Hamilton near Preston, in August 1648.

l. 188. *impregnable Clonmel*. Clonmel was attacked unsuccessfully by Cromwell, but the Irish subsequently evacuated it (May 1650). This was the last incident in Cromwell's Irish campaign.

l. 189. *And where . . . Fenwick scaled*. At the battle of the Dunes in June 1658, which preceded the occupation of Dunkirk, Lt.-Col. Roger Fenwick was mortally wounded in storming a sandhill. The day was a day of public prayer.

l. 190. *The sea between*. Although the sea was between.

115 ll. 191–2. *What man . . . Gibeon stayed.* Joshua 10: 12–14: 'And there was no day like that before it or after it, that the Lord harkened unto the voice of a man' (v. 14).

l. 194. *He conquered God . . . men.* See Genesis 32: 24–9.

ll. 201–2. *Friendship . . . name.* The Protector's branch of the Cromwell family was founded by his great-grandfather Richard Williams, nephew on his mother's side of Thomas Cromwell, Earl of Essex. His friendship with his uncle resulted in his being knighted and in his adoption of the surname Cromwell.

l. 203. *But within . . . all.* But while friendship attaches to one object, Cromwell's tenderness extended unto all.

116 l. 215. *cast.* Calculated; cf. l. 139; or perhaps 'diagnosed'.

l. 234. *Janus' double gate.* The *ianus geminus* in the Forum at Rome.

l. 242. *David.* 'And David danced before the Lord with all his might' (2 Samuel 6: 14).

l. 245. *Francisca.* Cromwell's youngest daughter, Frances.

117 l. 259. *feign.* Imagine.

l. 264. *honoured wreaths.* Oak garlands awarded for statesmanship.

ll. 269–70. *The tree . . . grew.* Cf. 'Eyes and Tears', ll. 5–6; also ll. 273–4.

l. 275. *seat.* Bodleian MS; 'state' Thompson.

l. 276. *Seeing how little . . . great.* Perhaps meaning: when they are high our eye diminishes their stature; it is when we see them at our own height that we understand their greatness.

l. 282. *Cynthia.* The moon.

l. 287. *pitch.* Height (as used in falconry).

118 l. 291. *at.* Conjectured by Margoliouth; 'yet' Bodleian MS.

l. 305. *Richard.* Richard Cromwell (1626–1712) was proclaimed Protector on the day of his father's death, but resigned the title in April 1659.

l. 317. *enchased.* Adorned.

119 *On Mr Milton's 'Paradise Lost'.* These verses, signed A.M., were prefixed to the second edition of *Paradise Lost* (1674) and reprinted in *1681*. Marvell is known to have supported Milton at the time of his greatest danger at the Restoration, and he also defended him in the second part of *The Rehearsal Transprosed* (1673).

Text from *1681*.

ll. 18–22. *Jealous I was . . . play.* Dryden asked and got Milton's permission to adapt *Paradise Lost* for the stage. The result was an 'opera' in rhymed couplets, called *The State of Innocence*, published in 1677, and probably never performed. Marvell obviously disliked Dryden, partly no doubt because he changed sides at the Restoration, and Dryden retaliated in the preface to *Religio Laici*.

119 l. 25. *that*. Bodleian MS; 'and' *1681*, Margoliouth.

l. 30. *detect*. Expose.

l. 33. *treat'st*. *1674*; 'treats' *1681*.

120 l. 39. *The bird . . . paradise*. The bird of paradise was supposed to have no feet and to remain in perpetual flight.

l. 43. *Tiresias*. The blind prophet of Thebes.

l. 45. *mightst*. *1674*, Bodleian MS; 'might' *1681*.

l. 46. *rhyme*. See Milton's adverse comments on rhyme in the prefatory remarks to *Paradise Lost*. Dryden's *State of Innocence* is mostly in heroic couplets. Milton is said to have given Dryden permission to 'tag' his verses.

l. 47. Town-Bays. Buckingham's *Rehearsal* (1672) presented Dryden as Bays, and the nickname stuck. See *The Rehearsal Transprosed, passim*.

l. 49. *bushy points*. Used for fastening hose; the ends were often tasselled. He means that whereas we use such frivolous ornaments only for 'fashion', such poets as Dryden employ them (viz., rhymes) as necessary to their halting verse.

ll. 51–2. *I too . . . commend*. Caught up in this fashion, I can't use the word 'praise' but must say 'commend' for the sake of the rhyme.

l. 54. *In number . . . measure*. Wisdom 11: 20: 'thou hast ordered all things in measure and number and weight.'

121 *Satires of the Reign of Charles II*.

*Clarendon's Housewarming*. First printed in 1667 with Denham's 'Directions to a Painter'. Reprinted in 1697. It also exists in various MSS. Margoliouth bases his text on *1667*, corrected by Bodleian MS Gough London 14, and our text is based on Margoliouth's. The poem was written in June–July 1667.

Edward Hyde, first Earl of Clarendon (1609–74), became a Member of Parliament in 1640. At first he opposed Charles I, then joined the royal forces at the outbreak of the Civil War in 1642 and thereafter remained a member of the court in exile. He was appointed Lord Chancellor in 1658, and was in effect chief minister from 1660 till his downfall in 1667. He was to die in exile in France.

l. 3. *three deluges*. Sent by Mars, Apollo, and Vulcan: the Dutch war, plague, and fire.

l. 4. *turn architect*. Clarendon's lavish house near St James's was begun in 1664. Built during war and plague, and at enormous expense, 'it raised a great outcry against him' (Gilbert Burnet, *History of my Own Time*).

l. 7. *brume*. Winter. Cf. 'brumal' in *The Rehearsal Transprosed* (ed. Smith), 229.

l. 13. *Rhodopis*. Conjectured by Margoliouth, Bodleian MS; 'Rhodope' *1667*. An Egyptian courtesan who built a pyramid out of her earnings.

121 ll. 15–16. *And wished . . . her quarry*. Alluding to the supposed pre-marital affairs of Anne Hyde, Clarendon's daughter.

l. 17. *Amphion*. Cf. 'First Anniversary', ll. 49–74.

l. 20. *Jew's-trump*. Jew's harp.

l. 21. *Virgil*. *Aeneid*, i. 367.

l. 22. *Poultney*. Sir William Poultney, among the original proprietors of the land granted to Clarendon in June 1664. *Did'*. Dido.

l. 24. *Hyde*. Clarendon's name, with pun on 'hide', a measure of land between 60 and 120 acres.

122 l. 26. *brickbat*. Bricks.

l. 27. *Denham*. Sir John Denham, poet. He went mad for a time in 1666.

l. 28. *And too much . . . chocolate*. See note on 'Last Instructions', l. 65 ff.

l. 30. *thong*. i.e. of bull's hide; another allusion to Dido (Margoliouth). *Caster*. a 'thong-caster' is the area of land you can encircle with a bull's hide cut into strips.

ll. 35–6. *What Joseph . . . hands*. Cf. Genesis 47: 13–26. Joseph bought all the land of Egypt for Pharaoh except that of the priests, which was held on a special tenure.

l. 41. *Scotch forts*. Built by Cromwell at Ayr, Leith, Perth, and Inverness. *Dunkirk*. Sold to France in 1662.

l. 43. *Tangier*. Part of the dowry of Charles II's wife Catherine of Braganza. *mold*. The harbour mole.

l. 44. *St Paul's*. Burnt down in 1666, when Clarendon House was nearly finished.

ll. 49–52. *Allens*. Sir Allen Apsley (1616–83) and Sir Allen Brodrick, Members of Parliament.

123 l. 57. *Worcester*. Clarendon rented a house from the marquess of Worcester in the Strand.

l. 60. *clustered of atom*. Lucretius, *De Rerum Natura*, ii. 394 ff. According to Epicurean physical science the atoms clung together as if by means of hooks (Margoliouth).

l. 63. *her that Rome once betrayed*. A reference to the story told in Livy, I. xi.

l. 65. *bishops . . . seal*. Clarendon was keeper of the Great Seal and no one had done more to restore episcopacy.

l. 66. *Sumners*. Lord, from several MSS; 'Sinners' Margoliouth. *farmers*. i.e. tax-farmers. *bankers*. Cf. 'Last Instructions', ll. 493-8. *patentees*. See 'Last Instructions', l. 258.

ll. 69–70. *Bulteale's . . . Kipp's*. Servants of Clarendon.

l. 71. *Act of Oblivion*. Act of pardon, etc., for those who had rebelled against Charles I and II, passed after the Restoration. It also confirmed sales of property made during the Interregnum.

123 l. 72. *benevolence.* A voluntary law. *snips.* Perquisites.

l. 73. *chimney contractors.* Collectors of Chimney Money. *smoked.* Made uncomfortable, or took note of.

ll. 74–6. *Nor would take . . . behind.* See note to 'Last Instructions', l. 258.

l. 78. *Bristol.* See note to 'Last Instructions', l. 933. Bristol went into hiding for two years after 1663.

l. 79. *St John.* Charles Paulet, son of the Marquess of Winchester, a scoundrel. The line seems to mean that St John will expiate his offence by providing lead to cover Clarendon's roof.

l. 82. *Worstenholm.* Sir John Worstenholm, a farmer of the customs.

l. 88. *Leslie's folly.* Dr John Leslie, Bishop of Orkney, built a palace so strong that it long resisted Cromwell's armies.

124 l. 90. *A lantern like Fawkes'.* Guy Fawkes was captured with a dark lantern and gunpowder in his possession.

l. 99. *handsel.* Inaugurate.

l. 100. *Buckingham's sacrifice.* Buckingham was arrested on a charge of high treason.

l. 103. *backside.* Back yard.

l. 108. *Tyburn.* There was a gallows there, in the north-east corner of Hyde Park.

l. 112. *St James's fair.* At the meeting of Parliament fixed for St James's Day, 25 July.

*The last Instructions to a Painter.* First published in *The Third Part of the Collection of Poems on Affairs of State* (1689), and reprinted in *Poems on Affairs of State* (1697). It is attributed to Marvell in all editions, but the attributions in these volumes are not, in themselves, to be relied upon.

The genre is sketched out in M. T. Osborne, *Advice-to-a-Painter Poems, 1633–1856* (Texas, 1949). In 1665 Edmund Waller published *Instructions to a Painter*, a panegyric on the Duke of York for his naval victory against the Dutch on 3 June 1665. This was quickly followed by second and third *Advices to a Painter*, sometimes attributed to Marvell, sometimes to Denham, mocking the tone of Waller's poems. A fourth and fifth followed. It seems that 'Last Instructions' put an end to this type of poem for the time being.

Margoliouth bases his text on *1689*. We follow Margoliouth for the most part, noting where we adopt readings different from his.

'Last Instructions' must have been written between June 1667, when the Dutch sailed up the Medway, and 29 November, when Clarendon fled to France, an event not mentioned by Marvell.

*London . . . 1667.* Bodleian MS; not in *1689* or Margoliouth.

ll. 1–2. *After two sittings . . . third time.* Three sittings were usual.

l. 6. *without a fleet.* The fleet was laid up in May 1667.

124 l. 7. *signpost.* Of an inn.

125 l. 9. *limn.* Draw.

l. 10. *alley-roof.* Roof of a covered bowling-alley.

l. 11. *Sketching in . . . tools.* The line is obscure. Editors take 'tools' to mean 'penises', but offer no evidence that 'antique masters' drew them on bowling-alley, or alehouse, roofs.

l. 14. *As the Indians . . . plumes.* Cf. 'Appleton House', l. 580 and note.

l. 15. *score out.* Sketch in outline. *compendious.* Minute. The idea seems to be that our fame, justly within small compass, should be enlarged, as Hooke enlarged the louse.

ll. 16–18. *With Hooke . . . staff.* A reference to R. Hooke's *Micrographia* (1665). One picture shows a louse on its back climbing along a single hair. The Comptroller of the Householder (then Lord Clifford of Chudleigh) bore a white staff.

l. 19. *pencil.* Brush.

ll. 21–6. *The painter . . . grin.* Pliny tells this story about Protagenes when he attempted to depict a heavily breathing dog. Unable to render the nature of froth—'the only mark he shot at'—despite his industry and skill, he hurled his sponge in anger at the painting and by chance achieved his desired end. *Natural History*, XXV. x. 36 (Donno).

l. 27. *perfect.* Accented on the first syllable.

l. 29. *St Albans.* Henry Jermyn, Earl of St Albans (d. 1684), pro-French ambassador to the French court.

l. 32. *salt.* Salacious.

l. 36. *Bacon.* Also bore the title St Albans.

l. 38. *treat.* Negotiate, and entertain.

ll. 39–40. *Draw no commission . . . supply.* The negotiations undertaken by St Albans were at first informal and unofficial, and Margoliouth attributes this failure to send a duly authorized ambassador with commission and seal to the Court's desire for a large grant from Parliament.

l. 41. *St James's lease.* Jermyn obtained a grant of land in Pall Mall in 1664 and planned St James's Square.

ll. 42–4. *Whose breeches . . . plenipotence.* 'Although he lacks an official commission, he will be able to produce ample evidence from his breeches. A gross insult is intended, whether this is an allusion to Jermyn's supposed affair with Queen Henrietta Maria (Lord and Donno) or to the size of his bottom (Legouis)' (Wilcher).

l. 45. *most Christian.* i.e. Louis XIV. *trepan.* Entrap.

l. 46. *St Germain, St Alban.* Playing on both parts of the ambassador's name (see note to l. 29).

126 l. 49. *her Highness.* Anne Hyde, Duchess of York.

126 l. 50. *Newcastle's wife*. Margaret Cavendish, poet and philosopher. She was widely held to have been very ridiculous. See Douglas Grant, *Margaret the First* (1957).

l. 51. *Archimedes*. He ran naked from his bath shouting 'eureka!' *put down*. outdo.

l. 52. *experiment upon the crown*. An attempt to secure the crown.

l. 53. *perfected*. Stressed on the first syllable. *engine*. Device.

l. 57. *Crowther*. Dr Joseph Crowther married Anne Hyde and the Duke of York, 3 September 1660.

l. 58. *Royal Society*. The Duke was a fellow of the Royal Society.

l. 60. *glassen*. Fragile. *D——*. *1689*; 'Dukes' Bodleian MS; 'Dildoes' MS note in British Library's copy of *1689*. Neither suggested reading seems to make much sense. *malleable*. (Stressed on the third syllable) resilient.

l. 62. *Wide mouth . . . proclaim*. Like a street-seller of asparagus.

l. 65–8. *Express her . . . cacao*. An allusion to the recent death of Lady Denham, the Duke of York's mistress, allegedly brought about by poison administered in a cup of chocolate. It was believed that china vessels would not endure contact with poison.

l. 72. *fawns*. The young of any animal.

l. 75. *Sidney's*. Henry Sidney (1641–1709), Groom of the Bedchamber to the Duke of York, was dismissed by the Duke in a fit of jealousy.

l. 76. *Denham's face*. See note to ll. 65–8.

l. 79. *Castlemaine*. Barbara Villiers (1640–1709), Countess of Castlemaine, created Duchess of Cleveland in 1670; mistress to Charles II.

127 l. 102. *Jermyn*. Henry Jermyn (1636–1708), nephew to St Albans.

l. 103–4. *Alexander . . . Apelles, give*. Alexander gave Campaspe to Appelles, who had fallen in love with her while painting her portrait (Pliny, *Natural History*, xxxv. 10).

l. 105. *pair of tables*. Folding board for backgammon.

l. 106. *the men*. Backgammon pieces.

l. 109. *trick-track*. A variety of backgammon.

l. 114. *Turner*. Sir Edward Turner (1617–76), Speaker of the House of Commons.

l. 116. *strike the die*. To cheat in throwing dice. In throwing for both court and country he arranges for the court to win, and takes his cut.

l. 120. *recreate*. Refresh by a change of occupation. Rubens was occasionally sent on diplomatic missions.

l. 121. *Cabal*. A committee for foreign affairs of the Privy Council.

l. 125. *fix to the revenue*. Parliament met on 21 September 1666. The sum fixed was £1,800,000. ('Revenue' is stressed on the second syllable.)

127 l. 126. *Goodrick.* Sir John Goodrick, MP for York. *Paston.* Sir Robert Paston, MP. He first moved the huge appropriation of £2,500,000 for the war.

l. 129. *Hyde's avarice.* See 'Clarendon's House-Warming'. *Bennet's luxury.* Henry Bennet (1618–85), the Earl of Arlington.

128 l. 136. *cassowar.* Cassowary; a type of ostrich which eats everything.

l. 138. *indented.* (1) furnished with teeth; (2) authorized by legal agreement.

l. 143. *Black Birch.* John Birch (1616–91), excise official.

l. 151. *wittols.* Complacent cuckolds.

l. 156. *Ashburnham.* John Ashburnham (1603–71) was believed to have betrayed Charles I in 1647 (see note to 'Horatian Ode', ll. 47–52).

l. 158. *word.* Password.

l. 160. *Steward.* Sir Nicholas Steward (1616–71), MP, one of the two Chamberlains of the Exchequer.

l. 162. *Wood.* Sir Henry Wood (1597–1671), Clerk of the Spicery to Charles I. As Clerk Comptroller for the Board of the Green Cloth he was responsible for order ('Cane') and provision ('Horn').

l. 167. *St Denis.* First Bishop of Paris; was beheaded and carried his head in his hands in pictures.

l. 168. *French martyrs.* Victims of syphilis.

l. 169. *as used.* As was customary.

l. 170. *Fox.* Sir Stephen Fox (1627–1716), Paymaster General.

129 l. 173. *Progers.* Edward Progers (1617–1714), MP; Gentleman of the Bedchamber.

l. 175. *Brouncker.* Henry Brouncker (?1627–88); Gentleman of the Bedchamber to the Duke of York.

l. 178. *to* teal *preferring* bull. A play on the name of John Bulteel. (teal=freshwater duck).

l. 180. *Wren.* Matthew Wren (1629–72), MP, secretary to Clarendon.

l. 181. *Charlton.* Sir John Charlton (1614–97), King's Sergeant. *coif.* White cap of office.

l. 182. *Mitre troop.* Lawyers associated with Mitre Court, one of the Inns of Court.

l. 186. *Finch.* Sir Heneage Finch (1621–82), Solicitor General. *Thurland.* Sir Edward Thurland (1624–85), solicitor to the Duke of York.

l. 187. *troop of privilege.* A reference to certain types of immunity to prosecution enjoyed by MPs.

l. 188. *Trelawney's.* Sir John Trelawney (1624–85), MP.

l. 193. *For . . . obeyed.* Sir Courtney Pool, MP, was a mover of the unpopular tax (chimney money) of 2*s.* (10p) on every hearth.

l. 195. *privateers.* MPs presenting private bills in their own interests.

129 ll. 197–8. *Before them Higgons . . . Countess . . . Act.* Sir Thomas Higgons (1624–91), MP, introduced a bill for the recovery of £4500 from the estate of his dead wife.

l. 199. *Sir Frederick and Sir Solomon.* Sir Frederick Hyde, MP, and Sir Solomon Swale, MP, members of the court party.

l. 200. *politics.* Politicians.

l. 203. *Carteret.* Sir George Carteret (d. 1680), Treasurer of the Navy, boasted he was worth £65,000.

l. 206. *Talbot.* Perhaps Sir Gilbert Talbot, MP.

l. 207. *Duncombe.* Sir John Duncombe, Commissioner of the Ordnance. *projectors.* Speculators.

l. 208. *Fitz-Hardinge.* Sir Charles Berkeley (1600–68), second Viscount Fitz-Hardinge. *Eaters Beef.* The yeomen of the Guard (Beefeaters).

l. 212. *Apsley.* Sir Alan Apsley (1616–83), Treasurer to the Duke of York's household, and MP. *Broderick.* Sir Allen Broderick, MP. See 'Clarendon's Housewarming', ll. 49–52.

130 l. 213. *Powell.* Sir Richard Powell, MP, Gentleman of the Horse to the Duchess of York.

l. 214. *French standard.* An allusion to the pox. *weltering.* Reeling.

l. 218. *Cornbury.* Henry Hyde, Lord Cornbury (1638–1709), Clarendon's eldest son.

l. 220. *Tothill Field.* At Westminster.

ll. 221–4. *Not the first . . . disperse.* George Monck, Duke of Albemarle (1608–70), had narrowly avoided losing his ship in a naval engagement with the Dutch, 'the sea-cod', in June 1666. Sir Thomas Tomkins had opposed the maintenance of the standing army that had been raised in the spring of 1667 to meet the threat from Holland. The fashionable footwear and headgear of the gallants of the standing army are glanced at contemptuously in the references to 'cork' and 'feather-men' (Wilcher). 'Cock' and 'cork' were similar in sound.

l. 225. *the two Coventrys.* Henry (1619–86) and Sir William (1627–86).

l. 226. *the other nought to lose.* Henry, without a job.

l. 230. *To fight . . . free.* Alluding to Henry's alleged cowardice at sea in 1665.

ll. 233–4. *They that . . . invade.* When the Dutch sailed up the Thames in June 1667, much of the English fleet was laid up in the expectation of peace, which Henry Coventry was helping to negotiate at Breda.

l. 239. *loose quarters.* Scattered billets, which are indefensible positions.

l. 245. *Strangeways.* Colonel Sir John Strangeways, opposition (country party) teller.

l. 249. *Cocles.* Horatio, the Roman soldier who alone defended a bridge against Porsena.

131 ll. 255–6. *Temple.* Sir Richard Temple, a leader of the country party (opposing the court). *conqueror/ Of Irish cattle.* The act against the importation of Irish cattle was passed in January 1667. *Solicitor.* Sir Heneage Finch (1621–82).

ll. 257–8. *Seymour.* Edward Seymour (1633–1708), MP; he attacked the monopolistic patent granted to the Corporation of the Canary company.

l. 259. *Whorwood.* Brome Whorwood, MP, active in the impeachment of Mordaunt (see following note).

l. 260. *Mordaunt.* John, Viscount Mordaunt (1627–75) was unsuccessfully impeached in 1666–7. William Taylor, an officer at Windsor, alleged that Mordaunt had imprisoned him because his daughter would not prostitute herself to him.

l. 261. *Williams.* Probably Colonel Henry Williams, MP, who spoke on financial matters in the House.

l. 262. *Lovelace.* John Lovelace (1638?-93), MP. *of chimney-men the cane.* Opponent of the hearth-tax.

ll. 263–4. *Old Waller . . . fight.* Edmund Waller, poet and political turncoat (and see headnote).

ll. 265–6. *Howard.* Sir Robert Howard (1626–98), brother-in-law of Dryden. *Montezumes.* A reference to the ranting hero of *The Indian Queen* (1664) by Howard and Dryden.

ll. 271–2. *Believes himself . . . can.* Cf. Virgil, *Aeneid*, v. 231.

ll. 275–6. *Such once Orlando . . . lance.* Cf. *Orlando Furioso*, ix. 68–9.

l. 280. *tobuck.* Tobacco.

l. 281. *recruit.* Reinforcement.

l. 287. *A gross of English gentry.* The country party, supporting the land tax rather than the excise.

132 l. 295. *battle.* Main body.

l. 296. *intervals.* The gaps in the ranks of this new force.

ll. 298–9. *Garway . . . Lyttleton . . . Lee.* William Garway, Sir Thomas Lyttleton, and Sir Thomas Lee all supported the land tax.

l. 301. *Sandys.* Colonel Samuel Sandys, MP.

l. 302. *St Dunstan.* St Dunstan served as Bishop of Worcester before becoming Archbishop of Canterbury in 959. Tempted while at work at a forge by Satan in the guise of a beautiful girl, the monk attacked the devil with a pair of pincers and forced him to flee.

l. 313. *seamen's clamour.* They rioted over arrears of pay.

ll. 327–8. *And the loved king . . . chide.* The king spoke on 18 January 1667. Cf. 'The King's Speech', p. 275.

l. 336. *The House prorogued.* On 8 February 1667.

133 l. 337. *Aeson*. The father of Jason; his youth was renewed by Medea's magic (Ovid, *Metamorphoses*, vii. 159–293).

l. 345. *sad tree*. The night jasmine of India; it loses its brightness by day.

l. 350. *Imprison parents*. See note to ll. 259–60.

l. 351. *The Irish herd . . . comes*. See note to ll. 255–6.

l. 352. *hecatombs*. Hundreds of cows.

l. 353. *Canary Patent*. See note to ll. 257–8.

l. 356. *fulminant*. Thundering.

ll. 357–8. *First Buckingham . . . rebel*. Buckingham, leading opponent of Clarendon, was arrested on trumped-up charges of high treason early in 1667.

l. 359. *the twelve Commons*. Twelve of the eighteen commissioners for the public accounts appointed in March 1667 were members of the Commons.

l. 360. *And roll . . . stone*. Perform an unending task.

l. 368. *To play for Flanders . . . lose*. The French aimed at a peace with England in order to be free to carry out their design on Flanders.

l. 370. *Holland's doors*. i.e. at Breda.

l. 371. *Cyclops*. One-eyed giants who forged thunderbolts for Zeus.

l. 374. *new sports*. From here to l. 394, Marvell describes a folk custom known as a 'skimmington ride' in which aggressive wives and their brow-beaten husbands were ridiculed by their neighbours.

134 l. 391. *quick*. Living. *effigy*. Stressed on the second syllable.

l. 398. *Candy*. Canvey Island, on the Essex coast.

l. 399. *Bab May*. Baptist May, Keeper of the Privy Purse.

l. 400. *so far off*. As Candia (Crete).

l. 406. *pilgrim Palmer*. Roger Palmer, Earl of Castlemaine. When his wife became the king's mistress he travelled in the near East.

l. 408. *Pasiphae's tomb*. Pasiphae of Crete, mother by a bull of the Minotaur. *bead*. Tear.

l. 409. *Morice*. Sir William Morice (1602–76), joint Secretary of State. *demonstrates*. Stressed on the second syllable. *the post*. i.e. by his knowledge of postal matters.

l. 415. *Bennet and May*. Bennet=Arlington. *May*. See l. 399.

135 l. 425. *The Bloodworth-Chancellor*. Sir Thomas Bloodworth was Lord Mayor of London during the year of the Great Fire. He seems to have had little effect, and the Chancellor's conduct is compared to his.

l. 431. *Dolman's disobedient*. Colonel Thomas Dolman was an Englishman commanding invading Dutch troops. Parliament ordered him to give himself up in 1665 but he refused.

l. 435. *prove*. Attempt.

135 l. 437. *De Witt . . . Ruyter*. Respectively, first minister and admiral of Holland.

l. 440. *undutiful*. i.e. as a nephew. Marvell, assuming St Albans was married to Henrietta Maria, would take him for Louis's uncle.

l. 442. *character*. Status as an ambassador.

l. 443. *gravelled*. Perplexed.

l. 444. *character*. Credentials.

l. 448. *Seneque*. i.e. Seneca; the French form may suggest deviousness.

l. 450. *Harry*. Henry Coventry, an English negotiator.

l. 454. *as the look adultery*. Matthew 5: 28.

l. 457. *Holles*. Sir Frescheville Holles (1642–72).

136 l. 460. infecta re. 'Without having done anything'.

l. 463. *shent*. Reproached.

l. 464. *eleventh commandment*. See ll. 453-4.(?) *commandment*. Stressed on the first syllable.

l. 471. *banned*. Cursed.

l. 480. *two-edged*. i.e. for use against an enemy, and for use at home.

ll. 487–8. *Hyde stamps . . . Myrmidons*. The Myrmidons were a race sprung from ants. Marvell combines this origin ('swarms') with memories of Pompey's boast in Plutarch that he had only to stamp on the ground of Italy to bring soldiers to him.

l. 493. *a proclamation*. 18 June 1667, to buoy up credit.

l. 494. banquiers banquerout. Bankrupt bankers.

l. 497. *Horse-leeches . . . vein*. Cf. Cleveland's 'Rebel Scot', ll. 83-5:

> Sure, England hath the haemorrhoids, and these
> On the north postern of the patient seize
> Like leeches . . .

*circling*. In mutual suction.

137 l. 503. *scrip*. Receipt for loan.

l. 510. *Monck in his shirt*. Albemarle just risen from bed.

l. 514. *Herb John*. A tasteless herb.

l. 532. *old*. He was 60.

138 l. 561. *Spragge*. Vice Admiral Sir Edward Spragge (d. 1673), commanding Sheerness.

139 l. 588. *captivate*. Capture.

l. 590. *recollect*. Gather.

l. 605. *Legge*. William Legge, Lieutenant-General of the Ordnance.

l. 607. *Upnor Castle*. About two miles below Chatham.

139 l. 611. *the* Royal Charles. Formerly the Naseby, it brought the king to Dover in 1660.

l. 615. *Admiral.* Flagship.

140 l. 629. *guards.* The regiment of Guards.

l. 631. *Daniel.* Sir Thomas Daniel, who commanded a company of foot guards which should have defended a ship that was destroyed.

l. 634. *shown.* Like a prize ox.

l. 636. *lac.* Crimson pigment.

l. 642. *Daniel . . . lion's den.* Daniel 6: 10–23.

l. 648. *Like Shadrack . . . Abednego.* Daniel 3: 13–27.

l. 649. *Douglas.* Archibald Douglas, who commanded a company of Lord George Douglas's Scotch regiment. See below, 'The loyal Scot'.

141 l. 666. *birding.* Shooting.

l. 667. *conjure.* Accent on the second syllable.

l. 682. *As the clear amber . . . does close.* Cf. Martial, iv. 32.

ll. 693–6. *Fortunate boy . . . Scot.* Cf. *Aeneid*, ix. 446–7 (of Nisus and Euryalus). Dryden's (later) translation runs:

> O happy pair! for if my verse can give
> Eternity your fame shall ever live,
> Fixed as the Capitol's foundation lies,
> And spread where'er the Roman eagle flies.

l. 695. *Oeta and Alcides.* Alcides (Hercules) was burnt on Mount Oeta.

l. 698. *The* Loyal London . . . *burns.* The ship *London* was blown up in 1665; the city was burnt in 1666; now a third *London* follows.

142 l. 712. *Our merchant-men . . . drown.* Ships were sunk in the Thames to prevent the Dutch sailing to London.

l. 715. *hole. 1697*; 'howl' *1689*, Bodleian MS, Margoliouth. 'Hole' means hold.

l. 719. *Gambo.* Gambia.

l. 722. *lest navigable.* Lest it should remain navigable.

ll. 735–6. *Such the feared . . . scorn.* Samson: Judges 16: 25.

143 ll. 761–4. *The court . . . Ruyter those.* Frances Stuart, who had married the Duke of Richmond, was the model for Britannia on coins. The motto *Quatuor* [sic] *maria Vindico* (I defend the four seas) appeared on farthings of the period.

l. 764. *those.* The four seas.

l. 765. *relish.* Gratify.

l. 767. *Pett.* Peter Pett (1610–72), Commissioner of the Navy at Chatham, was a scapegoat for the disaster.

143 l. 772. *Who treated . . . at Bergen.* In August 1665 a convoy of Dutch ships took refuge in Bergen. Sandwich wasted the opportunity to attack at once, and the Dutch escaped.

l. 775. *prevented.* Anticipated.

l. 784. *fanatic.* Sectarian; Pett had served under the Commonwealth.

144 l. 789. *one boat.* Laden with his own goods.

ll. 793–4. *Southampton dead . . . share.* Southampton, the Lord High Treasurer, died in May 1667. Sir John Duncombe took over some of his duties.

l. 797. *petre.* Gunpowder.

l. 799. *corn.* Grain of gunpowder.

ll. 801–2. *Who . . . least.* Duncombe's poverty and political inexperience are the butt of these lines.

l. 805. *brother.* Brother-in-law. *May.* Baptist May.

l. 811. *Sheldon.* Gilbert Sheldon (1598–1677), Archbishop of Canterbury.

l. 813. *With Boynton . . . Middleton.* Court beauties.

l. 817. *De. 1697*; [omitted] *1689*, Margoliouth.

l. 820. *Harry.* Henry Coventry, come from Breda.

l. 823. *job.* Hurry back on this instance.

l. 826. *Let them . . . again.* The House met on 25 July and was prorogued on 29 July.

l. 828. *vest.* The king designed the vest to out-do Paris in matters of fashion.

145 l. 837. *Turner.* The Speaker; see note to l. 114.

l. 854. *Expects.* Waits.

l. 855. *Ayton.* Sir John Ayton, Gentleman Usher of the Black Rod.

146 ll. 870–2. *But all . . . apron look.* The Speaker made money from private bills. Hence during their progress, he was hot and agitated like a cook.

l. 876. *Court-mushrooms.* Upstarts.

l. 882. *Norfolk.* James Norfolk, Sergeant-at-Arms.

ll. 883–4. *At night Chanticleer . . . Pertelotte.* Alluding to Chaucer's *Nun's Priest's Tale.*

l. 907. *starling.* Starting.

147 l. 913. *secure.* Overconfident.

l. 918. *Of grandsire Harry . . . sire.* Henri IV, who had been assassinated; Charles's father, Charles I, had been beheaded.

l. 933. *Bristol.* George Digby, Earl of Bristol, had attempted to impeach Clarendon in 1663. *Arlington.* Henry Bennet (l. 129), who had been in Bristol's service.

147 l. 935. *Who . . . betray*. Coventry resigned as secretary to Charles II's brother, the Duke of York.

l. 949. *tube*. Telescope.

148 l. 953. *trunk*. Nerve.

ll. 959–60. (*Kings . . . way*). Charles escaped in disguise after his defeat at the battle of Worcester.

l. 980. *Burrowing . . . store*. Pliny, *Natural History*, viii. 43.

149 *The loyal Scot*. In June 1667 the Dutch fleet under De Ruyter sailed up the Medway, on their return burning the *Royal Oak* 'and in her . . . Douglas . . . [who] had received orders to defend his ship, which he did with the utmost resolution; but having none to retire, he chose to burn with her, rather than live to be reproached with having deserted his command' (Kelliher).

The title plays on that of Cleveland's 'The Rebel Scot', which is often echoed in Marvell's poem.

Margoliouth identified three strata in the poem: '(1) Lines 15–62, the death of Douglas, originally part of "The Last Instructions" [ll. 649-96], and, therefore, written in 1667. (2) Lines 1–14 and 274–85, the Cleveland framework, and ll. 63–86 and 236–73, which assert the essential unity of England and Scotland. The most probable date . . . of these portions is 1669–70, when . . . Parliamentary union . . . was . . . mooted . . . (3) Lines 87–253, an anti-Prelatical tirade, including (ll. 178–85) an English version of [Marvell's] "Bludius et Corona".' Blood's attempt on the crown took place in May 1671.

There are various early printed versions: a short version in Charles Gildon's *Chorus Poetarum* (1694), ascribed to Marvell, and a slightly longer version in *Poems on Affairs of State* (1697), and the poem also exists in an undated, but probably late seventeenth-century, pamphlet in the Free Library of Philadelphia. Manuscripts give longer versions: the Bodleian MS, and Bodleian MS Douce 257, and BL MS Sloane 655, being among the most complete. Margoliouth bases his text on Bodleian MS Douce 257, and we follow Margoliouth for the most part.

*By Cleveland's Ghost*. Sloane 655; [omitted] Douce 357, Margoliouth.

l. 9. *tumour*. Bodleian MS; 'humour' Sloane 655, Margoliouth.

150 l. 39. *glories*. Most other versions; 'glory' Douce 357. 'honours' Bodleian MS.

l. 42. *Monck*. George Monck, Duke of Albemarle.

l. 63. *Skip-saddles*. The text may be corrupt at this point. If Marvell wrote 'Skip-saddles' we do not know what it means.

l. 64. *Galloway*. Scotch horse.

l. 67. *Curtius*. To save his country, Curtius leaped into the chasm that had suddenly opened up in the forum.

150 l. 71. *Corinthian metal.* Bronze.

ll. 73–4. *Shall fix . . . before.* These lines are omitted in Bodleian MS Douce 357, *1689*, and *1694*.

151 l. 80. *That whosoever . . . kill.* After this line *1697* has:

> Will you the Tweed that sullen bounder call
> Of soil, of wit, of manners, and of all?
> Why draw you not as well the thrifty line
> From Thames, from Humber, or at least the Tyne?
> So may we the state corpulence redress,
> And little England, when we please, make less.

(Bodleian MS and *1694* are similar.)

l. 86. *influence.* Astrological.

l. 88. *Holy Island.* Lindisfarne.

l. 93. *Sales.* St François de Sales, Bishop of Geneva.

l. 94. *Burnet.* Alexander Burnet, anti-puritan Archbishop of Glasgow. Lauderdale (John Maitland, Duke of) followed a policy of toleration. This compelled him to resign in 1669.

l. 95. *For Becket's sake.* Thomas à Becket is said to have cursed the men of Strood (near Rochester) for cutting off his horses' tails; their children were to be born with horsetails.

l. 96. *pacify.* Reconcile. *prayers.* The Book of Common Prayer.

l. 97. *Or to . . . chairs.* After this line the Bodleian MS and *1697* have:

> Though kingdoms join, yet church will kirk oppose
> The mitre still divides, the crown does close.
> As in Rogation-Week they whip us round,
> To keep in mind the Scotch and English bound.
> What the ocean binds is by the bishop rent,
> Their sees make islands in our continent.
> Nature in vain us in one land compiles,
> If the cathedral still will have its aisles.

l. 100. *surcingle.* Girdle.

l. 101. *zone.* (1) geographical sense; (2) girdle.

l. 106. *hocus.* A juggler.

l. 109. *Pharoah.* Exodus 7; parodying Cleveland, 'Rebel Scot', 63–4:

> Had Cain been Scot, God would have changed his doom,
> Not forced him wander, but confined him home.

l. 113. *musk, even.* Bodleian MS, Philadelphia pamphlet; 'dung the' Douce 257.

l. 115. *rennet.* Also, a bawd.

152 l. 116. *(Lord) are.* Bodleian MS; 'are Lord' Douce 257; 'are "Lord" ' Margoliouth.

152 l. 118. *forsook*. Bodleian MS, Philadelphia pamphlet, Margoliouth; 'mistook' Douce 257, Margoliouth (1st edn.).

l. 135. *Aaron cast . . . calcines*. Moses, holding the temporal power, checks the mischiefs done by Aaron, the high priest (Margoliouth), *calcines*. Burns to ashes.

l. 141. *ambigu*. A meal at which meat and pudding are served together.

l. 143. in commendam. Beneficed.

l. 145. *These Templar Lords . . . Knights*. These spiritual lords exceed those who were members of the medieval religious and military order founded to protect the Holy Sepulchre in Jerusalem; they came to be noted not only for their wealth, but also for their hard drinking (Donno).

l. 146. *baron bishop*. Lord Bishop.

l. 147. *Leviathan . . . behemoth*. Two monstrous beasts commonly linked.

l. 150. *flamen*. Heathen priest.

l. 151. *mitre*. (1) headgear of bishop; (2) oriental headdress. *four*. Disyllabic.

153 l. 158. *Seth's pillars*. 'The descendants of Seth preserved the knowledge that had been acquired on pillars of brick and of stone' (Donno). Also alluding to Seth Ward, Bishop of Salisbury, enthusiastic persecutor of dissenters.

ll. 158–65. *Seth's pillars . . . Snow*. These lines play on the names of various bishops.

l. 164. *Abbot . . . doe*. Bodleian MS; not in Douce 257.

l. 170. *congruous dress*. i.e. black and white.

l. 178. *daring Blood*. Colonel Thomas Blood. He was deprived of some lands in Ireland at the Restoration, and, in return, attempted to steal the crown, sceptre, and orb from the Tower of London on 9 May 1671. The attempt had been carefully prepared for three weeks beforehand. Blood, in clerical dress of 'cassock, surcingle, and gown', had made friends with the keeper, Edwards. When the attempt was made, Edwards was bound and stabbed but not mortally. Blood and his three accomplices were disturbed by the unexpected return of Edwards' son, but it is possible that they would have got away with the regalia if they had killed Edwards and so prevented him from giving information to their pursuers.

Blood was examined by the king in person, was pardoned, and had his lands restored (Margoliouth).

l. 180. *surcingle*. (Here) belt for cassock.

154 l. 202. *moot*. Argue.

l. 204. *Bishops even*. Bodleian MS; 'Even Bishops' Douce 257.

l. 205. jus divinum. Divine right.

ll. 212–15. *To conform's . . . indifferent*. This passage is hopelessly corrupt in the various witnesses. After l. 215, the Bodleian MS reads:

New oaths 'tis necessary to invent,
To give new taxes is indifferent.

154 ll. 230–2. *Luxury . . . Id-*. Sloane 655; Douce 257 has

Luxury, malice, pride and superstition
Oppression, avarice, and ambition,
Sloth, and all vice that did abound.

155 l. 246. *shibboleth*. Judges 12: 6.

l. 261. *cross and pile*. Heads and tails.

l. 264. *atone*. Reconcile.

l. 269. *comb-case*. Comb without honey.

l. 276. *My former satire*. 'The Rebel Scot'.

l. 277. *The hare's head . . . set*. Proverbial; 'tit for tat'.

157 *The Rehearsal Transprosed*. Marvell gives an account of his motives for writing this work below (pp. 271–2). In it he answers three books by Samuel Parker:

(1) *A Discourse of Ecclesiastical Policy* (1670);
(2) *A Defence and Continuation of the Ecclesiastical Policy in a Letter to the Author of the 'Friendly Debate'* (1671);
(3) *A Preface showing what Grounds there are for fears and jealousies of Popery, prefixed to the 'Vindication' of Bishop Bramhall* (1672).

However, theological disputation is not the main interest of this work today; and for precise reference to these and other books we refer the reader to the edition by D. I. B. Smith.

The title and much of the method is given by the Duke of Buckingham's popular farce *The Rehearsal*:

*Bays*: Why, thus, sir; nothing so easy when understood: I take a book in my hand, either at home or elsewhere, for that's all one, if there be any wit in't, as there is no book but has some, I transverse it; that is, if it be prose put it into verse . . . and if it be verse, put it into prose.
*Johnson*: Methinks, Mr Bays, that putting verse into prose should be called transprosing.

Text from the edition by D. I. B. Smith.

*The author*. Samuel Parker (1640–88), Archdeacon of Canterbury. An Anglican cleric on the make. Rochester portrayed him soon after this in 'Tunbridge Wells', ll. 68 ff.

Listening, I found the cob of all this rabble
Pert Bays, with his importance comfortable.
He, being raised to an archdeaconry
By trampling on religion, liberty,
Was grown too great, and looked too fat and jolly,
To be disturbed with care and melancholy,
Though Marvell has enough exposed his folly.

He drank to carry off some old remains
His lazy dull distemper left in 's veins.
Let him drink on, but 'tis not a whole flood
Can give sufficient sweetness to his blood
To make his nature or his manners good.

157 *peevish*. Perverse, stupid.

chi lava . . . sapone. 'He who washes the ass's head wastes the soap.'

*'S'il . . . offendi'*. Spoken by Amarillis in Guarini's *Il Pastor Fido* (1590), III. iv. 19–24. The Italian really means 'If it is so sweet to sin, yet so necessary not to, O nature too imperfect in rebelling against law! O law too severe in offending nature.'

158 *Bishop Bramhall's*. John Bramhall (1594–1663) was Archbishop of Armagh, and chaplain to Strafford in Ireland in 1632.

*bookseller*. Publisher.

*trick up*. Dress.

*coif*. A close-fitting cap covering top, back, and sides of head.

*bull's head*. False frizzled hair.

*indite*. Write.

*Bear-garden*. Bear-baiting, proscribed in 1642, was officially revived at the Restoration. It took place on the south Bank of the Thames.

*nick a juncture of affairs*. Hit a critical moment.

*malapert*. Presumptuous.

*save them from hanging*. Referring to benefit of clergy = exemption from trial by a secular court. By reciting the 'neck-verse' a felon could demonstrate that he was a clerk and escape execution.

Imprimatur. Permission to print (Lat. 'let it be printed').

*conventicles*. The meeting of dissenters for religious worship, forbidden by the Conventicles Act of 1664.

*mere*. Only.

*sponges*. Inking pads.

159 *B. and L.* Sir John Berkenhead (1617–79), and Sir Roger L'Estrange (1616–1704), both busy in censoring books for the government.

*Cadmus . . . sowed*. 'I know they [books] are as lively, and as variously productive, as those fabulous dragon's teeth; and being more sown up and down, may chance to spring up armed men.' Milton, *Areopagitica* (1644). And see Ovid, *Metamorphoses*, iii. 104–13.

*bulky Dutchman*. Laurence Koster (*c.*1370–1440), printer.

*syntagms*. Systematically arranged treatises; orderly arranged statements.

*J.O.* John Owen (1616–83), Vice-Chancellor of Oxford under Cromwell and a prolific theological controversialist. Friend of Marvell.

159 *Malmesbury*. Prosperous weaving town in Wiltshire, famous as the birthplace of Thomas Hobbes.

*railers*. Abusers, scoffers.

160 *'Was . . . importunity'*. A parody of Virgil, *Eclogues*, viii. 49–50.

*A.C. and James Collins*. London publishers. 'A.C.' = Andrew Clarke.

*'Put . . . marquess'*. 'Put up thy wife's trumpery, good noble marquis', 'The Session of the Poets' (1668), l. 90. The marquess was William Cavendish, Duke of Newcastle, who is represented as drawing the works of his wife Margaret, poet and philosopher, from his breeches.

*Bays*. The main character in Buckingham's farce *The Rehearsal* (1672), based partially on Dryden, whose heroic plays are frequently parodied. The name 'bays' (see note to 'The Garden', l. 2, p. 47) is an allusion to Dryden's appointment as Poet Laureate in 1668. Marvell hijacks the name (Bays) to stand for Parker.

*''Tis no matter . . . see't'*. Cf. *The Rehearsal*, I. i. 64–6.

*'that could not guess . . . love'*. Ibid., I. I. 146–7.

161 *homely*. Familiarly.

*Volscius*. A character in ibid., III. v.

*'whose mother . . . wall'*. Ibid., l. 50.

*'Go on . . . none'*. Ibid., ll. 95–7.

*'For as . . . neither'*. Ibid., ll. 100–9.

*Declaration of Indulgence*. 15 March 1672.

*dog-days*. July–August.

*'the season . . . undone'*. Parodying Davenant's *Gondibert* (1651), I. ii. 42.

*Ephemerides*. A book of prognostication.

*Sidrophel*. William Lilly (1602–81) a notorious astrologer, perhaps portrayed by Samuel Butler in *Hudibras*, ii. 3.

*jade*. Exhaust.

volo nolo. 'I want, I don't want'.

162 *'he had prevailed . . . book'*. Owen, *Truth and Innocence Vindicated*, 4.

*crupper*. A leather strap to prevent the saddle on a horse slipping forward; or the sense may be the rump of a horse.

*push-pin*. 'A trivial occupation'.

*procatarctical cause*. The immediate cause.

*Roscius*. Roman comedian, friend of Cicero, who defended him in a lawsuit.

*Lacy's*. John Lacy (d. 1681), dramatist and player, who played Bays in *The Rehearsal*.

163 *tuant*. Biting.

163 *Irrefragable Doctor*. Alexander of Hales (d. 1245), celebrated medieval theologian.

*recreate*. Console.

*jump*. Agree.

164 Grand Cyrus. A novel by Madeleine de Scudéry (1649–53), translated into English in 1653.

Cassandra. By Coste de la Calprenède, translated into English in 1652.

Knight of the Sun. By Ortunez de Calahorra (tr. 1579).

King Arthur. Malory's *Morte Darthur* (1485).

*Bishop of Cologne*. Henry Maximilian (1621–88).

*Bishop of Strasbourg*. Franz Egon Fürstenberg (1625–82). 'Both men were aggressive supporters of Louis XIV' (Smith).

*Bishop of Münster*. Christoph Bernard von Galen (1606–78). In 1672 he was fighting for Louis XIV (Smith).

*Tories*. Irish papists, or outlaws.

*'Down by . . . lice'*. Apparently from John Ogilby's 'Character of a Trooper'.

*pendets*. A pendet Brahmin is a learned Hindu versed in Sanskrit. The word developed into 'pundit' in the eighteenth century.

*history of the Mogol*. By François Bernier (1648, tr. 1671).

165 Legend of Captain Jones. By David Lloyd (1648); Grosart quotes from the title-page: 'relating his adventures at sea, combat with a mighty bear, furious battle with 36 men, sea-fight, combat with Badaher Cham, a giant of the race of Og, etc.'

De Virtutibus Sancti Patricii. 'On the Virtues of St Patrick'. Saint Patrick (373–463) brought Christianity to Ireland.

*writer . . . Debates*. Simon Patrick (1626–1707), Bishop of Ely, author of *A Friendly Debate* (1669).

*Secundinus*. Fifth-century Bishop of Armagh.

*Jocelyn*. He wrote *Vita S. Patricci* ('The Life of St Patrick').

*Odo, Bishop of Bayeux*. Eleventh-century Earl of Kent, who fought with a mace instead of a sword.

166 *Earl of Strafford*. Thomas Wentworth (1593–1641), Lord Deputy of Ireland. He was beheaded in 1641.

*Bishop Ussher*. James Ussher (1581–1656), Archbishop of Armagh.

De Primordiis Ecclesiae Britannicae. 'On the early History of the British Church'.

*Pelagius*. A fifth-century British theologian.

*Grub-street*. Near Moorfields in the east end of London, now Milton Street. Proverbial for hack-writers.

166 *Irish Rebellion*. In 1641, in Ulster.

167 *Alexander's architect*. Deinocrates.

*gravelled*. Perplexed.

*Isthmus of Peloponnesus*. Isthmus of Corinth. The idea occurred to Julius Caesar, and Nero, among others.

168 *rebating*. Blunting.

*conned him thanks for*. Expressed thanks.

*Austin*. St Augustine (d. 604), first Archbishop of Canterbury.

169 *'manage . . . extent'*. Parker, *Preface*.

*Clarencieux*. An English dukedom: the second king of arms in England (*OED*).

*conference of Worcester House*. In 1660, when Presbyterians and Anglicans met to settle differences in religion, abortively.

170 *Arminian*. The doctrines of Jacob Arminius (1560–1609), Dutch Protestant theologian, opposed to Calvin.

171 *Grotius*. Hugo Grotius (De Groot) (1583–1645), celebrated Dutch writer on jurisprudence.

Gazette. i.e. information from *The London Gazette*.

172 *'prefers . . . together'*. *The Rehearsal*, IV. i. 108–10.

*Drawcansir*. A character in *The Rehearsal*, parodying Dryden's Almanzor in *The Conquest of Granada*.

*cymar*. A loose-fitting woman's robe or bishop's gown.

*prolocutor*. Spokesman.

synodical individuum. A one-man synod.

*fifth council*. Protestants generally acknowledged only four councils: Nicaea (325), Constantinople (381), Ephesus (431), and Chalcedon (451).

*those two*. 'Sergius and Abdullah: "A Nestorian monk of Constantinople and a paynime [pagan] Jew" Purchas, *Purchas his Pilgrimage* (1613), p. 200' (Smith).

173 Hudibras. Burlesque poem by Samuel Butler, published 1663–78.

*whifflers*. Triflers.

*oilmen*. Sellers of oil.

*Baudius*. Dominicus Baudius (1516–1613) was a Flemish scholar.

*concentre*. Draw into a centre.

opprobrium academiae. 'The shame of the university'.

pestis ecclesiae. 'The plague of the Church'.

*Mr L.* Sir Roger L'Estrange, press censor.

174 *Mr Calvin*. 'Master' Jean Calvin (1509–64), French Protestant theologian.

*Dodona's Grove*. The seat of Zeus.

175 *Mistress Mopsa*. The dim daughter of Miso in Sidney's *Arcadia* (1590). She tells stories which are very silly and inconsequential.

*Helvetian Passage*. Caesar, *De Bello Gallica*, i. 6.

*altar-wise*. The position of the communion-table was a subject of much heat in the seventeenth century.

176 *roguing*. Acting like a rogue.

*sold Mr Bays a bargain*. To sell a bargain is to make a fool of someone by replying coarsely to a question.

The Rehearsal. v. i. 299.

*'reasons . . . blackberries'*. *1 Henry IV*, II. iv. 242–3.

*city of 'roaring lions'*. Lyon (still pronounced 'lions' in English, and spelled 'Lyons').

*Millecantons*. 'The result of a thousand cantons like those few of Switzerland' (Grosart).

177 *nuts to Mother Midnight*. 'Nuts'=goodies. Mother Midnight meant a midwife, and sometimes a bawd. The phrase seems to be proverbial.

*fits and girds*. Fits and starts.

ipse dixit. An unsupported statement ('He himself said it').

178 Sunt . . . ineptiae. Calvin to Francford, 15 February 1555. 'There are in that book certain tolerable absurdities.'

*make good cheer*. Be well entertained.

*pall-mall*. Ball game a little like croquet.

*arbalet*. Cross-bow.

*court boule*. A form of bowls.

*fort of St Katherine*. Near Geneva.

escalade. Mounting, scaling (Fr.). Geneva was attacked in 1602.

*Sancho*. Cervantes, *Don Quixote*, ii, ch. 42.

*Bedlam or Hoxton*. Madhouses.

179 *'Before . . . holla'*. *The Rehearsal*, v. i. 89–90.

*quadrature*. Squaring.

*nobleman*. Gilbert Sheldon (1598–1677), Archbishop of Canterbury.

*clapped in*. Applied himself with briskness.

*drolling upon*. Making sport of.

*would take*. Would be popular.

180 *tippet*. A band of silk with two ends hanging, worn by clergymen; here, standing for clergymen.

*shifted himself*. Dressed.

*Maudlin de la Croix*. Sixteenth-century levitating abbess.

180 *speculate*. Inspect.

*baby*. Reflection.

*the King's or Duke's house*. The two London playhouses.

181 *King of France*. Charles VI (1368–1422).

*Cabal*. Cabbala; esoteric interpretation.

182 *by-word*. Nickname, epithet of scorn.

*Bishop Prideaux*. John Prideaux (1578–1650), Bishop of Worcester.

*professor of divinity*. At Oxford.

cujus . . . te doctorem creo. 'Of which . . . I create you Doctor'.

*promotion*. The act of conferring a doctorate.

Rationale. Anthony Sparrow (1612–85), Bishop of Norwich, *Rationale upon the Book of Common Prayer* (1657).

*Mr B*. Richard Baxter (1615–91), Presbyterian. Bramhall in his *Vindication* wrote against him.

*bullace*. A wild plum.

peccavi. 'I have sinned'.

183 *'Art . . . misery'*. From an anonymous paraphrase of 1 Samuel 28: 8–20, entitled 'In Guilty Night' or 'The Witch of Endor'. Marvell could have heard the work when a student at Trinity College, Cambridge. Parker in the role of Saul desperately conjures up the ghost of Samuel (Bramhall) (Smith).

*proverb*. 'Confess and be hanged.'

*Now, one* . . . Ghibilinus es. The incident took place in 1300. The Pope was Boniface VIII, the Ghibelline the Franciscan Archbishop of Genoa. 'Remember that you are dust and to dust will return' . . . 'that you are Ghibelline.'

184 *as Gonzales . . . ganzas*. F. Godwin, *The Man in the Moon* (1638). *ganzas*. Wild swan.

*disintricated*. Disentangled.

*gantlope*. A punishment in which the culprit had to run stripped to the waist between two rows of men who struck at him with sticks or knotted cords (later, 'gauntlet').

185 *sow in Arcadia*. Varro, *Rerum Rusticarum libri tres*, ii. 4.

*criss-cross row*. The alphabet, from the figure of the cross formerly attached to it in horn books.

*Mr Bales*. Thomas Bales, Commissioner of buildings, highways, etc., and Magistrate for Westminster.

186 *blue-John*. After-wort; the second run of beer.

io . . . paean. 'Hurrah!' ('Paean' was a surname of Apollo.)

*Io*. Ovid, *Metamorphoses*, i. 569–748.

186 *ragout*. Stew.

*Fayal*. One of the Azores, formerly volcanic.

187 *that fellow's*. 'Richardson, the famous fire-eater' (Smith).

*aspic*. Asp; small venomous serpent.

*cowage*. The stinging hairs of the pod of a tropical plant.

*genial bed*. Marriage bed.

*Tom Triplet*. Dr Thomas Triplet (1603–70), prebendary of Westminster.

*Dr Gill*. Alexander Gill (1565–1635), High Master of St Paul's School.

*Priscian*. Roman grammarian.

188 *jerking*. Whipping.

*Osbolston . . . Dr Busby*. Teachers at Westminster School.

*J.O.* Owen was vice-chancellor of Oxford 1652–8. Parker was up at Wadham College during this time.

*cogent*. Peremptory.

189 *holy brotherhood*. 'Santa Hermandad', in Spain the name of a combination formed to resist the exactions of the nobles in 1476. It was subsequently given police functions. Marvell may also be referring to the Jesuits here (Smith).

*renegade*. Apostate.

*intelligencer*. Spy, informer.

190 *resemble*. Compare.

*politic engine*. Probably the notorious informer John Pouler.

*sent over a Declaration*. 14 April 1660, from Breda.

*Act . . . Indemnity*. 29 August 1660.

*late combustions*. The civil war.

191 *minister of state*. Edward Hyde (1609–74), Earl of Clarendon.

quod scripsi scripsi. John 19: 22: 'What I have written, I have written.'

192 sui juris. 'Under their own law'.

193 *enabled*. Authorized.

*confine upon*. Adjoin.

*dictature*. Dictatorial manner.

194 Gondibert. By William Davenant (1651).

Leviathan. Thomas Hobbes, *Leviathan* (1651).

*fetch*. Stratagem.

praemunire. 'The writ for a person accused of asserting Papal jurisdiction in England, and therefore denying the ecclesiastical supremacy of the sovereign' (Smith).

194 non-obstante. A dispensation ('not opposing').

jejunium Cecilianum. Jejunium='fast'; in 1562 William Cecil, Lord Burghley, as a measure to help the navy, proposed making Wednesday, as well as Friday, a meatless day.

*lightly*. Casually.

195 *like the ass in the fable*. Aesop, 24.

*ramp*. Raise the forepaws in the air.

*forbear*. Put up with.

*his book*. Grotius, *De Imperio Summarum Potestatum Circa Sacra* (Paris, 1647).

*Erastian*. Follower of Thomas Erastus (1524–83), Swiss theologian, held to have advocated the supremacy of the state over the Church.

196 *answerer*. J. Owen, *Truth and Innocence Vindicated* (1669), 108.

jus divinum. 'Divine authority'.

*'gravest . . . flatterers'*. Jonson, *The Alchemist*, II. ii. 59.

Hunc . . . tutus. *Satires*, II. i. 20. 'If you flatter him the wrong way, he kicks back on all sides.'

*trepanned*. Cheated, conned.

*projectors for concealed lands*. 'Concealed land' is that held privily from the king by a person having no title to it. 'Projectors' informed against the holders of such lands (Smith).

*prowling*. Plundering.

197 *conversation*. Behaviour, mode of life.

*trepan*. Inveigle.

*peach*. Inform against.

198 *cuts*. Trims his discourse.

*casual divinity*. Casuistry.

*hallow*. Make holy.

199 'Ma . . . nume'. Guarini, *Il Pastor Fido*, III. iv. 28–9.

'Quaesitum . . . verpos'. Juvenal, xiv. 104: 'To bring only the circumcised to the desired fountain.'

'Nullum . . . prudentia.' Juvenal, x. 365–6 (slightly misquoted).

*school-divines . . . systematical*. The distinction seems to be between teaching divines and theoretical ones.

*come off*. Succeed.

200 *rebellion . . . witchcraft*. 1 Samuel 15: 23.

*aggravates upon*. Lays charges against.

saevior . . . orbem. Juvenal, vi. 292–3: 'More savage than any foe, Luxury oppresses, and punishes a conquered world.'

201 *Sybarites*. Lovers of luxury. For dancing horses see Athenaeus, *Deipnosophista*, xii. 520.

*Julian*. (331–63), Roman Emperor. 'The Apostate'.

*acted*. Actuated.

*mure up*. Wall up.

*Bartholomew register*. The Act of Uniformity came into force on St Bartholomew's Day, 24 August 1662.

*March licenses*. Licenses to preach, issued after the Declaration of Indulgence, 15 March 1672.

*breaks no square*. Does no harm.

202 *one neck*. Caligula, Roman Emperor; Suetonius, *Caligula*, xxx.

203 ciurma. Galley slaves.

*Sir John Baptist Dutel*. Of French extraction, a Knight of Malta.

*the tired . . . suffering*. Tertullian, *Ad Scapulam*.

non tibi . . . Regi. Henlyn, *Cosmography* (1669), i. 93. 'Not to you, but to Peter . . . to the king.'

*entremets*. Snacks between courses.

204 non legit. 'He does not read'.

*Saint Dominic*. Dominic de Guzman (1170–1221), founder of the Dominican order.

*Synesius*. Greek writer, Bishop of Ptolemais (d. *c*.430).

arcanum. Secret.

*experiments*. Experience.

*opiniastre*. Stubborn in opinion.

205 *imp*. A young shoot of a plant, a sucker.

*disvalise*. Strip someone of their luggage.

*scurvy disease*. Syphilis.

*nodes*. Inflamed swellings.

*buboes*. Abscesses in glandular parts of body.

206 *Mas Johns*. Contemptuous for Presbyterian preachers.

*archbishop*. Alexander Burnet (1614–84), Archbishop of Glasgow, 'noted for his severity towards Presbyterians' (Smith).

*a book too of J.O.'s*. *Truth and Innocence Vindicated* (1669).

*all the teeth*. Like Richard III.

207 Jus Divinum. Divine Right.

*old excellent*. i.e. long expert.

*great prelate*. John Cosin (1594–1672), Bishop of Durham.

*'look to't . . . do't'*. *The Rehearsal*, I. i. 397.

207 *in* querpo. Naked (Sp.).

208 sui juris. 'Of one's own right'.

209 *remonstrate to*. State a grievance.

*remonstrants*. Referring to the Arminian party at the Synod of Dort (1617).

*ingenuity*. Ingenuousness.

*makes indentures*. Zig-zags, doubles.

210 *Sardanapalus*. Luxurious king of the Assyrian empire.

*Justine*. Marcus Justinianus Justinus, Roman historian.

*teeming*. Pregnancy.

*pounces*. Claws, talons.

*tokens of loyalty*. 'The nonconformists lent large sums of money to Charles II' (Smith).

211 *first book*. *Discourse*.

*prosper*. Successful.

*marched up Holborn*. To Tyburn.

*ear-mark*. Mark of ownership.

*claw it off*. Get rid of it.

*answerer*. J. Owen, *Truth and Innocence Vindicated*.

213 Sortes Virgilianae. 'Virgilian auguries': a method of divination by opening Virgil at random.

214 *quetch*. Utter a sound.

*lock*. A grapple in wrestling.

*counterscarp*. In fortification, an outer wall supporting a covered way.

*scold*. A woman of ribald speech.

*divaricated*. Stretched apart.

*amain*. Vehemently.

*insulted over*. Triumphed over.

215 *augurate*. Inaugurate.

*Tiberius . . . letter*. Suetonius, *Tiberius*, lxvii.

Dii deaeque me perdant. 'May the gods and goddesses destroy me.'

216 *John a Nokes . . . John a Styles*. Fictional names for the parties to a legal action.

*rencounters*. Skirmishes.

*drolling*. Jesting.

*Verulam*. Francis Bacon.

217 *Tarlton*. Richard Talton (d. 1588), jester.

effigies. Portrait.

*vouchers*. Proofs.

217 *mum*. i.e. silence.

*sleeps upon both ears*. Enjoys undisturbed sleep.

218 *Bedlam*. A madhouse.

*'I am . . . do't'*. *The Rehearsal*, I. i. 371–97.

*burning the ships at Chatham*. See 'The Last Instructions', above.

*conflagration*. The 'Great Fire' of London, 1666.

*yearly . . . solemnity*. On 10 October 1666 a general fast was ordered.

*Diana's temple*. It was said that St Paul's Cathedral was erected on the site of a temple dedicated to Diana.

*Socinian*. Doctrines derived from Faustus Socinius (1539–1604) who denied the divinity of Christ.

219 *tagging*. Supplying rhymes.

*supererogate*. Speak beyond necessity.

*Doctor Thorndike*. Herbert Thorndike (1598–1672), clergyman.

*Richard Montague*. Bishop of Norwich (1577–1641).

*Mr Hooker*. Richard Hooker (1554–1600), theologian.

220 *Mr Hales*. John Hales (1584–1656), theologian.

Treatise of Schism. Hales's book, probably written in 1636, was published in 1642.

221 Cautissimi . . . feceris. Pliny, *Epistles*, I. xviii. 5. 'The precept of every cautious person—what you have doubts about, don't do.'

*second Council of Nice*. 787, about the veneration of holy images.

pectus praeparatum. Horace, *Odes*, II. x. 15. 'His breast (mind) prepared.'

222 cardinalism, nepotism, putanism. Respectively, the institution of cardinals, the practice of granting favours to nephews, and the trade of a whore.

223 *unprejudicate*. Unprejudiced.

*you are adreamed*. You have been visited by a dream.

*'words left betwixt the sheets'*. Juvenal, *Satires*, vi. 195.

*Macedo*. Ferdinand de Macedo, an unreliable witness in a recent case before the Council concerning Sir John Branston.

224 *assassinates*. Assassinations (here figurative).

*have reason*. Are right.

*to conquer . . . whole armies*. *The Rehearsal*, v. i.

Evangelical . . . Unity. By John Owen (1672).

*dexterity in sinking*. *The Merry Wives of Windsor*, III. v. 9–11.

225 *shibboleth*. Hebrew word used by Jephthah as a test-word to sort out Ephraimites (Judges 12: 4–6).

225 *collar of* -nesses. A play on the shirt of Nessus and the 'collar of S's', an ornamental chain consisting of a series of S's, worn by certain dignitaries in England.

226 *add or diminish in the scripture.* Revelation 22: 18–19.

227 adepti. Initiates.

*Elias.* Elijah, whose return will announce the arrival of the Messiah.

*barber's basin . . . Mambrino's helmet. Don Quixote* iii, ch. 7.

*panel.* A cloth under the saddle.

*furniture.* Decoration.

228 *Saint Paul.* Acts 24: 25.

*'Book of Martyrs'.* John Foxe, *The Acts and Monuments of these latter days* (1563).

*Scotch history.* 'Persecution in Scotland' in Foxe.

*great cardinal . . . Wishart.* David Beaton (1494–1546), Cardinal Archbishop of St Andrews, and George Wishart (1513–46), Scotch reformer.

*Campanella.* Tommaso Campanella (1568–1639), Italian philosopher.

superciliums. Eyebrows. (Also, 'supercilium' = 'pride'.)

*modern author lately dead.* This could be Sir John Denham (d. 1669), Davenant (d. 1668), or Cowley (d. 1667). In view of 'Appleton House', l. 340, it could be a reference to Marvell himself.

*that of the lie.* For the variety of lies Grosart aptly compares *As You Like It*, V. iv. 78–83.

229 *brothers of the blade.* Gentlemen who debate honour and duelling.

hoc est verum, hoc est falsum. 'That is true . . . false'.

quod restat probandum. 'What remains to be proved'.

*'Villain . . . I trow.' The Rehearsal*, V. i. 216–19.

*'They fly . . . the lie.'* Ibid, III. i. 242.

*rapper.* An arrant lie, a whopper.

*the strumpet and the beast.* The Whore of Babylon, and the Beast in Revelation (the Roman Catholic Church is meant).

*Elagabalus.* (205–22), Roman Emperor, 'abandoned to the grossest pleasures with ungoverned fury' (Gibbon).

*Bishop Bonner.* Edward Bonner (1500–69), Bishop of London. A notorious persecutor.

*clary.* Drink of wine and honey.

230 *Doctor Bayly's.* Thomas Bayly (d. 1657), royal clergyman, who wrote about herbs.

*Heliodorus Bishop of Trissa.* See Socrates [a Constantinopolitan lawyer], *Ecclesiastical History*, v. 22.

230 *ordination dinner*. The Jesuit Christopher Holywood put it about that the Anglican succession originated at the Nag's Head Tavern.

*excusation dinner*. 'Excusation' is the act of excusing. Marvell's precise meaning here is obscure.

*porridge*. Dissenters called the Anglican service this.

phaesianum . . . muluellum. '[Who ate] pheasant as others ate cod'.

*Mr Cartwright*. The actor who played Thunder in *The Rehearsal*.

*Nonconformist Cartwright*. Thomas Cartwright (1535–1603), puritan divine.

*waxen . . . matter*. The first matter, or *prima materia*, on which God imprinted the forms.

*gig*. A whipping-top.

*the Wheel of Fortune*. A revolving wheel with sections indicating chances taken or bets placed.

*Sir Edmund Godfrey*. (1621–78), JP for Westminster.

231 *vitiate*. Deflower.

*run up to the wall by an angel*. Numbers 22: 25.

*picked*. Choice.

232 *pudder*. Turmoil.

Signa . . . appellantur. 'Signs relating to divine matters are to be called sacraments.'

*Augustulus*. 'Little Augustus'.

Quum . . . emendes. Horace, *Epistles*, II. i. 1–3. 'Since you carry so many weighty affairs on your shoulders, strengthening Rome's defences, promoting decent behaviour, reforming our laws . . . '

233 Ep. 5ta ad Marcellinum. Augustine's *Fifth Letter to Marcellinus*.

Nimis . . . appellantur. 'It would take too long to argue about the variety of signs, which when they pertain to divine things are called sacraments.'

*mousled*. Pulled about roughly.

234 pro hac vice. 'For his office'.

terra incognita. 'Unknown land'.

*baffled*. Dishonoured.

235 curtana. Ornamental sword worn at the coronation.

*Solomon's sword*. 1 Kings 3: 24.

*Scanderbag*. Scanderbeg, George Castriot (1403–67), Albanian national hero.

*stounded*. Stupefied.

*prelate*. John Cosin (1594–1672), Bishop of Durham, 'whose *Private Devotions* (1627) mentioned Seven Sacraments rather than five' (Smith).

*Roman Emperor*. Caligula; Suetonius, *Caligula*, xliv.

236 *innovated upon*. Revolted against.

insana laurus. 'Laurel which maddens'.

237 Hic jacet . . . precare. 'Here lies the body of Herbert Thorndike, Prebendary of this church, who when living sought out with prayer and study the true doctrine and mode of the Reformed Church . . . Reader, pray for his peace and blessed resurrection in Christ.'

238 mutatis mutandis. 'After making the necessary changes'.

*tap-lash*. The 'lashings' or washings of casks or glasses; the dregs of liquor.

239 *Augustus Caesar*. Suetonius, *Augustus*, lxxxv.

homousians . . . homoiousians. The Arian heresy, about the nature of the Trinity, turned on the terms: 'of the same nature', 'of a similar nature'.

240 *'Let . . . order'*. 1 Corinthians 14: 40.

*quit cost*. Give a return on investment.

241 *tilting*. propping up (?).

*'cloak . . . parchments'*. 2 Timothy 4: 13.

*beaten*. Well-worn.

*bran*. Sort.

242 Unum . . . deesse. A line from a fragment of a poem by Julius Caesar on Terence. 'I torment myself and grieve that you lack this one thing.'

*'Politic Would-be's'*. As in Jonson's *Volpone*.

vestra clementia. 'Your mercy'.

243 *father*. Adoptive father, Julius Caesar.

244 *one of your Roman emperors*. Caligula; Suetonius, *Caligula*, lviii.

*the word*. The password.

*Parliament of Poland*. 'The authority of the king was so limited by the General Diet in Poland . . . that he was virtually helpless' (Smith).

*tyrant*. Ivan IV (1530–84).

io . . . bestie. 'I don't want to rule over beasts.'

*sturdy Swiss*. William Tell.

*unhoopable*. Not capable of being contained.

245 *5° Eliz. of the Wednesday fast*. See p. 194.

246 *rochet*. A clerical vestment of linen.

*scourges of heaven*. Plague, fire, the Dutch.

248 *Robert Parker*. (1564–1614) Puritan divine.

de Cruce. 'Concerning the cross'.

*Archbishop . . . Parker*. Mathew Parker (1504–75).

de . . . Britannicae. 'Concerning the antiquities of the British Church'.

249 *glorious*. Ostentatious.

250 *gird*. Strike.

*fetch*. Trick, ploy.

*crimination*. Accusation.

252 *shrewdly*. Maliciously.

253 *as if . . . November*. 15 March signified the Declaration of Indulgence to tender consciences, while 5 November was the Gunpowder Plot.

*harbergeon*. Mail jacket.

*Beza*. Theodore Beza (1519–1604), Calvinist theologian.

*bandoleer*. Gun belt worn across the shoulder.

*Keckerman*. Bartholomeus Keckerman (1571–1609), German theologian.

254 *Doctor Patrick*. Simon Patrick, author of *The Friendly Debate*.

*Father Patrick*. Father Patrick MaGinn, one of the Queen's confessors.

255 *leasings*. Falsehoods.

*laystall*. Dungheap.

sanbenitas. Black garment forcibly worn by impenitents while being murdered by the Inquisition.

256 jus patronatus. 'The sum of rights of a patron over his freedmen'.

*Petra . . . Rudolpho*. An allusion to Gregory VII's duplicity. Having humbled and absolved Henry IV, Gregory sent a crown to Rudolph of Swabia, inscribed with the words Marvell quotes (Smith).

eo nomine. 'As such'.

257 *Abbot*. George Abbot (1562–1633), an opponent of Laud.

*Archbishop Laud*. William Laud (1573–1645), Archbishop of Canterbury.

*fate than he met with*. He was beheaded.

*impositions of money*. The 'forced loans' of 1626–7.

*ordinaries*. Taverns.

*pieces*. Gold coins worth just over a pound.

*half a crown*. 2*s*. 6*d*, now 12½p.

258 sui juris. 'Under their own law'.

*pickthankness*. The quality of being a pickthank, i.e. an informer, flatterer, etc.

*There was . . .* Abbot's *Narrative*.

*Doctor Pierce*. William Pierce (1580–1670), Bishop of Bath and Wells.

259 *St Matthew*. 22: 21.

*Public Readers*. Teachers appointed to read and expound to students.

*the marrying . . . to the Lady R*. On 26 December 1605, Laud married Charles Blount, Earl of Devonshire, to the divorced wife of Lord Rich.

260 *Doctor Worral.* (1589–1639), Canon of St Paul's.

*hand over head.* Without delay (?).

*counsel at the Temple.* John Selden (1584–1654).

meum . . . tuum. Distinctions of property ('mine . . . yours').

*Montague's Arminian book.* Richard Montague, *Appello Caesarem* (1625).

*Doctor Manwaring.* Smith quotes: 'censured by the Lords in Parliament, and perpetually disabled from future Ecclesiastical Preferments'.

261 *aggravated.* Enlarged.

*Turks . . . Abiram.* Numbers 16: 1.

*Theudas and Judas.* Acts 5: 36 7.

*that of the Maccabees.* The Maccabees rebelled against the Romans in the 2nd century BC.

jure divino. 'By divine law'.

jure regio. 'By royal law'.

*'letters of reprisal'.* Official warrants authorizing an aggrieved subject to exact forcible reparation from the subjects of another state.

262 *magnificate.* Magnify.

*remembrancer.* One engaged to remind someone of something.

263 *Barnevelt.* Johan van Olden Barenveldt (1547–1619). Grand pensionary of Holland, a republican, and an Arminian. Following the Synod of Dort, he was arrested and executed (Smith).

265 *bead-roll.* A list of those who are to be mentioned at prayers; hence any (long) list of names.

*Declaration.* Of Breda, 1660.

*Conference.* The Savoy Conference, 25 March 1661.

266 *one memorable day.* The Ejection of the two thousand in 1662.

267 *trinkle.* Intrigue.

*superfoetation.* Second conception; unnecessary accumulation.

270 *close.* Secret, intimate.

271 *All* . . . pork. Livy, *The Roman History*, tr. Philemon Holland, 1600, p. 916. Marvell elaborates the point in the Second Part.

*'five-mile Act'.* The Act made it penal for any Nonconformist minister who had not taken the oath of non-resistance to go within five miles of a town where he had earlier preached.

*King Phys . . . King Ush.* Characters in *The Rehearsal.*

*the poet.* Davenant.

*Bishop Davenant.* John Davenant (1576–1641), Bishop of Salisbury; 'a moderate Calvinist' (Smith).

*'For . . . lost'. Gondibert* (1651), II. vi. 85.

272 *'praise . . . minds'. Gondibert*, st. 84.

*Its utmost . . . palace tear*. Ibid., st. 87.

*Astragon*. A character in *Gondibert*.

*the* Rehearsal. IV. i.

*Aretino*. Pietro Aretino (1492–1556), Italian poet.

275 *The King's Speech*. Understandably, this mock-speech, one of Marvell's most politically audacious performances, was not printed until 1704, in Part 3 of *Poems on Affairs of State*. Our version is from BL MS Add. 34362, as given in Bradbrook and Thomas, *Andrew Marvell*, 125–7.

*my Lord Treasurer*. Thomas Osborne, Earl of Danby.

*reformado*. A military term, applied to officers waiting for a place. Here, it probably means 'ex-mistresses'.

*your religion*. As Wilcher points out, Marvell hints that Charles was already what he undoubtedly became, a Roman Catholic.

276 *proclamation*. Against dissenters and Roman Catholics.

*Lord Lauderdale*. John Maitland (1616–82), Earl of Lauderdale and ruler of Scotland.

*George*. The Earl of Northumberland (b. 1674), Charles's youngest son by Barbara Villiers, Duchess of Cleveland.

*Portsmouth*. Louise de Keroualle (1649–1734), Duchess of Portsmouth, the king's mistress since 1670.

*Lord of Pembroke*. Philip Herbert (1653–83), Earl of Pembroke, a brute.

*Lord of Inchiquin*. Murrough O'Brien (1614–74), Earl of Inchiquin.

*Lord Vaughan*. John Vaughan, Irish peer.

*Crew*. Nathaniel Crew, Bishop of Durham.

*Brideoak*. Ralph Brideoak (1613–78), chaplain to the king.

*Lord St John*. Charles Paulet, Lord St John.

*Mistress Hyde and Emerton*. Referring to the topical scandal. Bridget Hyde, 12-year-old step-daughter of Sir Robert Viner, had been secretly married to a Mr Emerton while Danby was negotiating a match between her and his son. Danby was later accused of putting pressure on the officiating clergyman to deny that the marriage had taken place.

# FURTHER READING

(The place of publication is London unless otherwise specified.)

EDITIONS

*The Complete Work in Verse and Prose of Alexander Marvell*, ed. Alexander B. Grosart (4 vols., 1872–5).

*The Poems and Letters of Andrew Marvell*, ed. H. M. Margoliouth (2 vols., Oxford, 1927; 3rd edn. revised by Pierre Legouis and E. E. Duncan-Jones, 2 vols., Oxford, 1971).

*The Selected Poems of Marvell*, ed. Frank Kermode (New York, 1967).

*Andrew Marvell: Complete Poetry*, ed. George de F. Lord (New York, 1968; London, 1984).

*Andrew Marvell, The Complete Poems*, ed. Elizabeth Story Donno (1972).

*Andrew Marvell: Selected Poetry and Prose*, ed. Robert Wilcher (1986).

*'The Rehearsal Transpros'd' and 'The Rehearsal Transpros'd: The Second Part'*, ed. D. I. B. Smith (Oxford, 1971).

BIOGRAPHIES

Kelliher, Hilton, *Andrew Marvell: Poet and Politician 1621–78* (1978).

Legouis, Pierre, *Andrew Marvell: Poet, Puritan, Patriot* (Oxford, 1965).

REFERENCE

Collins, Dan S., *Andrew Marvell: A Reference Guide* (Boston, Mass., 1981).

Guffey, George R., *A Concordance to the English Poems of Andrew Marvell* (Chapel Hill, N.C., 1974).

CRITICAL

Bradbrook, M. C., and Lloyd Thomas, M. G., *Andrew Marvell* (Cambridge, 1940).

Brett, R. L. (ed.), *Andrew Marvell: Essays on the Tercentenary of his Death* (Oxford, 1979).

Carey, John (ed.), *Andrew Marvell: A Critical Anthology* (1969).

Chernaik, Warren L., *The Poet's Time: Politics and Religion in the Work of Andrew Marvell* (Cambridge, 1983).

Colie, Rosalie L., *'My Ecchoing Song': Andrew Marvell's Poetry of Criticism* (Princeton, N.J., 1970).

Craze, Michael, *The Life and Lyrics of Andrew Marvell* (1979).

Duncan-Jones, E. E., 'Marvell: A Great Master of Words', *Proceedings of the British Academy*, 61, *1975* (Oxford, 1976), pp. 267–90.

Eliot, T. S., 'Andrew Marvell', in *Selected Essays* (3rd edn., 1951).

Empson, William, *Some Versions of Pastoral* (1935).

Friedenrich, Kenneth, (ed.), *Tercentenary Essays in Honour of Andrew Marvell* (Hamden, Conn., 1977).

Friedman, Donald F., *Marvell's Pastoral Art* (1970).
Hodge, R. I. V., *Foreshortened Time: Andrew Marvell and Seventeenth Century Revolutions* (Cambridge, 1978).
Leishman, J. B., *The Art of Marvell's Poetry* (1966).
Lord, George de F. (ed.), *Andrew Marvell: A Collection of Critical Essays* (Englewood Cliffs, N.J., 1968).
Patrides, C. A. (ed.), *Approaches to Marvell: The York Tercentenary Essays* (1978).
Patterson, Annabel M., *Marvell and the Civic Crown* (Princeton, N.J., 1978).
Scoular, Kitty W., *Natural Magic: Studies in the Presentation of Nature in English Poetry from Spenser to Marvell* (Oxford, 1965).
Stocker, Margarita, *Apocalyptic Marvell: The Second Coming in Seventeenth Century Poetry* (Brighton, 1986).
Wallace, John M., *Destiny his Choice: The Loyalism of Andrew Marvell* (Cambridge, 1968).
Wilcher, Robert, *Andrew Marvell* (Cambridge, 1985).
Wilding, Michael, *Dragons Teeth: Literature in the English Revolution* (Oxford, 1987).

# INDEX OF FIRST LINES